HPBooks

METAL FABRICATOR'S Handbook

by Ron and Sue Fournier

About the Authors

Ron Fournier's career as a metal fabricator spans over 25 years. He began with Holman and Moody in 1964, and since that time racing greats such as Roger Penske, A.J. Foyt, Kar Kraft and Bob Sharp Racing have all utilized Ron's unique skills to transform metal into various components for their championship-winning race cars. In the mid-Seventies, Ron founded Race Craft, which soon developed a nationwide reputation as one of the finest metal fabrication shops in the country. Today, he supervises the development of automotive prototype sheet metal projects for the Troy Pioneer Development Center in Auburn Hills, Michigan.

Ron's wife, Susan, is the reason Ron's years of fabrication experience can be translated into written form. After receiving a B.A. from Michigan State University, Sue went on to graduate work in English at West Chester State University in Pennsylvania and at the University of Houston. Her publishing credits include a number of poems and a high school English textbook.

Ron and Sue have also produced a sequel to the *Metal Fabricator's Handbook*, HPBooks' *Sheet Metal Handbook.* They frequently write technical articles for various car enthusiast publications. It has been an unusual, yet productive joint effort. Over the years, Sue has learned a great deal aboutt metal fabrication, while Ron's spelling has improved.

Interior photos by Ron Fournier unless otherwise noted.

Front Cover: The author installing a fabricated aluminum firewall on his roadster. All other metal components were fabricated by the author using the techniques in this book. Photo by Michael Lutfy.

HPBooks are published by
The Berkley Publishing Group,
A division of Penguin Putnam Inc.
375 Hudson Street, New York, New York 10014
ISBN 0-89586-870-9

The Penguin Putnam Inc. World Wide Web site address is
http://www.penguinputnam.com
Library of Congress Catalog Number: 90-61805
Printed in U.S.A.

25 24 23 22 21 20

NOTICE: The information in this book is true and complete to the best of our knowledge. All recommendations are without guarantees on the part of the authors or the publisher. Parts and procedures may not be legal for sale or use in some states; check local laws. The authors and publisher disclaim all liability incurred in connection with use of this information.

CONTENTS

Exhaust headers, page 150.

Sheet-metal interiors, page 160.

Acknowledgments

Special thanks to Hydrocraft, Inc., for their cooperation in letting me photograph projects as I worked. Bob Sharp Racing and Mecca, Inc., allowed me to take some of the photos necessary to explain technical ideas. Eckold Machine Tools and Sykes-Pickavant supplied photos of their products. Thanks!

Metal shaping using the English wheel, page 85.

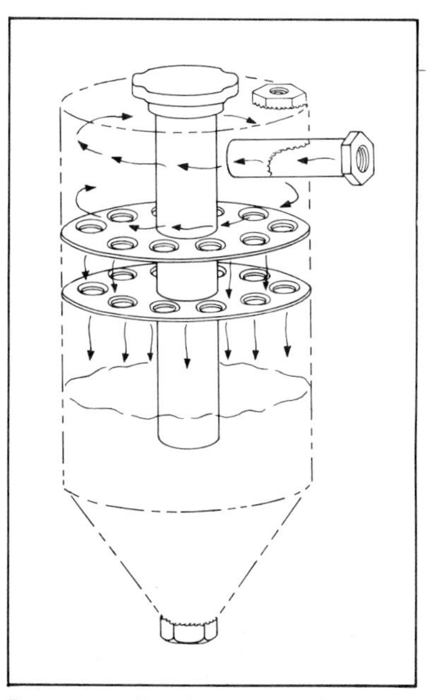

Dry-sump oil tanks, page 143.

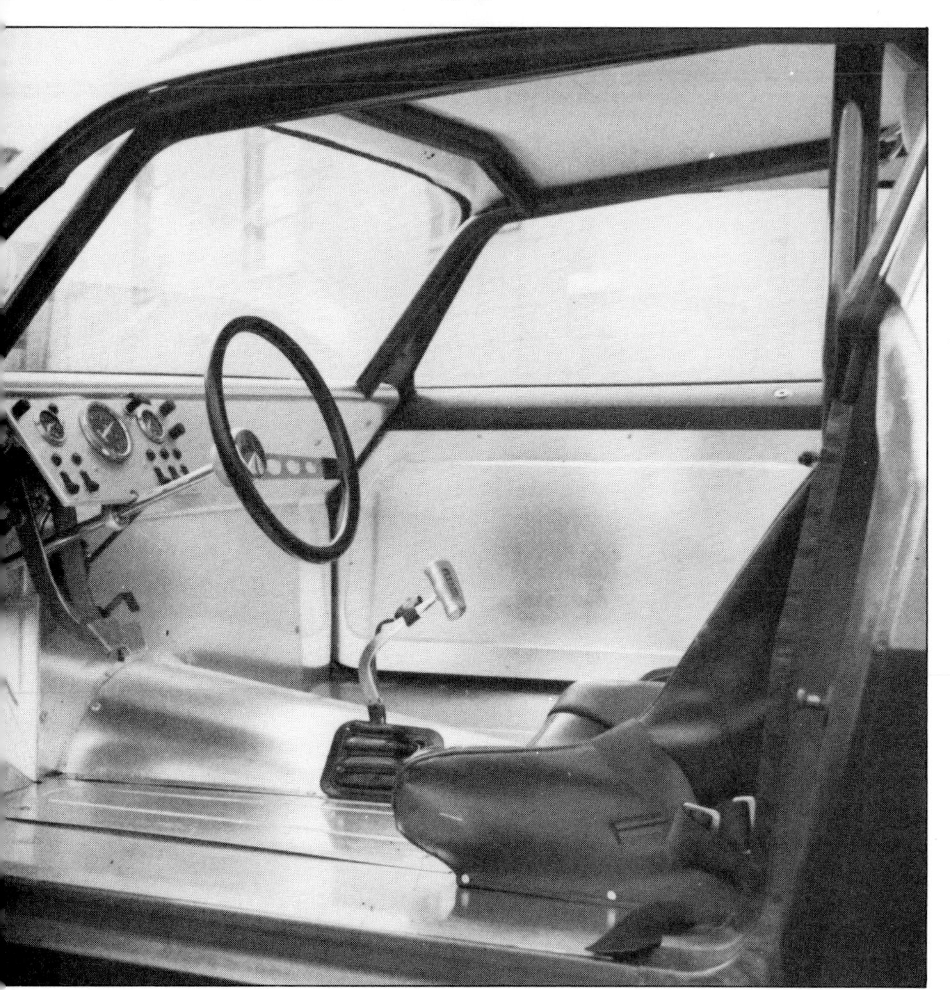

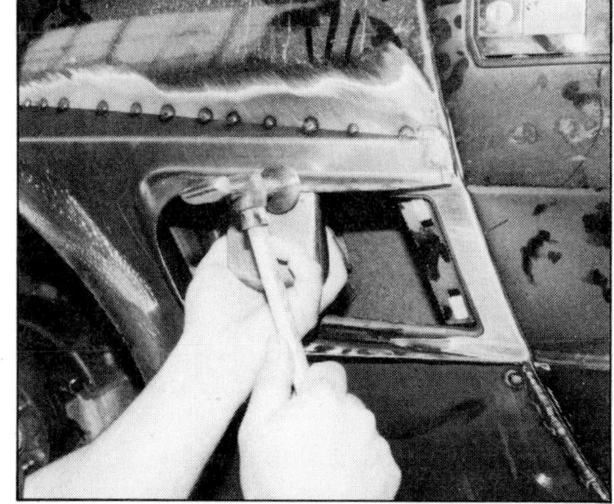

Hammers and dollies, pages 10 and 12.

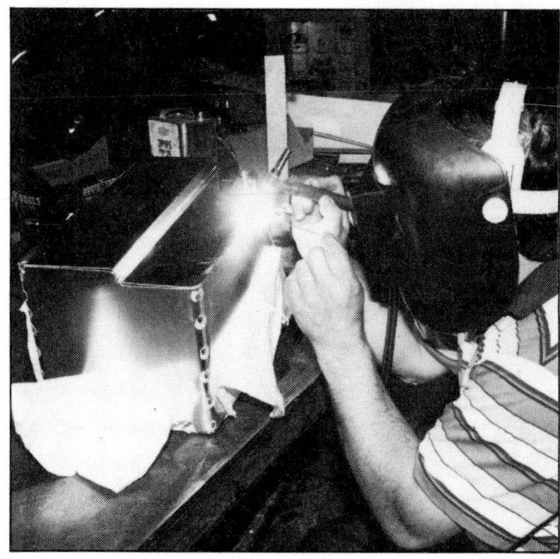

TIG-welding, page 68.

INTRODUCTION

The quality of a car's metal fabrication can determine its level of performance and appearance. It can be the difference between an outstanding custom or race car and an also-ran.

Many people appreciate custom metal fabrication, but few attempt it. This, I think, is because no one ever showed them how. Most guys ended their metalwork career right after getting a C-minus on their handmade ashtray in high-school metal shop. From that point, how things such as headers or fender flares were fabricated remained a mystery.

Here's where you can learn how to build those tricky parts. I'll fill you in on how to achieve good-looking, structurally sound metal parts. Clear explanations and lots of pictures and drawings will carry you a long way. You can start by making parts which require minimal experience and ability, then progress to parts requiring a higher level of skill as your confidence and ability grows. You can learn how to shape metal. You can build exhaust headers. You may even try a complete sheet-metal interior. It's up to you.

I show how to recognize quality construction . . . and exactly where and what to look for in a custom job. Whether you are a customer or a builder, you need this information. High-quality touches aren't necessarily complicated, but they add up to a better-looking, more-durable vehicle. I'll show you where mistakes can happen and how to avoid them.

All types of metalworking equipment and how to use that equipment, are fully discussed. I'll start with a description of the tools you'll need to fabricate a complete part or assembly, then show how it is installed. You'll see how to make those tricky compound curves from a flat sheet by hand or by machine. I will tell you how to weld—both gas and electric—and how to recognize good and bad welding.

You'll have the information necessary to go from a raw idea to a fully built fuel, oil or water tank. How to make the necessary measurements, then transfer these measurements into working patterns is shown. You'll have the information needed to determine the correct metal to use. Maybe your car needs roll bars. You'll see how they must work, and where and how to mount them.

If you like racy-looking spoilers or hood scoops, I show how to build those. How to make fender flares carries custom sheet-metal work further. You'll see how to do it all.

I've included a lot of drawings. I find that one good drawing can be more help than a pile of words. Many handy formulas are also included.

So, jump in! It's all waiting for you. I've had fun applying my years as a custom metal fabricator to this book. I think you'll enjoy pulling some ideas out of it. You may even end up doing some classy projects.

Pictured between Roger Penske, kneeling, and Mark Donohue is the author, Ron Fournier, with the Donohue Indy-Car crew. The occasion is the 1970 Indy 500. Much of the metalwork on this car is the product of Ron's skills. The words in this book are the product of Sue Fournier's skills, Ron's wife. Photo courtesy of the Indianpolis Motor Speedway.

SUNOCO
Special
66
GOODYEAR
Ford
CHAMPION

HAND TOOLS

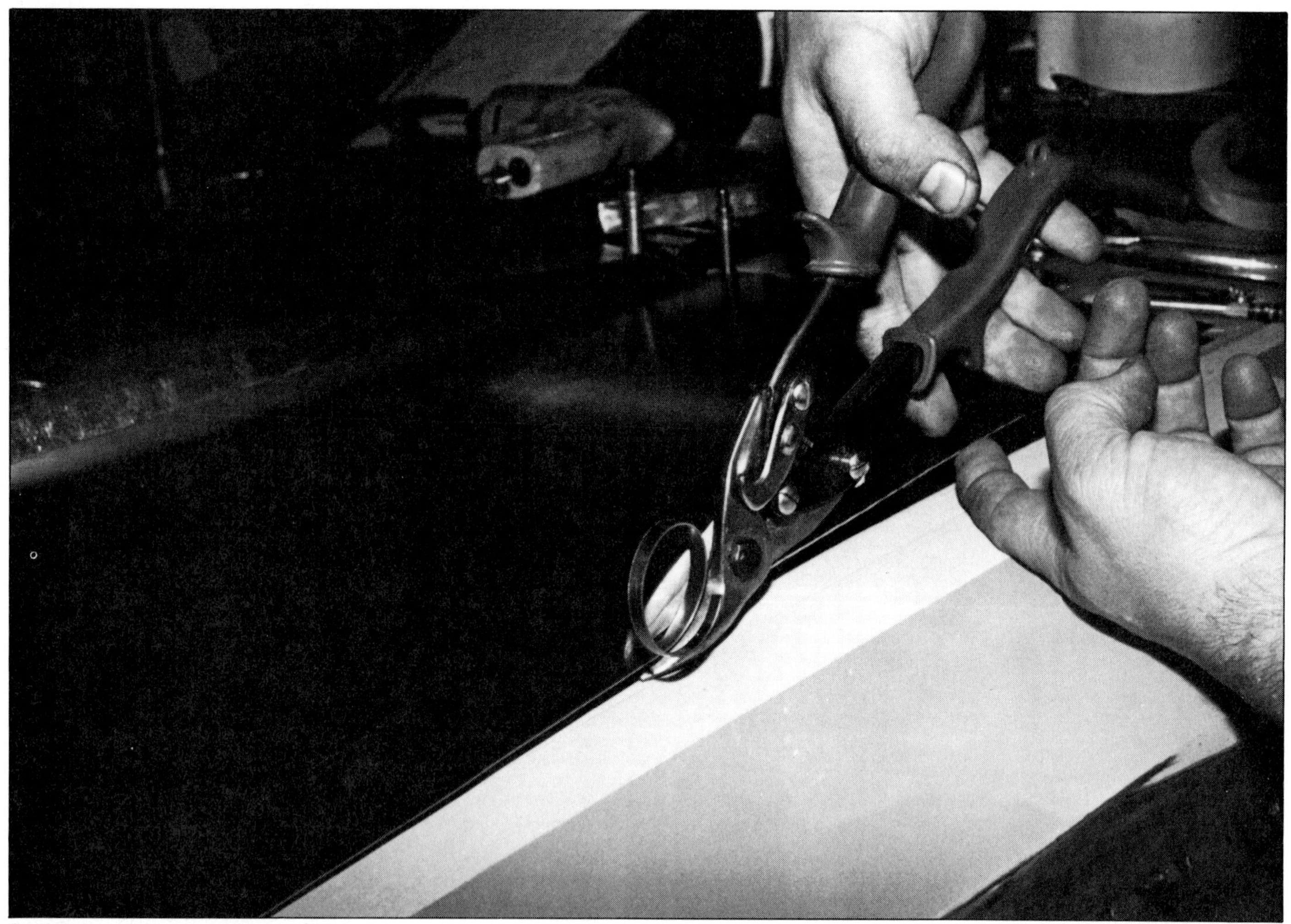

Basic hand tools are essential to accomplish any metal-working project. Trimming the edge of a hood panel with left-cut aviation snips.

Before you can build a car, or jump into any metal-fabrication project, you must have the tools and knowledge needed to "make the project happen." By choosing your tools well and developing your skills you'll find projects go easier. And the final product will be better in function and appearance. It is also worth taking time to select proper tools because the investment will be sizable. So, choose tools that will do the job best for the most reasonable cost.

Metalworking tools are separated into four groups. First, the basic hand tools are a must. There are some optional hand tools. Second, power tools are needed to form and install some parts. Third, large equipment is primarily used for cutting, bending and shaping. Fourth, welding equipment is needed. This chapter covers the first group—basic hand tools and optional hand tools.

BASIC HAND TOOLS

Basic hand tools are essential. They need not be the most expensive, but they must be very dependable. I'll discuss some of the brands I'm familiar with and make suggestions. Remember: Even with the best tools, the skill and respect with which they are used are most important. You'll gain skill as you experiment on your own projects by using similar projects in this book as a guide.

I've used all the basic hand tools on a daily basis for over 20 years for every kind of car-build work. I've learned from this to buy well-known brand-name tools simply because they are durable. Also, stick to basics. It is silly to buy tools that are beyond your capabilities.

Don't be a mad scientist and experiment on your car or a friend's car. If you must experiment with a new tool or technique, dream up a project for this purpose only. If the project ends up being best suited for the trash can,

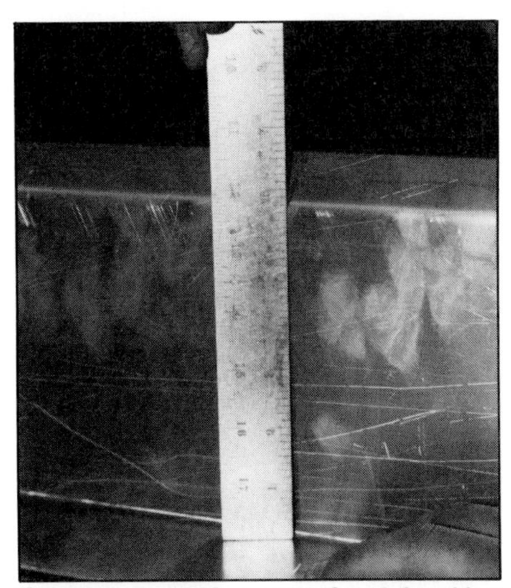

They are so much a part of metalworking that I can't imagine being without an assortment of various-length steel rules.

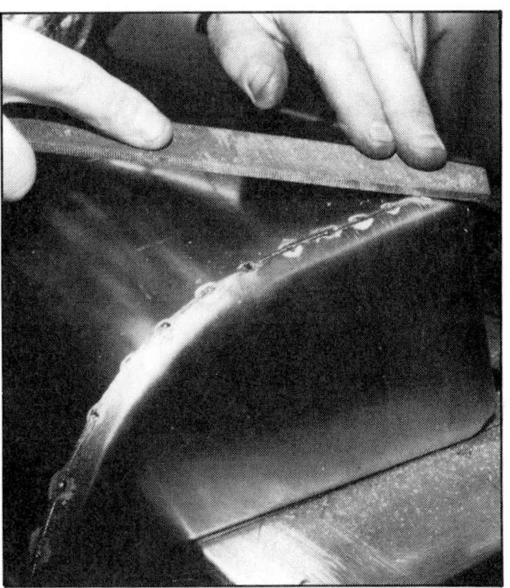

One of the hundred or so uses I have for a file is knocking the tops off of tack-welds. This lets me get a smoother final bead.

One of the many uses of a hacksaw: cutting exhaust pipes. Although it doesn't replace the bandsaw, a hacksaw sure is handy—and it's portable.

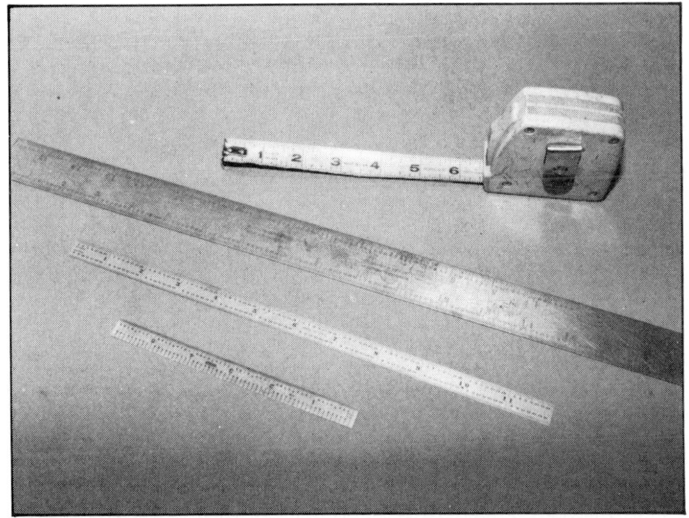

Measuring tools I use most often: 10-ft retractable steel tape; an 18-in. rigid steel rule; a 12-in. flex rule; and a 6-in. flex rule.

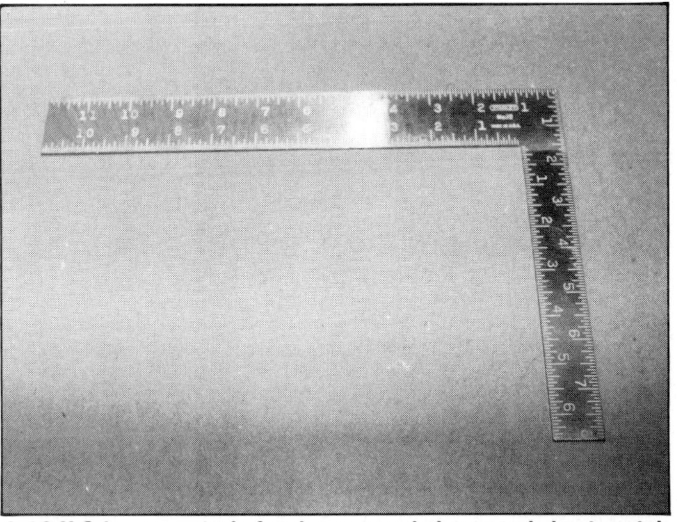

A 12 X 8-in. carpenter's framing square helps speed sheet-metal layout work. It's easy to read at a glance. I use this and larger ones.

you haven't destroyed your best friend's pride and joy or your friendship—and you will have learned something. Try again.

MEASURING

Don't underrate the value of a good ruler. It's the one basic tool you'll use in all metalwork. You can't make anything right without accurate measurements.

Measuring devices come in all shapes and sizes. They are also graduated in a variety of ways: millimeter/centimeter or inch/foot.

Recommendations—A *retractable flexible-steel tape measure* is handy because it is easily carried. I use mine constantly. Not only does it make straight measurements, a flexible tape is great for measuring around curves, bends or edges. Several manufacturers make a 10-ft flexible-steel tape measure. Choose a reasonably priced tape that's at least 1/2-in. wide. Narrow tapes are hard to read. And they wear out quickly.

Steel rulers—Steel rulers, ranging in length from 6 in. to 48 in., are readily available. Their widths range from 1/2 in. to 2 in. I use a variety of rulers. Short, narrow ones work well on small pieces. Long, wide rules are handy for the big pieces. Choose an assortment of sturdy rulers.

Squares—A *common square*—a special L-shape steel ruler—is a great timesaver. Marking or checking a corner for squareness is the main use of a square. It can also be used to measure two sides of a 90° angle simultaneously.

Common squares come in varying sizes. An 8 X 12-in. common square is the one I use most often. Larger squares are handy. If you can afford it, add one to your collection.

Several manufacturers offer good common squares. Tool suppliers usually carry a range from which you can choose.

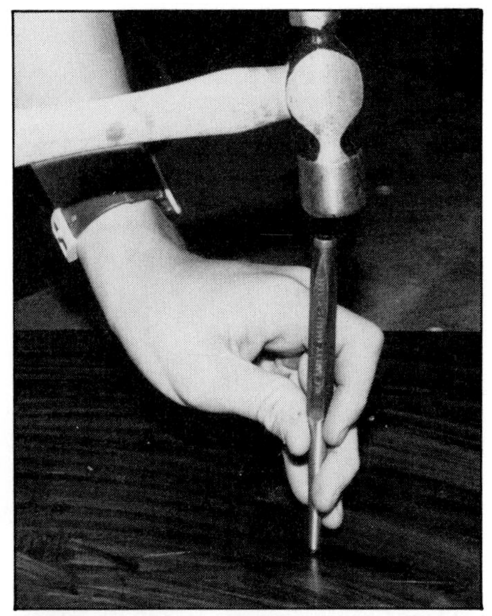

Center punching a metal blank. You'll need an assortment of various size center punches. Most tool companies sell four- to six-piece sets, and I recommend you purchase one of those. Photo by Michael Lutfy.

A good combination square is essential. Always use the factory straightedge to square off from whenever possible. Here, I'm using the 45-deg. side of the combination square to make a triangular bracket. Photo by Michael Lutfy.

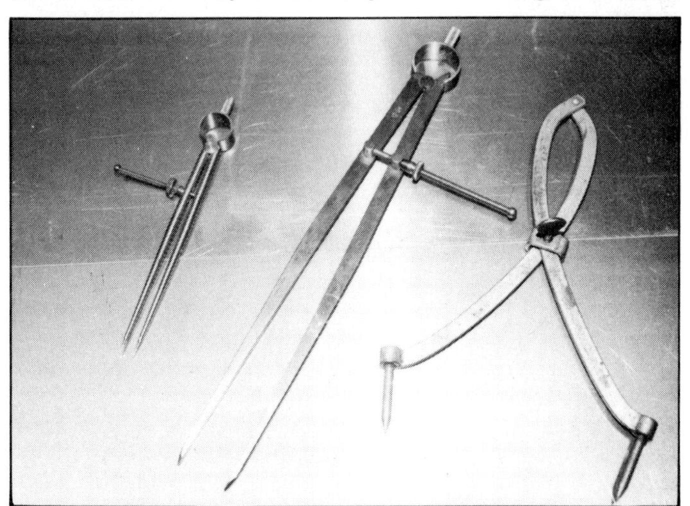

Dividers come in many sizes. I keep these in constant use.

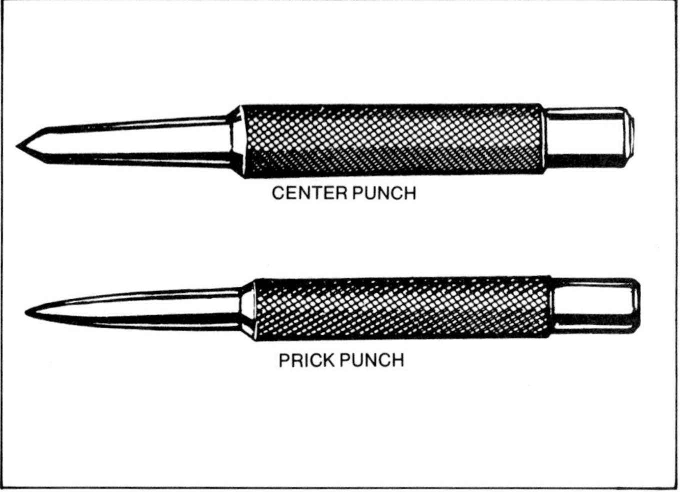

There is a difference. In addition to center punches, you'll also need a prick punch for marking metal.

Combination Square—A *combination square* is a two-piece measuring device. The *head* slides on a steel ruler. A clamping screw engages a groove in the ruler to hold the head at any measurement. Angles of 45° and 90° are made with the head. As a bonus, a bubble level is included in the head.

Combination-square sets come with two other attachments. A *protractor head*—a device for measuring angles—is used on the ruler to mark and measure all angles. The *center square*, in combination with the ruler, is mainly used to locate the center of a round section.

Recommendation—Definitely consider the combination-square set as necessary equipment for your toolbox. Sears' Craftsman set is a terrific value for the money.

Center Punch—Like a chisel, a center punch is a small hardened-steel tool, but with a very sharp, conical point. A center punch looks like a large nail or steel stake. The main use for a center punch is to make a mark, or small indentation, for drawing a circle or arc or to drill a hole.

To make a center-punch mark, strike the upper end of the punch with a hammer while holding the pointed end square against the surface you're marking. The resulting indentation in the metal will help locate the drill or compass. The drill won't be as likely to "walk off" target as it spins. A center-punch tip ground sharp at a 60° angle works best.

Recommendations—Center punches come in various sizes. General, Sears and Snap-on offer good center-punch sets with an assortment of sizes.

Dividers—Another tool you must have is a set of *dividers*. A divider is basically a compass with two arms, but with steel points at both ends rather than one. The steel arms are joined by a pivot at the end opposite the points. An adjusting screw or locking device bridges the legs to allow them to be spread, or opened, and held in a specific position for marking

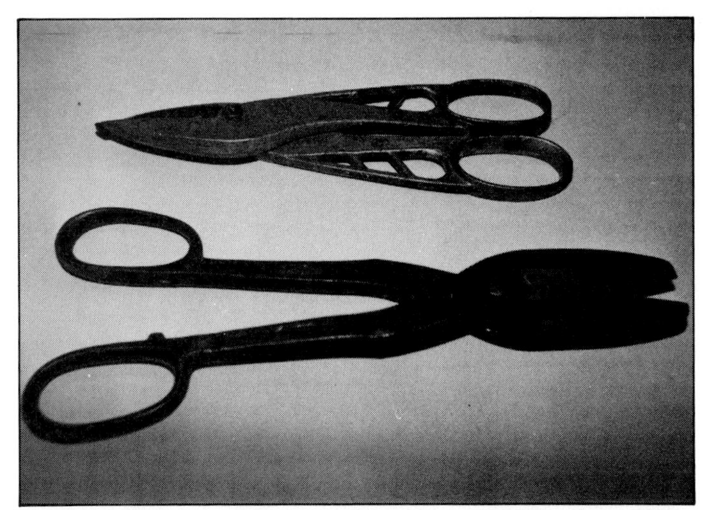

Tinner's snips are good for making long straight cuts on thin metal—such as 0.040-in. aluminum. I also use mine to cut cardboard patterns.

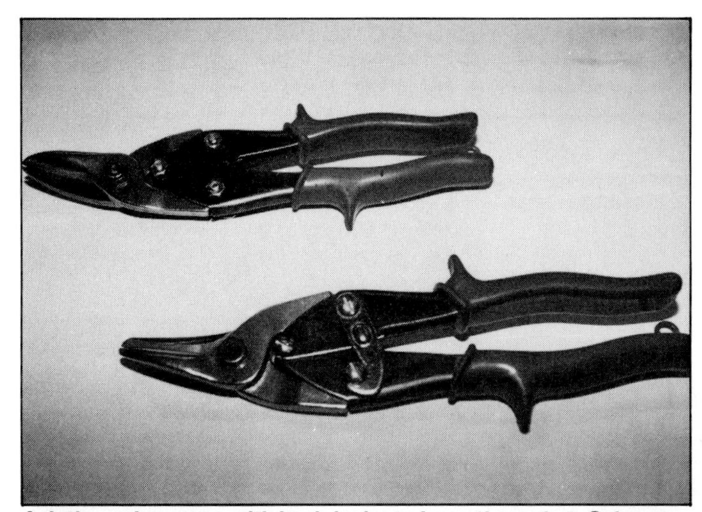

Aviation snips come with hook locks to keep them shut. Snips are marked RIGHT or LEFT. Right-cut snips at top have green handles. Left-cut snips at bottom have red handles. Not shown are straight-cut snips that have yellow handles.

Right-cut aviation snips are for cutting to the right of a curved line, not for the right hand.

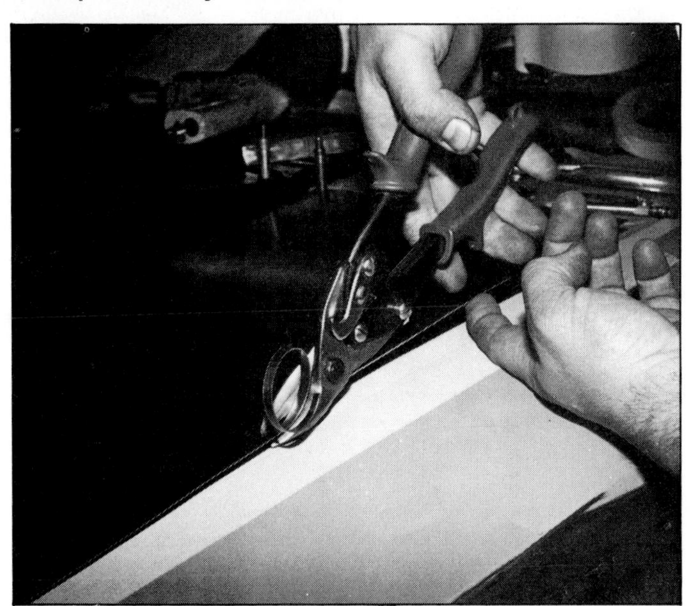

Left-cut aviation snips cut away material to the left of a curved line.

distances or drawing arcs.

Dividers are mostly used to make arcs and circles by lightly scratching the metal rather than drawing as is done with a lead or ink compass. They are also used for transferring dimensions from a ruler or from a part to the metal being worked.

To draw an circle or arc with a pair of dividers, first mark the center of the arc with a center punch. After setting the radius, position one point of the dividers in the center-punch mark. With the other point *scribe*—or scratch—the arc by rotating the dividers. Lean the dividers slightly in the direction of rotation and apply a slight downward pressure. For best results the divider points must be sharp.

Recommendations—My collection of dividers ranges in size from 3 to 12 in. I recommend you start with a 6- or 8-in. set. This would suit most of your purposes. And you can add dividers as you need them. General, Sears and Starrett make good dividers.

METAL SNIPS

After the part has been laid out, the first step is to cut the metal into pieces. The kind of cut determines what cutting tool must be used. A straight cut requires only the use of *standard tinner's snips.* They look like a pair of common scissors, but with long handles and short blades.

Tinner's Snips—Tinner's snips come in several sizes. I suggest a medium-size set of tinner's snips with about a 12-in. overall length for your toolbox. There are also *straight aviation snips.* Don't choose them for straight cuts because tinner's snips will do the job better with more leverage, and with longer, cleaner cuts.

Aviation Snips—A small, curved, intricate cut requires the use of *aviation snips.* A standard in the metalworking industry, aviation snips are so named because they were developed in the early days of the aircraft industry. There was a need for compact, easy-to-handle snips. Aviation snips filled this need.

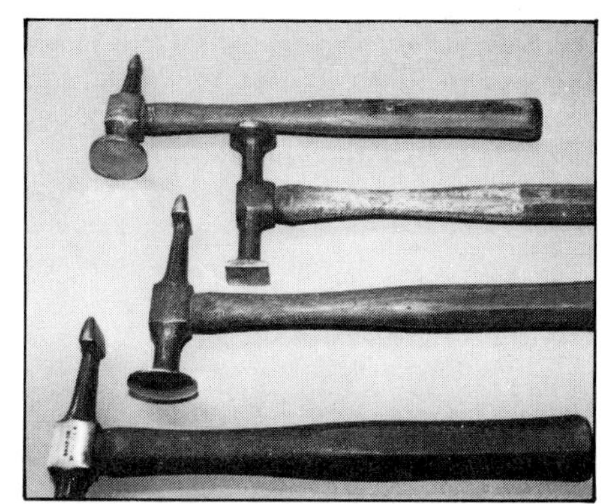

Here's are a few of my body hammers. Each has its own special application.

Seems like the Proto 1427 body hammer is always in my hand. It's versatile and I'm always reaching for it. Note the replaceable fiberglass handle.

Unlike common tinner's snips with a single pivot, aviation snips have *compound leverage*—multiplied leverage—for increased shearing power. Fine *serrations*—teeth—on the cutting edges of both jaws provide good control when cutting. See the accompanying photo.

In addition to straight snips, two other types of aviation snips are available: *right-hand* and *left-hand.* They sound like snips for left- or right-handed aircraft metalworkers. They are neither. Right-hand aviation snips *remove metal to the right for cutting a curved line.* Left-hand aviation snips are, you guessed it, snips used to cut away sheet metal to the left.

Special snips for cutting curves to the right or left are necessary because, unlike paper, metal is heavy and stiff. The material being removed doesn't easily move out of the way of the handles as you make the turn. Consequently, special snips are needed. The action of the jaws bends or raises the section being removed out of the way. This section is on the right for right-hand snips and on the left for left-hand snips. So you'll need both right- and left-hand aviation snips to follow a curved line that changes direction.

Care of Snips—Aviation snips are expensive. Avoid cutting through welds or metal with scaly surfaces. This will ruin the snips or rapidly accelerate blade wear. Do not use them to cut steel thicker than 0.063 in. Continuous use on stainless steel will quickly dull the blades.

Save your snips by using a bandsaw for cutting thick steel or any stainless steel. To prolong their life you should always keep them clean and well lubricated with light oil. I wear out about three sets of snips a year, even though I give them the utmost care and respect.

Recommendations—Aviation snips are available from several manufacturers. I've found Wiss snips to be the most durable and expensive. Craftsman, Snap-on's Blue Point and

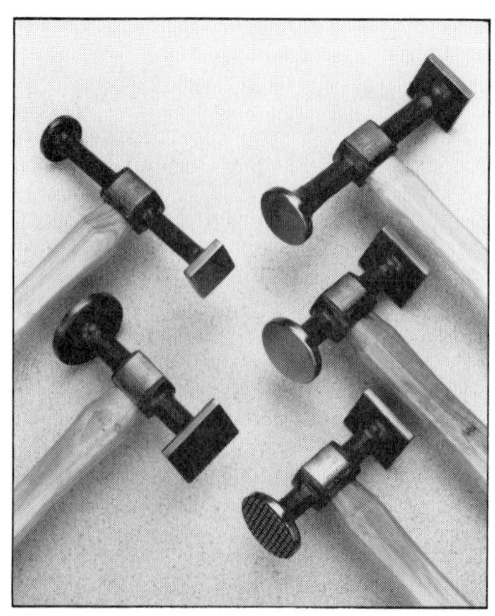

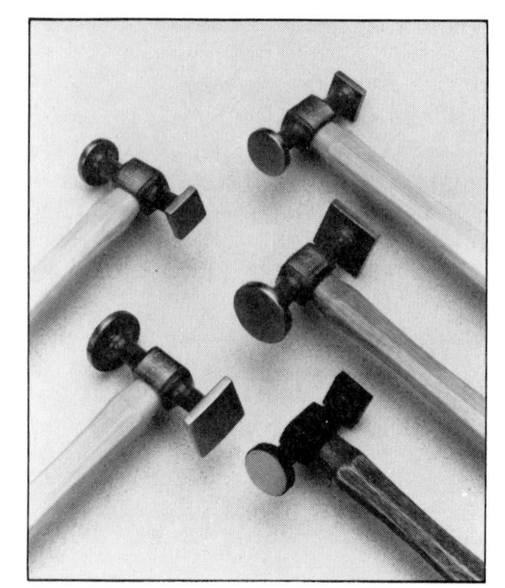

This assortment of body hammers is made by Sykes-Pickavant, a British tool manufacturer. They are exceptionally well made and will last a lifetime if taken care of properly. They are distributed in the United States by U.S. Industrial Tool and Supply Co. For the address, see the Supplier's Index on page 174. Courtesy Sykes-Pickavant.

Century offer satisfactory quality for general use.

HAMMERS

Body Hammers—There are three basic body hammers you'll use all of the time. First is a *dinging hammer*, a body hammer with both ends of the head ending in a full flat surface. It is used for flattening. Both ends work equally well. The dinging hammer is available in two sizes. I suggest starting with the smaller size.

Second, there is a *pick-hammer* style body hammer. One end of the head is flat like the dinging hammer, and one end is pointed like a pickax. It's used to bring up low spots in sheet metal. Pick hammers come with different-length *picks* of varying sharpness.

Third, a body hammer with a curved face, sometimes called a *shallow-domed face*, is needed to form curved surfaces. I use the shallow-domed hammer with the *cross-peened* end most often. I find the combination of the shallow-domed and cross-peened ends allows me to work a continuous curve into a smaller V-shape most easily. A cross-peened/ shallow-domed body hammer is often needed to put the finishing touch on a part. My favorite is Proto's 1427 cross-peened body hammer, shown on page 10.

Sometimes I feel like all I need is my shallow-domed/cross-peened body hammer. But then I soon find myself searching for a special hammer, one with a particular shape for a specific application. Regardless, you can begin with the three basic hammers I just discussed. As your skills increase you'll develop a need for a larger selection of hammers.

Let's take a look at some of the specialized hammers I use that make certain types of metalwork easier.

Additional Body Hammers—A body hammer with a *large-domed face* is used to stretch metal down into a *bag*, or into hardwood. I show this on page 82. This process is called *hollowing*. Hollowing is a handy way to stretch a deep curve in either steel or aluminum. A large domed-face hammer will be a good addition to your hammer collection once you have become proficient with the three basic body hammers.

Another group of body hammers you may find interesting as your skill grows is the *curved-face* hammers. This hammer face is cylindrical in shape. The faces come in different radii and the curved heads are usually 90° from one another. They are terrific for forming concave surfaces.

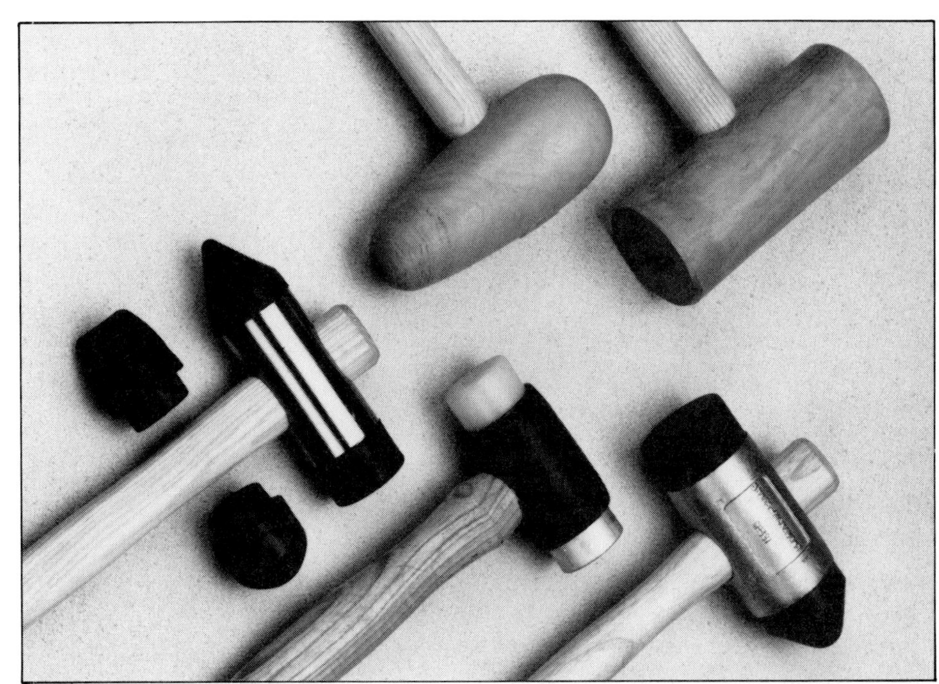

A good selection of mallets is a must. Sykes-Pickavant offers some of the best around. If you use them carefully, they won't mar the metal surface of your project. Courtesy Sykes-Pickavant.

I take a lot of kidding about the number of hammers I've acquired over the years. Yet I find not one collects dust, and I wouldn't want to part with any of them.

Recommendations—Body hammers are available from Snap-on, Mac, Vulcan, Craftsman, Proto and Sykes-Pickavant. Not all manufacturers carry all the same styles, however. For instance, Proto's 1427 hammer is sold, but not manufactured by, Snap-on. In general, I think Sykes-Pickavant offers the best value, especially in terms of cost and selection.

Mallets—Like hammers, mallets are used to form metal. However, mallet faces can be reshaped to tailor their form to a particular job. Another advantage is that mallets are less likely to mar metal than hammers because of their softer material. They are often used to *shrink* metal.

Rawhide mallets have a head made of hard rolled rawhide and a wood handle. The only way a mallet can be used to stretch metal is by striking the metal over something soft—such as a *shot bag*, or a *sand bag*. I will describe these next.

Wooden mallets and rawhide mallets serve much the same function. Rawhide mallets are heavier than wood. Some are even weighted.

A shot-filled leather bag is useful for stretching metal. You may have to go to a leather shop to have one custom-made. Shot bags go under the metal to absorb hammer blows. By conforming to the shape of the metal, bags allow metal to stretch evenly. Photo by Michael Lutfy.

Wooden mallets are lighter and made from very hard woods. Both types of mallets need to be lightly oil-treated so they'll stay in good condition. You'll grow into using mallets as you gain experience with hammers.

Sometimes mallets are used to shrink metal. To shrink with a mallet, you must use the mallet on metal placed on a hard surface—such as a metal work table. This technique is called *cold shrinking*. I'll describe this process when discussing metal-shaping methods.

SHOT BAGS

Shot bags are hand-made leather sacks filled with #9 birdshot. Just as you need a hammer or mallet to strike

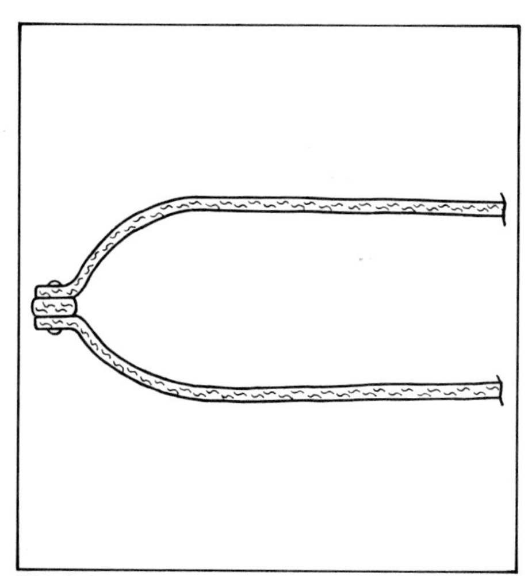
Thick leather ring between the seam of my shot bag has survived for 10 years. The seam is stronger than if the leather halves were just sewn together.

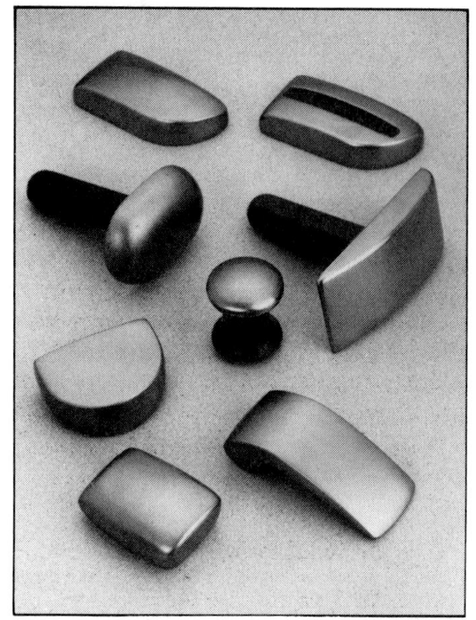
All of these dollies are manufactured by Sykes-Pickavant. I especially like the dollies with post handles. This feature makes them easier to grasp and use. Courtesy Sykes-Pickavant.

My favorite dollies: toe, heel, and an egg dolly. The names describe the shapes. Surfaces should be polished smooth because any irregularities transfer to the metal being formed. Keep each dolly in an old sock to protect it.

the metal, you need a surface *beneath* the metal to help form it. This is what shot bags are for.

Shot bags provide a surface into which you can pound and evenly stretch metals. A shot bag absorbs the impact of a hammer according to how densely the bag is filled with shot. A loosely filled circular shot bag, about 2-in. thick and with a 14-in. diameter is ideal for *slapping* out gradual shallow shapes over large areas. Slapping is discussed beginning on the following page. A densely filled rectangular shot bag, about 2-in. thick, measuring 6 X 10 in. is especially good for hollowing smaller deep areas.

There is an old metalworker's story about the guy who could hand-form an entire car body using only sheet metal, a hammer and an old tree stump. Shot bags are so effective that I think they must have been inspired by the tree stump in the story.

The first shot bag I used belonged to someone else. It worked so well I wanted some of my own. I ended up at a saddle maker's—an old-timer in Texas. He laughed, then he told me he would use thick, soft leather. The old-timer showed me special stitching he thought would never tear under years of pounding. He said if I kept the shot bag well oiled it would have a long life. He was right. The shot bags he made for me are still in good condition after years of constant use.

A shot bag isn't something you can pick up at your nearest hardware store. You'll have to do as I did. Find someone to make it. Maybe you'll be luckier than I was. I did the obvious and inquired at several shoe-repair shops, but to no avail. They all refused, claiming they lacked suitable equipment to do the job. You might find a leatherworker who will try.

Sand bags are freebie tools you can make yourself. Start at your bank. See if they'll give you several canvas coin bags. Some banks may charge a small fee for these. I got mine free.

With the canvas bags in hand, fill them with clean, dry sand. Leave enough canvas to allow for a double-fold at the top. I used a heavy needle and carpet thread to sew the top shut. Bingo—sand bag.

I frequently use sand bags to weight down large, awkward aluminum panels I can't clamp to the workbench. Occasionally I use one like a shot bag for forming metal. Sand bags are cheaper and nearly as useful. But unlike a good shot bag, a sand bag will split or leak sand if it's used very often for forming. Otherwise they'll last pretty well. You might try using a sand bag for forming a piece just to see how it holds up for you.

DOLLIES

Dollies are usually made of solid cast iron, formed into shapes and polished to a smooth finish. They are used to back up one side of the metal being worked on as a body hammer strikes the opposite side to form three-dimensional shapes.

Try to think of a dolly as a hand-held anvil, moved by hand and manipulated in coordination with the hammer to form the desired shape. Dollies give you control when shape is important. They are the necessary partners of the body hammers. The shape and weight of the dolly encourages the metal to form in a specific way under the impact of a hammer.

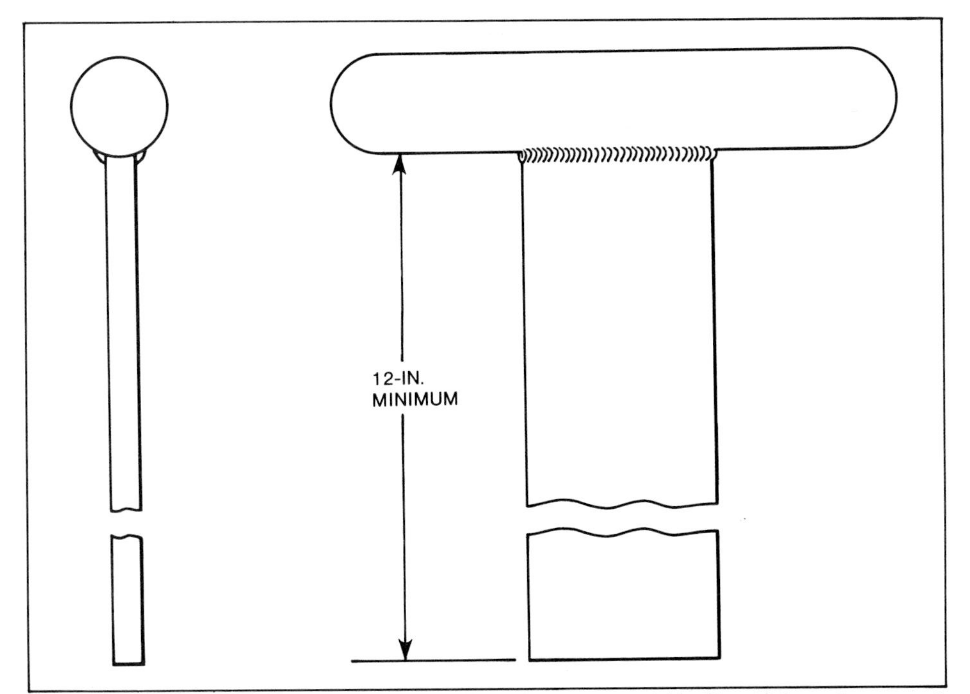

Like apple pie, T-dollies are homemade and just about as good. You'll want to make several different sizes using the same basic construction.

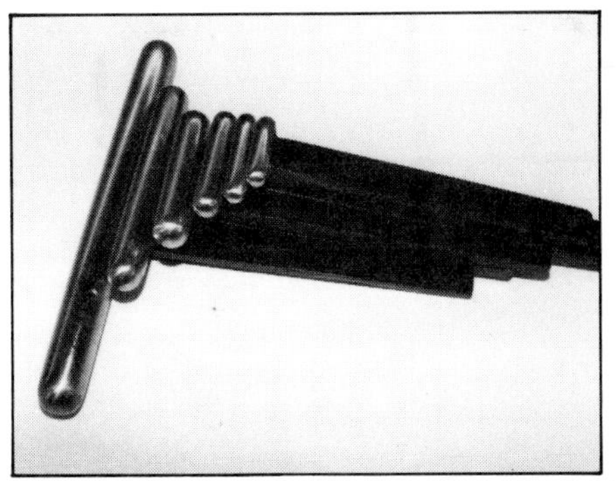

Some of the T-dollies I've made over the years. They can be hand held or secured in a vise.

T-dolly secured in a vise makes a good, firm working surface to form sheet metal. The T-dolly, like other body tools, should be kept polished and scratch free.

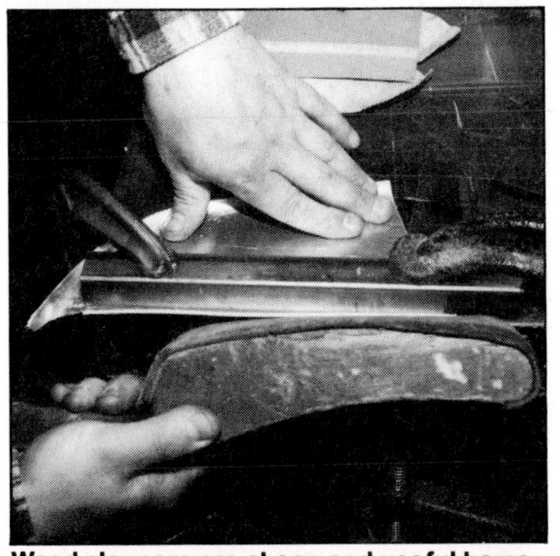

Wood slappers are cheap and useful homemade tools. What more could you ask? They should be covered with leather when used to work aluminum.

Heel, egg, and universal dollies provide good beginning shapes for your toolbox. Note the photo of these dollies. With these three dollies you can form a surprising number of curves, vees, bends, and shapes.

Bending *tight angles* is a common use for the straight side of the *heel dolly.* By tight angles I mean bends with little or no radius. The opposite side of the heel dolly is excellent for similar tight bends on a curved shape.

The *egg dolly* gets its name from its egg-like profile. It's an excellent dolly for metal shaping. An egg dolly combines a large rounded end, a sweeping curved top and a smaller rounded end. Its flat bottom provides a fourth convenient surface. Because these four surfaces offer versatility, the egg dolly is frequently used.

The *universal dolly* belongs in your toolbox because of it has a wide variety of usable shapes: sharp edges, flat surfaces, and several curved ends. Two attractive features of the dolly are its ease of holding and its relatively heavy weight. Weight is an advantage when working with steel.

T-dollies are great homemade tools which are fabricated from two pieces. A T-dolly, which looks like a capital T, consists of the dolly and a *post.* The dolly portion, the top part of the T, is made from a round bar of cold-rolled steel cut to the desired length. Each end of the bar is ground to a spherical shape and polished.

Used for holding the dolly, the post, or the vertical part of the T, should be a minimum of 12-in. long. Cut it from 1/4-in.-thick, 2-in.-wide, flat hot-rolled steel. Weld the post to the dolly as pictured.

The T-dolly is held in a vise while being used. The 12-in.-long post gives you good distance for clamping the T-dolly and leaving adequate working area. They can be used for many shaping jobs. I started out with two T-dollies and quickly found them very useful. They are so easy to make I frequently make a new one for a particular project. I commonly use a T-dolly for rounding tank edges that will be joined later by welding.

Care of Dollies—Dollies must be kept clean and oiled to prevent rust. Avoid hitting a dolly directly. A hammer will leave an indentation that will transfer to metal being worked. This goes for all imperfections. So keep your dollies sanded to a very smooth finish. Just as indentations in a dolly telegraph into metal being worked, smoothness is also transferred.

SLAPPERS

Wooden slappers are used like a hammer or mallet to shape metal. They are made from a hard wood, such as oak, and sometimes covered on the slapping side with leather. Unlike a hammer or mallet, a slapper has several curved surfaces for forming the desired curve in the metal being worked.

You can make your own slappers. It's not difficult. Draw a pattern on cardboard similar to that pictured. After cutting out the cardboard, transfer the pattern to a 2-in. wood block.

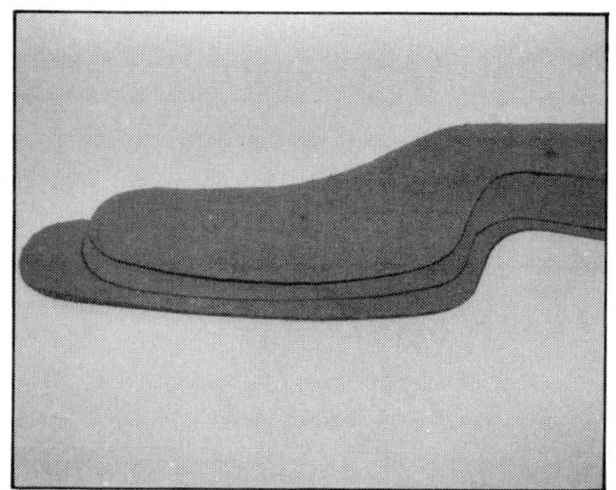

Make three full-size cardboard slapper patterns about about 10-in. long. Three different curves will give you a good start.

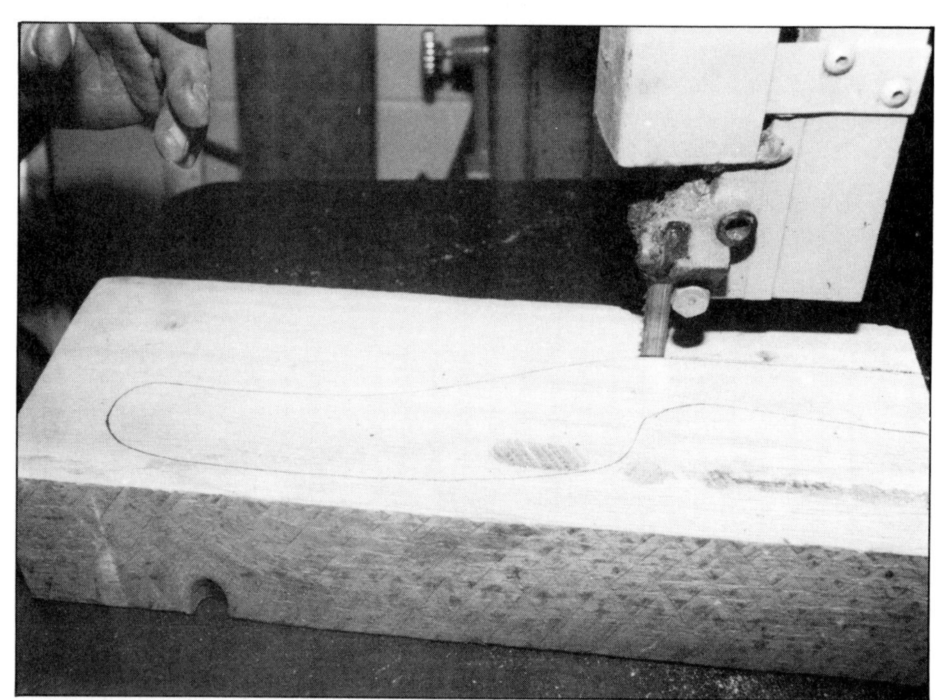

Cut out a 1-1/2—2-in.-thick hardwood blank using your pattern as a guide. Make sure you sand it smooth, all around, especially the handle. You don't want a handful of splinters. See page 84 for more details on slapper construction.

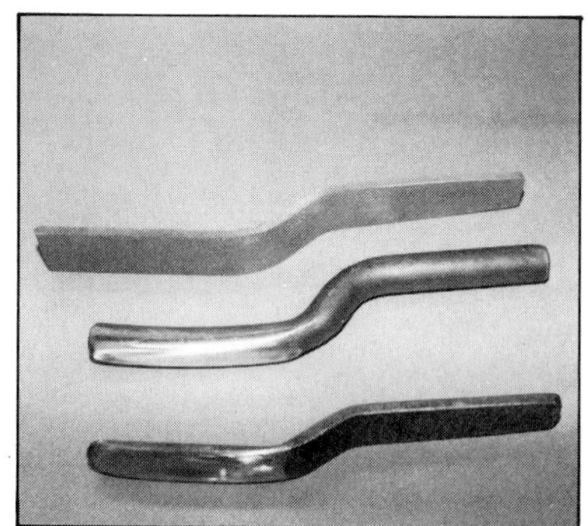

Steel slappers are used to smooth metal after forming and welding. The two lower slappers are homemade. The top slapper is a shrinking slapper made by Snap-on. You can make a slapper from an old file like the lower one. After bending, grind the teeth off to make it smooth.

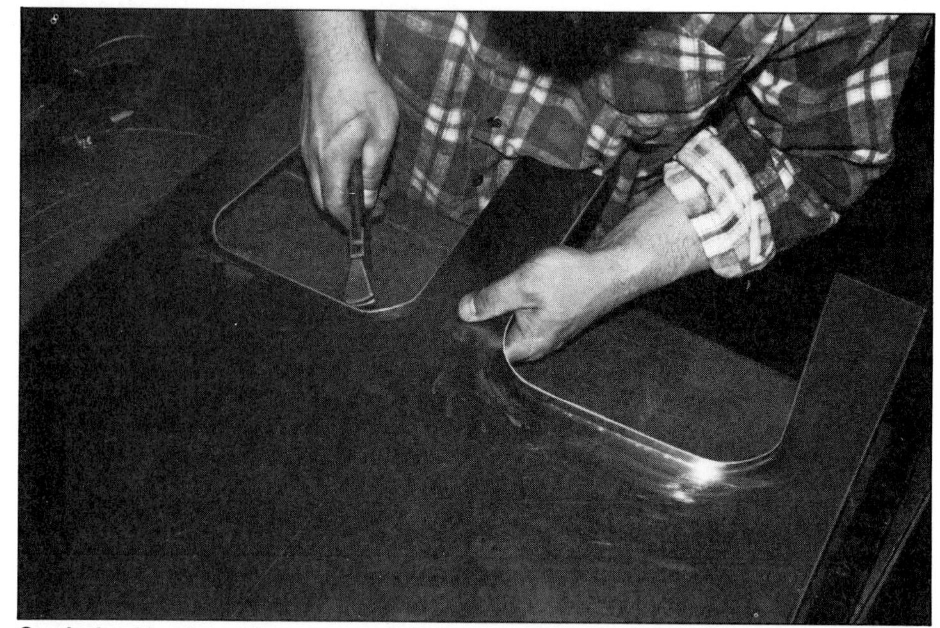

One-inch-wide glass pliers are great for bending flanges. They come with straight-jaw edges, but you can reshape them by grinding like I've done with these. A curved jaw will bend a flange around a curve. Make *sure* the jaws have no sharp edges or burrs.

Any hardwood such as oak or rock maple makes a good slapper. Cut out the slapper, attach a leather cover if necessary, and you are ready to slap away. Leather covers are needed when working with aluminum. Steel can be slapped with bare wood. Note the nearby photo of slapper construction.

Slappers in the sizes and shapes indicated by the cardboard patterns in the top photo have a wide range of possible uses. Different curved slappers do the job in different ways. You'll have a better idea of what best suits your needs after making and using your own slappers.

Steel slappers are used for *planishing*, or to make metal level and smooth by pounding. Planishing is done to a welded or shaped area. Slappers let you strike the metal in a completely different way than you can with a hammer or mallet.

Planishing is done by lightly striking the upper surface of the metal with a slapper. At the same time, a long dolly is held against the underside of the metal, *parallel* to the slapper's face. Lots and lots of soft slaps are better than a few hard ones. Skill, not muscle, is needed to planish metal successfully. Think of steel slappers and long dollies as tools for blending and smoothing an uneven shape by using many gentle pats.

GLASS PLIERS

Glass pliers are a tool from the *glazier's* trade. They are available from glass-supply houses. I use glass pliers to bend metal in small areas. They are very good for working small metal pieces. I consider them a very small hand-held sheet-metal *brake*. Brakes are discussed on page 32.

Recommendations—Glass pliers have no serrations, so they won't scratch the work. Glass pliers come

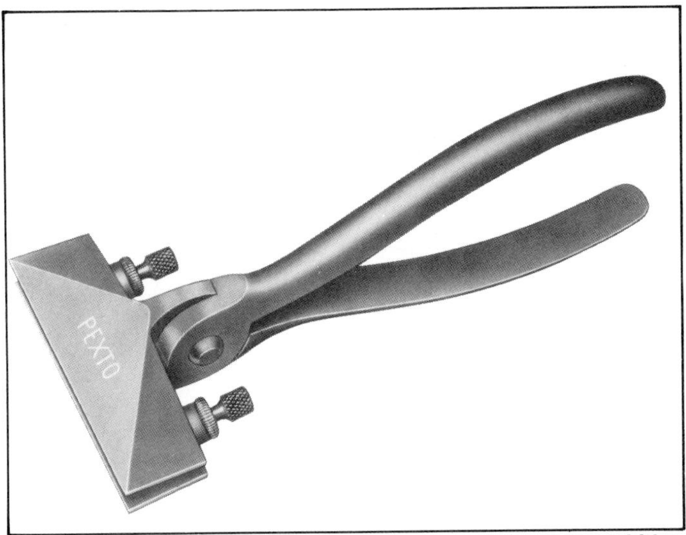

Hand seamers are a bargain. Pexto has a variety of jaw widths available. Photo courtesy of Peck, Stow & Wilcox, Co.

Vise-Grips are a high-quality, frequently used tool. They come in lots of styles and sizes. Here are two holding a spoiler panel in place while another part is being developed.

You'll run into situations where you'll need more C-clamps. They keep things in place while a project is being assembled. They were very helpful during the construction of this Indy-Car tub.

with several jaw widths. I recommend the 1-in.-wide jaws as most useful.

HAND SEAMER

The heating-and-cooling industry has provided a tool similar to glass pliers. *Hand seamers,* similar to but unlike glass pliers, are intended for sheet-metal use. Their jaws are wider—up to 6 in.—and some feature compound-leverage handles. Like glass pliers, the hand seamer is also a small, portable sheet-metal brake for small bending jobs. Hand seamers are carried by sheet-metal-tool supply houses. Two popular brand names are Pexto and Malco. The Malco S-2 hand seamer with 3-1/4-in.-wide jaws is my favorite.

C-CLAMPS

A C-clamp is a tool for securing metal pieces in place while they are being welded, formed or assembled. The name originates from its shape—the frame of a C-clamp looks like a C. A C-clamp has a screw that bridges its opening. This opening is adjusted by a screw which is turned and tightened by hand.

Some C-clamps have steel ball-and-socket swivels at the end of the screw. This offers a better grip on uneven surfaces. Quality C-clamps have drop-forged, heat-treated steel frames.

I use C-clamps with openings from 1 in. to 10 in. It's not uncommon to use several C-clamps on one job. Easy to use, C-clamps speed up the work and are like extra hands. They also help control warpage during welding.

Care of C-Clamps—C-clamp screw threads must be oiled to protect them. Arc welding near a C-clamp will ruin it. The weld spatter sticks to the screw threads, causing the screw to jam. Overtightening a C-clamp will bend the frame and cause misalignment.

VISE-GRIP PLIERS

Vise-Grip pliers are for clamping things. They are adjustable, locking pliers and should be in every metal fabricator's toolbox.

Vise-Grip pliers have a variety of uses. They can clamp, lock, twist or hold metal. They come in several

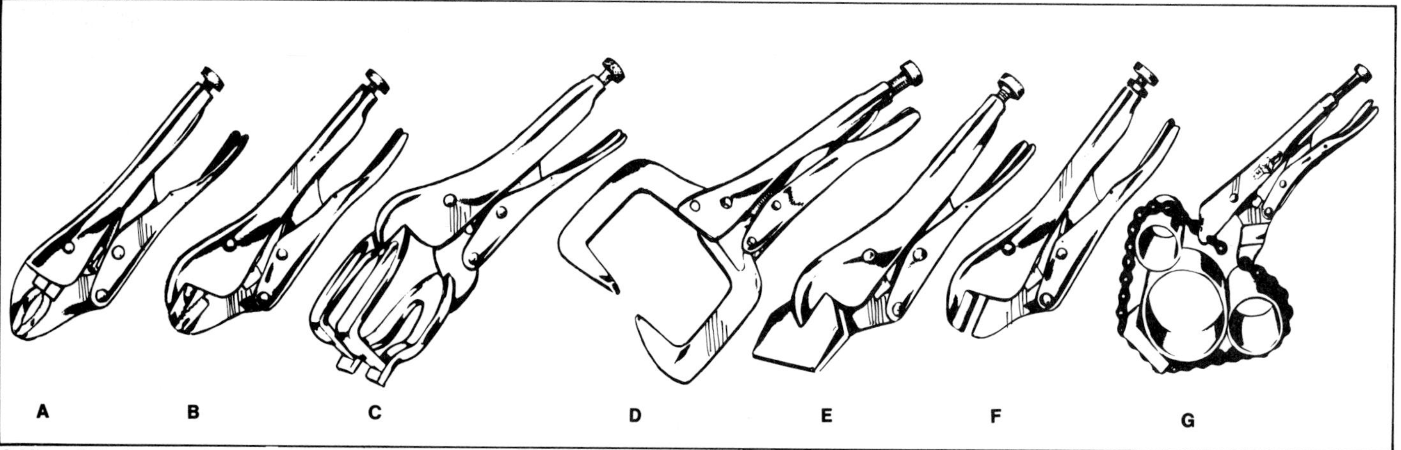

A Vise-Grip for nearly every clamping need. A: wire cutter; B: wire cutter; C: welding clamp; D: C-clamp; E: bending tool; F: hose/tubing pinch-off; G: chain wrench. Drawing courtesy of Snap-on Tools Corporation.

Properly cared for, hole saws last a long time. This hole saw is cutting a hole through 0.063-in. steel sheet. Clamp small pieces tightly when using a hole saw. Note the plywood underneath to prevent damaging the drill-press table.

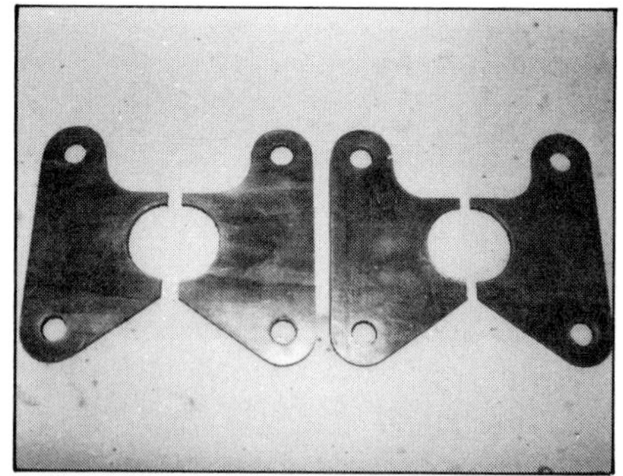

Hole saws are time and work savers. By grouping these brackets together and cutting radii with a hole saw, I saved bandsaw time. One center hole was used for a pair of brackets. The parts are street-rod front-axle brackets made of 3/16-in. mild steel.

styles. The most common Vise-Grip plier is the straight-jaw style. However, I find three other types more useful for metal fabricating; the C-clamp, sheet-metal-tool and welding clamp styles. I use these three styles more often than any other type.

I find the sheet-metal-style Vise-Grip handy because it is similar to the hand seamer. It has the added locking feature, plus it can be adjusted for different material thicknesses. These features make bending metal easier. If your budget is limited, a pair of sheet-metal-style Vise-Grips could substitute nicely for hand seamers.

Care of Vise-Grips—Because Vise-Grips depend on an internal spring to operate properly, heat will damage their usefulness. If the spring is damaged, it can be replaced to restore the Vise-Grips to good health. The adjusting screw in the handle should be oiled.

HOLE SAWS

Hole saws are a must. Basically a saw blade bent in a circle, the hole saw is a cutting tool which can be used in an electric drill or drill press for cutting round holes. It has a *pilot* drill in the center. The steel hole saw will cut aluminum, steel, and fiberglass and wood.

Many sizes are available. You shouldn't have any trouble finding the particular size hole saw you need. Buy individual hole saws according to the holes you want to make. For example, I use a 3-in.-diameter hole saw to make holes for installing Stewart-Warner gages.

Care of Hole Saws—Overheating a hole saw is a common mistake. Overheating anneals or softens the saw teeth, making it useless. On the other hand, keeping a hole saw cool and well-oiled will keep it useful a long time.

KNOCK-OUT PUNCHES

Knock-out punches, also known as *chassis punches*, are simple three-piece round-hole punches. A knock-out punch cuts a clean, accurate hole in steel up to 0.090-in. thick and in aluminum up to 0.125-in. thick.

A knock-out punch consists of a *punch, die* and *screw*. The punch does the actual cutting. The die backs up the punch on the opposite side of the material being worked. The screw draws the punch and die together to cut, or *punch,* the desired hole.

Before using the knock-out punch, you must first drill a *pilot hole*—a guide hole—of the same diameter as the punch screw. The pilot hole must have a common center with the hole you're going to punch. After the pilot hole is drilled, the die is placed over the pilot hole with the screw inserted through the hole. The die is placed on the opposite side of the material and

Greenlee radio chassis punches cut accurate, burr-free holes in sheet metal. The smallest punch cuts a 1/2-in. hole: The large punch cuts a 2-1/2-in. hole. All you'll need is a pilot hole for the bolt.

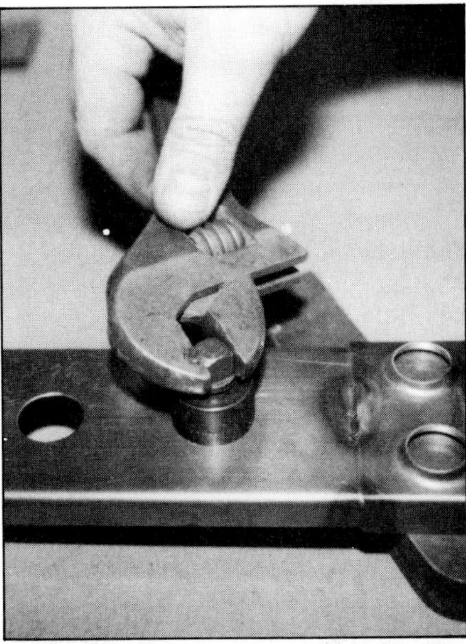

Tightening the chassis-punch bolt. The punches work equally well on aluminum or steel. Mine are still sharp after years of use. Note the cleanly punched hole at left.

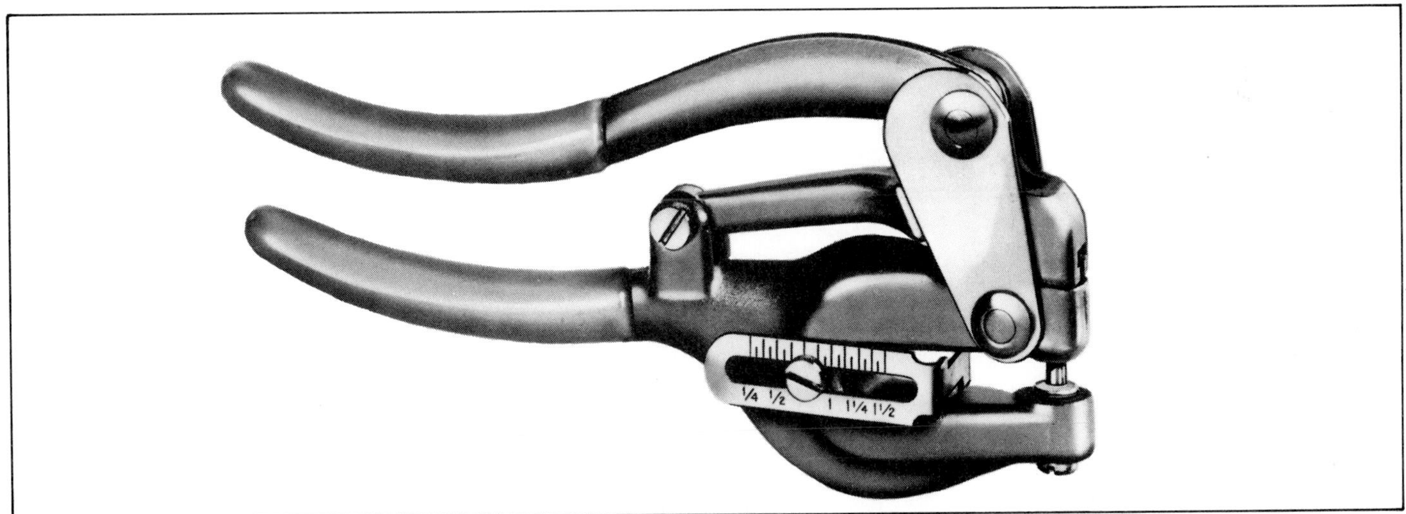

Whitney-Jensen 5 Junior hand-lever punch is almost a necessity. It's very strong and durable, and it saves lots of time. A pilot hole is not needed with this punch. Photo courtesy Roper-Whitney Inc.

the screw is threaded in. The screw is then tightened to punch the hole.

The screw and cutting edge of a knock-out punch should be oiled with each use. Punch sizes range from 1/2-in. to 5-1/8-in. diameter. They are expensive, putting them in the professional or well-heeled-amateur category.

HAND-LEVER PUNCH

The hand-lever punch operates on the same principle as the knock-out punch, only the die is driven by a hand lever. Although it's not capable of punching as large a hole, the hand-lever punch is much quicker. Punch and die *sets* are interchangeable to allow punching holes of different

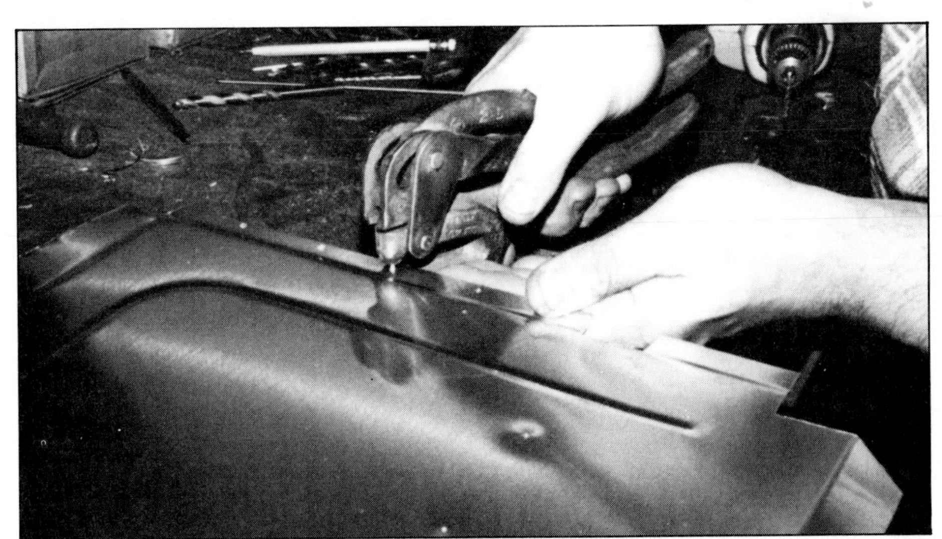

Using the Whitney-Jensen 5 Junior to punch burr-free rivet holes along an edge of this interior panel.

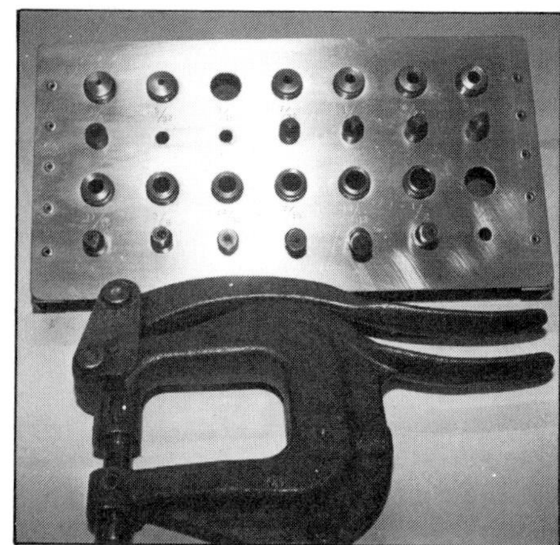

The dies that came with my Whitney-Jensen XX punch were not enough. I purchased several others and made this aluminum holding board for storing all of them.

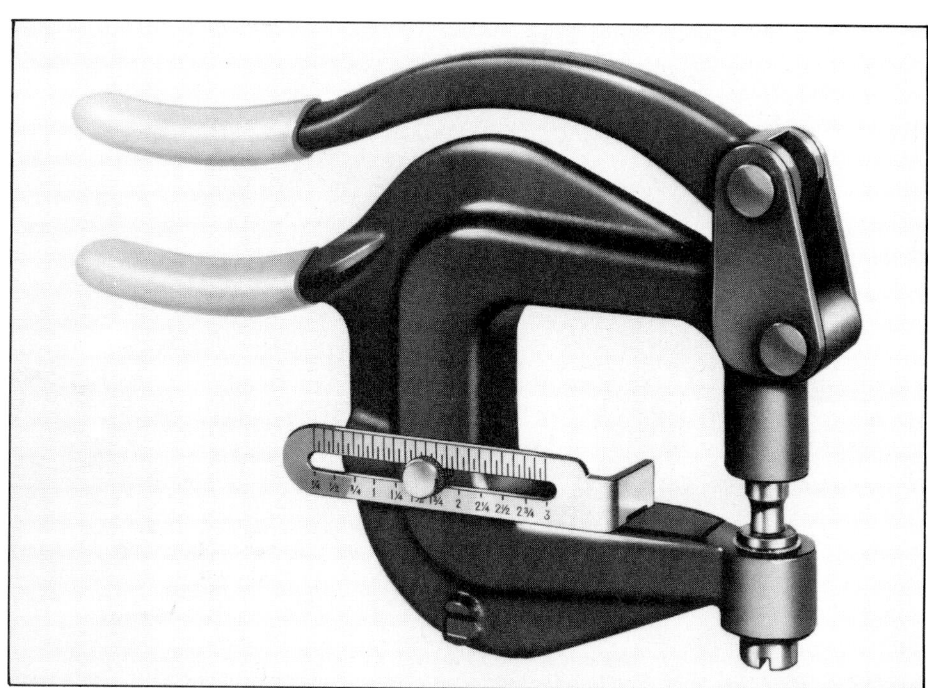

Whitney-Jensen XX punch gives extra throat depth and a wide range of punches. It's worth the cost. Note the back gage, or stop, for punching holes a set distance from an edge. Photo courtesy Roper-Whitney Inc.

Punching holes in this flanged radiator bracket was no problem because of the large throat of the XX punch.

Using Clecos to hold the skin of this wing in place while riveting and bonding. Upon removal, Clecos go straight into the can of lacquer thinner to remove adhesive.

sizes. A pilot hole is not required with the hand-lever punch.

Referring to the accompanying photos, the punch and die of a hand-lever punch face one another in the ends of a C-frame. The punch is slipped over the edge of the sheet metal. How far from the edge a hole can be punched is limited by the *throat depth* of the C-frame. When the levers are squeezed, the work is clamped between the punch and die. Additional lever force punches the hole.

A hand-lever punch saves time and gives a quality look to a project. An adjustable *back stop* can be used to simplify punching a series of holes equidistant from an edge.

For instance, rather than drilling a series of 1/8-in. rivet holes 1/2-in. from the edge after laying them out and center-punching them, the hand-lever punch will do this job in much less time. All you need to do is lay out the *pitch* of the holes—the distance between them—and set the back stop, or gage, to 1/2-in. In addition, punched holes require no *deburring.*

Deburring is the removal of uneven or jagged material or *burrs* from the edge or corner of a metal piece such as that produced when a drill breaks through the material being drilled. No burrs means a lot less finish work on any major project.

Recommendations—The Whitney-Jensen 5 Junior punch is my choice. The Whitney-Jensen 5 Junior comes in a metal box with seven punch-and-die sets—3/32, 1/8, 5/32, 3/16, 7/32, 1/4, and 9/32 in. It can punch up to a 9/32-in.-diameter hole through 16-gage (0.0625-in.) sheet

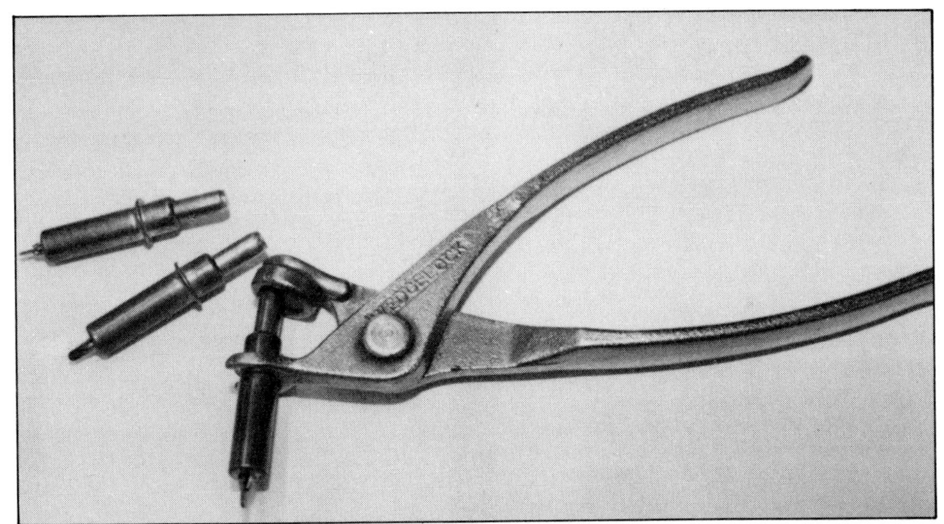

Peanuts and Clecos have something in common—you never get enough of them. Like portable rivets, Clecos do a great job holding parts in place for fitting or riveting. Cleco clamp, shown at right, is a variation on the Cleco: It doesn't need a hole for clamping. Photo at right by Tom Monroe.

metal about 1-1/2 in. from the edge.

The XX deep-throat Whitney-Jensen hand punch is also available. It not only can do what the 5 Junior can, it has a much deeper throat for reaching in about 3-1/2 in. from an edge. It also has a selection of dies up to 17/32-in. diameter. The disadvantage is that the Whitney-Jensen XX deep-throat hand punch is more expensive than the 5 Junior.

JOIN THE PIECES

Once you've cut, punched and hammered on several pieces of metal that will combine to make one assembly, you will need a way of joining those pieces. The popular items used to join metal parts are *Clecos* and *Pop rivets,* or *blind rivets.* Of course, welding is a joining technique too, but I talk about it in the welding section beginning on page 56. Solid rivets, as opposed to Pop rivets, are discussed in the next chapter beginning on page 24. Let's look at Clecos first.

Clecos—Clecos, used for *temporary* joining of parts before permanent joining by riveting or welding, are another great metalworker's aid derived from the aircraft industry. Consequently they're available from aircraft supply houses. I use them to fit and refit a piece I'm fabricating. Clecos hold a part in place for exact fitting without the bother of tack-welding or riveting. This makes them especially handy for assembling complicated multipanel structures such as a *monocoque tub*—an Indy or formula car's main structure—or holding a single panel in place for precise fitting.

Cleco operation is simple. An internal spring retracts and expands *prongs* that extend from the body of the Cleco. A *plunger* extends from the opposite end of the Cleco body. Compressing the spring with *Cleco pliers* via the plunger extends and contracts the *prongs* so they'll slip through two matching holes in the parts you wish to join. Releasing the spring with the Cleco pliers causes the prongs to retract and expand, pulling the parts snugly together.

A collection of various-sized Clecos should be part of your metalworking supplies. Regardless of their size, Clecos operate the same.

There is one quick way of distinguishing the size of a Cleco—color. Common sizes include: 3/32 in., silver; 1/8 in., copper; 5/32 in., black; and 3/16 in., brass.

Cleco Pliers—To install and remove Clecos, you'll need Cleco pliers. Cleco-pliers jaws fit under a collar on the Cleco body and over its plunger. Squeezing the pliers extends and contracts the Cleco prongs so the Cleco can be installed or removed.

One pair of Cleco pliers will handle the four sizes of Clecos normally used in car fabrication. All Cleco pliers are similar. The only real difference is their compactness from one brand to the next. Some are better for getting into tight spots. Cleco pliers can usually be found in aircraft-surplus stores.

Care of Clecos—Both Cleco pliers and Clecos are long lasting, provided they are properly cared for. Three things will ruin Clecos. They rust if they aren't kept clean and dry. Heat

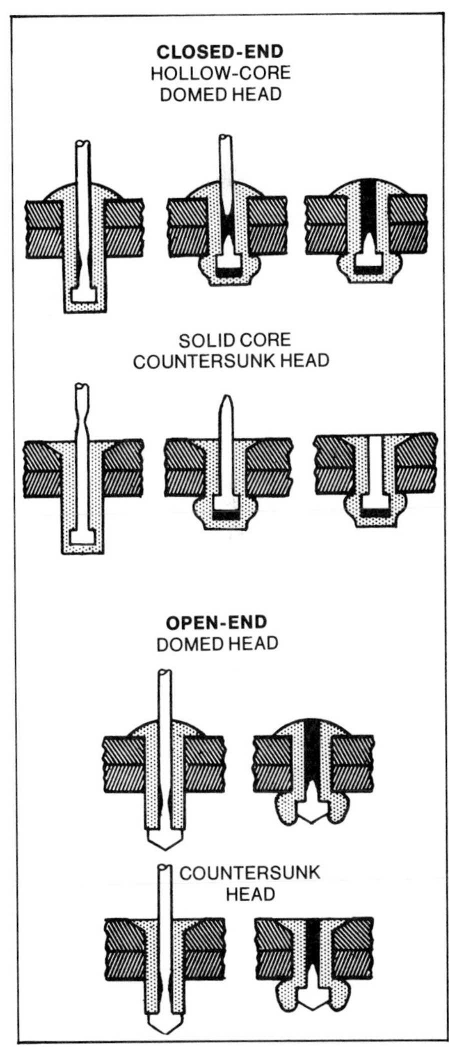

Pop rivets are hollow with a mandrel, or nail inside: Some Pop rivets have closed ends and some have open ends. There are also different heads: domed and countersunk. Countersunk heads install flush with the surface. In all cases, the Pop-rivet gun breaks off the nail after it expands and tightens the rivet in its hole.

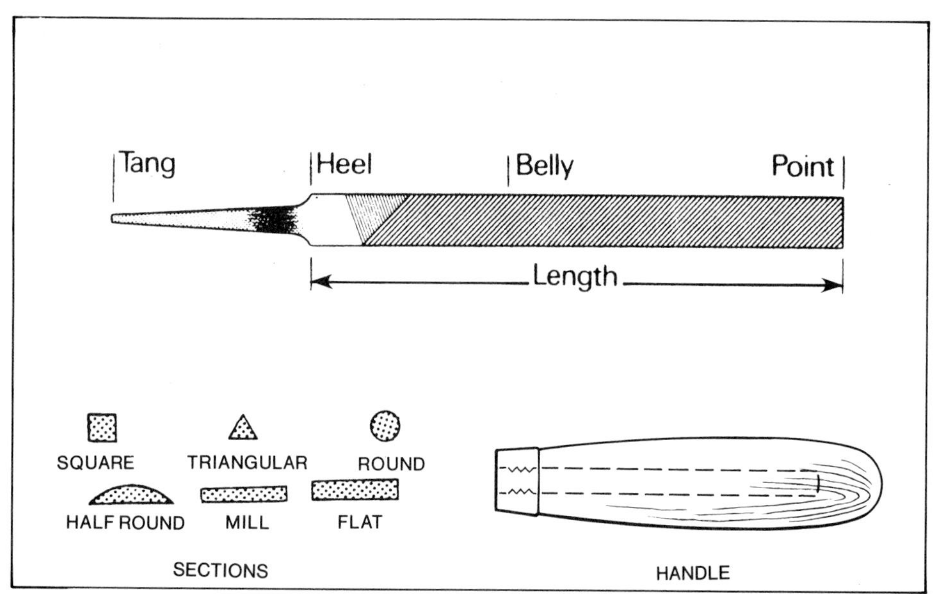

A file and its terms. Get an assortment of files. You'll find the second-cut half-round file the most useful. Use a handle on each of your files to prevent putting a hole in the palm of your hand. The handle should screw on so it will stay put. Screw-on handles are reusable.

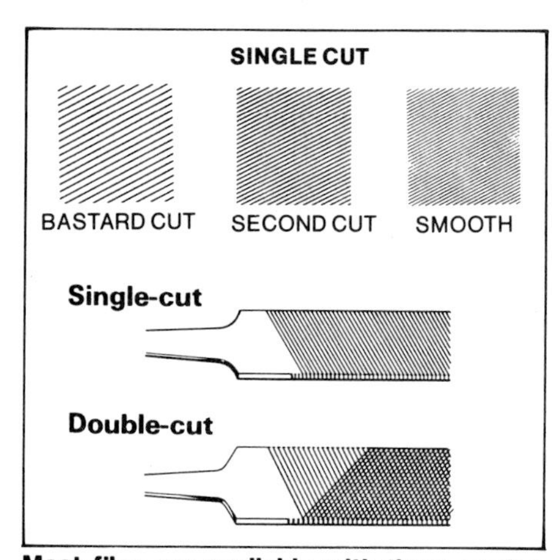

Most files are available with three grades of coarseness: bastard-cut, second-cut and smooth-cut in single- or double-cut. Bastard-cut file removes the most material with each pass; smooth-cut removes the least. Double-cut file is used with more pressure than the single-cut to remove material faster from the workpiece.

Files come in all styles and shapes. Half-round, flat or mill and round files are the ones you'll need most. Photo and drawings on this page courtesy of Nicholson File Company.

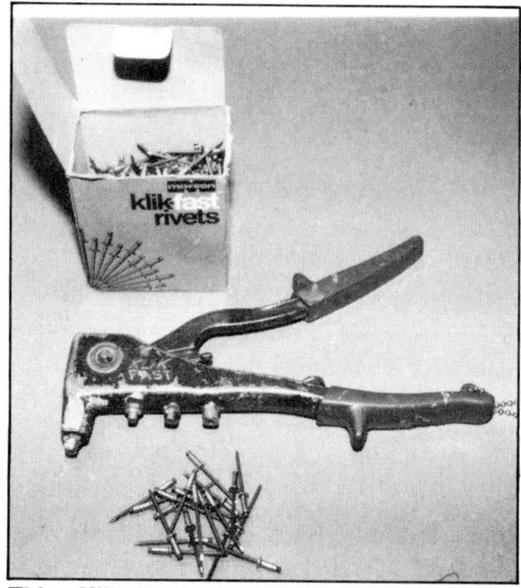

This Klik-Fast rivet gun will *pull* four different-diameter rivets. It has replaceable jaws.

from welding too close will weaken or destroy a Cleco's spring. Don't weld closer than 1 in. Structural adhesive, such as that used for monocoque and wing construction, will render Clecos useless. To avoid adhesive damage, drop removed Clecos into a can of solvent as each Cleco is removed.

Pop rivets—I've used *Pop rivets*—blind rivets—to join or attach everything from an aluminum splatter shield for my mother-in-law's kitchen range to a driver's footrest in a Trans-Am race car. Pop rivets are reliable and easy to use. A main advantage of using a Pop rivet is only one side of the joint needs access. This is where the term *blind* rivet comes from.

Pop rivets consist of two parts: the *mandrel*—also called the *nail*—and the *body.* The nail fits inside the body. When the rivet nail is *pulled,* it expands the rivet body, pulling together and locking the two pieces being joined. In the process the nail is broken off.

The *size* of a Pop rivet is determined by its body diameter. *Grip length*—length of the rivet body measured from the underside the rivet head to the end of the body—determines the total material thickness that can be joined. Minimum grip length must be equal to the combined thickness of the metal pieces *plus* an amount equal to the rivet diameter.

Pop rivets are readily available in two materials: aluminum and plated steel. Although less common, stainless-steel and copper rivets are available for special applications. I never use aluminum rivets simply because steel Pop rivets are stronger. Aluminum Pop rivets tend to shake loose and lose their *grip*, or clamping force.

Of the many sizes available, I use 3/32-, 1/8-, 5/32- and 3/16-in.-diameter Pop rivets most often. Pop rivets are also available with *countersunk* heads. The rivet head installs flush, or even, with the metal surface when installed in a *countersunk* hole—the end of the rivet hole is angled to provide a recess for the rivet head. A final choice is *open-end* or *closed-end* Pop rivets. What distinguishes the closed-end rivet from the open-end style is the end of the body opposite the rivet head is closed. The nail doesn't project from this end of the rivet body. When used with sealer at the rivet head, the closed-end rivet will not leak when used with a liquid container or tank.

When buying a Pop-rivet kit, you automatically receive a good selection of various size Pop rivets. Nice, huh? Unfortunately many of the rivets will be aluminum.

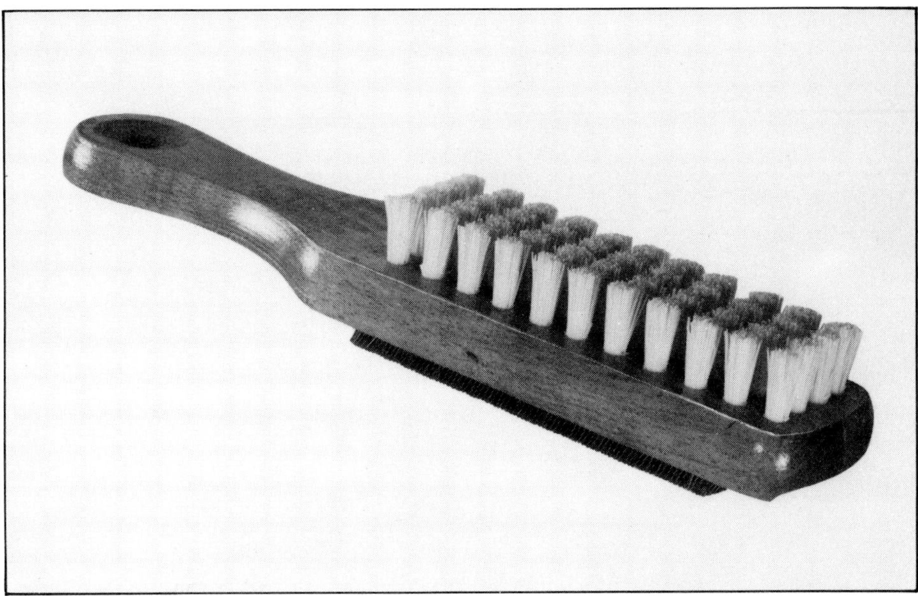

A file loaded with wax or aluminum won't cut well. Short wire bristles on the bottom-side of this file card and the softer bristles on top will clean out the grooves easily. Photo courtesy of Nicholson File Company.

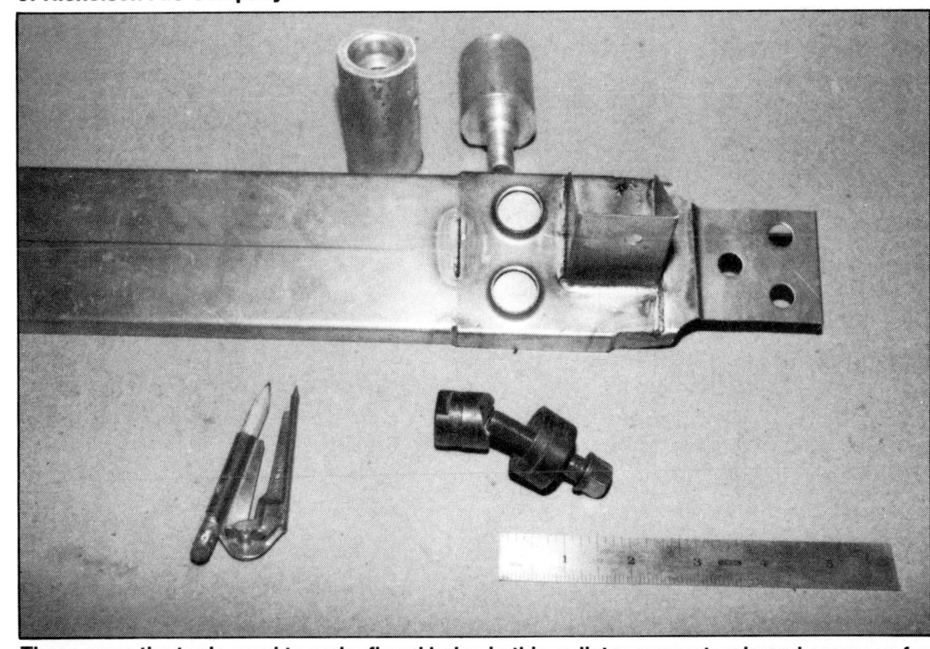

These were the tools used to make flared holes in this radiator support: rule and compass for marking (left); punch for making the hole (right); and a flaring or belling tool (top).

Pop-rivet Tools—There are hand-operated and pneumatic Pop-rivet tools. Both do a good job, are reliable and easy to use. Hand tools are available practically everywhere at a moderate cost. Air-powered Pop-rivet tools are available from industrial supply houses. They have the advantage of being almost effortless to operate, letting you pull—or install—large numbers of rivets quickly. Depending upon how much you use Pop rivets and the size of your wallet, you may want to have the air model as well as the hand model. You'll find a description of the pneumatic Pop-rivet tool in Chapter 2. You should know about them if you do a lot of riveting.

Regardless of the type of Pop-riveter you get, make sure it has interchangeable nose pieces for using different-diameter rivets.

FILES

Newly fabricated parts should be properly finished for a smooth appearance and a good fit. A very useful group of tools for doing this job is files. Files are classified by their *length, cut* and *shape.*

Length is the distance from the file *shoulder*—where the *tang* begins—to the end of the file. The tang is the pointed end of the file on which a wooden handle can be installed. The file should be long enough so it's easy to work with.

Shape is the shape of a file's *cross-section.* It may be round, half-round or flat, triangular or square. And most of these shapes are tapered as well.

Cut determines how much metal will be removed by a file on a single pass over the metal. A *bastard-cut* file will remove a lot of metal, but it leaves a rough, coarse finish. A *second-cut* file removes less metal, but leaves a smoother finish. Therefore,

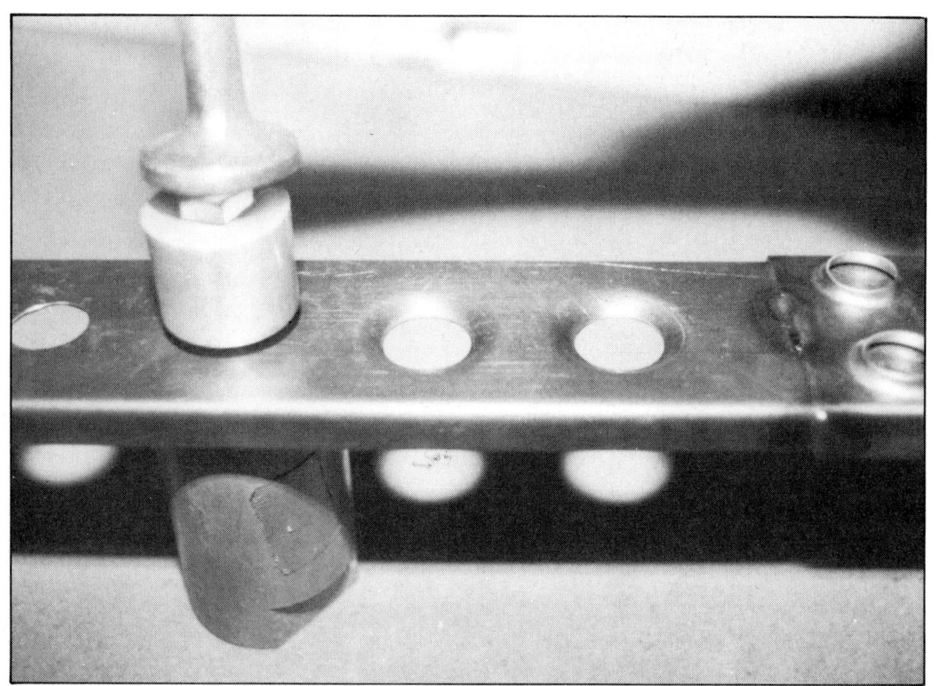
Make your own flaring, or belling, tool. It is worth the trouble. The tool is easy to use and adds a professional touch to parts. The upper die slips through the hole in the metal into the lower die. Hammer on the top die to form the flare.

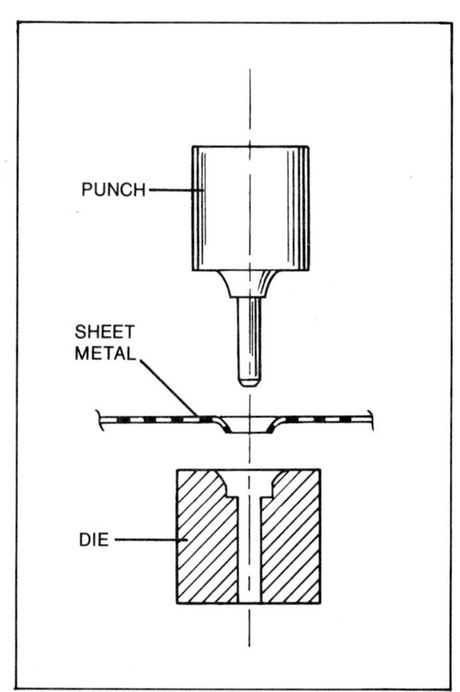

Typical construction of a belling, or flaring, tool. Machine the punch and die from aluminum on a lathe according the size holes you want to flare.

Flaring does more than make parts look good. Flaring allows a part to be made lighter without loss of strength. This is true whether it's on a radiator bracket or formula-car tub.

the use of one cut or another depends on the speed of removing material and the finish desired.

The cut of a file is always stamped on the tang end for identification. Although there are several other cuts of files available, I use the bastard cut and the second cut most of the time.

The basic metal file I rely on most is the 6-in. half-round file in second cut. You will need this and several general use files such as *round,* or *rat-tail,* and *mill,* or *flat,* files in bastard and second cuts. You may want to add other shapes and cuts later, but these are good starters. They aren't expensive and last a long time if properly cared for.

Care of Files—Caring for files is easy, but necessary to preserve their usefulness. Don't just toss them in the bottom drawer of your tool cabinet where they are free to bang against each other and your other tools. Keep them dry and separated. Rust loves moisture and hates files. Store files *separately* from each other so they won't dull one another by constantly coming into contact. I hang my files to keep them apart and handy. You may want to use this method to store your files.

Never use a file without a handle. Without a handle, the odds are you'll puncture the palm of your hand with the tang of the file. It's sharp. You not only don't want blood all over that nice aluminum part, you'll have to go to the doctor for a tetanus shot. Tetanus shots hurt and will make your metalworking arm swollen and sore.

File Card—File teeth tend to *load*—become clogged—when filing soft metals such as aluminum. A good way to keep a file clean is with a *file card.* A file card is a wire brush with short bristles. File cards are available at hardware and tool stores.

FLARING TOOL

Many times a fabricator puts holes in a part to lighten it. A *flaring tool*—sometimes called a *belling tool*—is used to flare, or turn the edges of a hole down smoothly. This not only gives the part a professional look, it also strengthens it.

A flaring tool is a tool you'll have to make by hand. It consists of an upper and lower half, or a punch and die, each turned on a lathe.

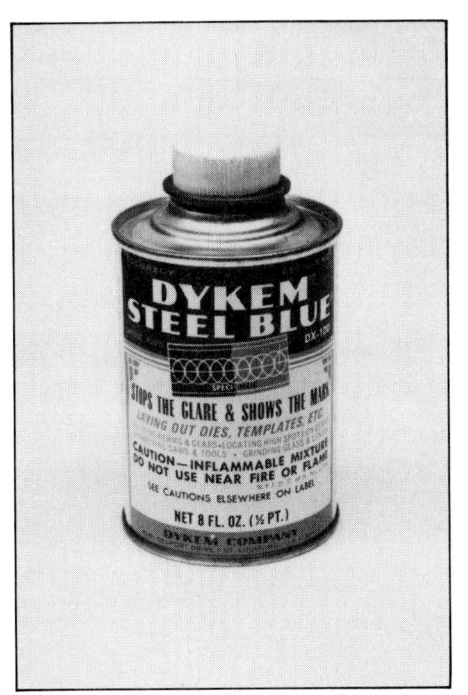

Machinist's blue or layout dye. Dye is applied with a brush attached to the underside of the lid. Dye also comes in aerosol cans. Photo by Tom Monroe.

Extensive use of flare holes on this formula car makes it stronger, lighter and better looking.

The hole that's to be flared governs the size of the flaring tool. Use the accompanying drawing as a guide to scale your flaring tool up or down to the desired size. Use aluminum for flaring tools to minimize scratches.

Miscellaneous Tools—Some miscellaneous items you'll need for doing metalwork aren't expensive, but they are important to do professional quality work.

You'll need soft white and silver pencils, a metal scriber and a common pencil compass for marking metal. Layout dye, Dykem, prussian blue, machinist's blue, or whatever you prefer calling it, stains metal so scribed lines stand out clearly. Fine-line, felt-tip permanent markers are handy for marking exhaust pipes or cardboard patterns. Lightweight cardboard—the kind that comes in shirts from the cleaners—is used for patterns. You'll need sturdy 8-in. scissors to cut out cardboard patterns. Safety glasses and gloves are a necessity if you're fond of eyes and fingers.

Good eye protection is a must. Many tools throw chips or sparks. Safety goggles belong on everyone near these tools. Photo courtesy of Norton Co.

STORAGE

Your hand tools and miscellaneous supplies and power tools that are discussed later will need a place to call home. I assume you have a garage, shop or some place to work. It needs some utility shelves, maybe a metal locker or closet. A locked tool-storage room or box adequate for your tools will keep them from "growing legs." I've noticed my tools can be pretty footloose if they aren't locked up.

If your working area is going to be your home away from home, you'll probably wish it had a wash-up area, too. In a pinch, make do with whatever you've got. Michelangelo worked upside down on a rickety wooden scaffold in a huge, unheated room when he was painting the Sistine Chapel. It turned out "OK."

Basic hand tools will get you started. In addition, maybe you'll add some of the more sophisticated hand tools. I know you'll want some power tools such as an electric drill, disc grinder, and saber saw. I'll talk about those tools next.

POWER HAND TOOLS

2

Cup-type grinding stone on a disc grinder can remove a lot of metal in a hurry. It does a great job removing cutting-torch slag. Full face protection must be worn.

Power hand tools are valuable assets to a metalworker. They do jobs which are either impossible to do or very difficult to do as well without them. You must have some power hand tools, and may want others to save muscle, time or aggravation.

Although I have my favorite sizes and styles of power hand tools, I encourage you to browse in tool stores, supply houses and tool catalogs to choose your own. Keep up on what is available and what is new. Most of the power hand tools I use are old familiar ones, yet I enjoy "checking out" any new versions being offered.

In this chapter I concentrate on the power hand tools I use most often. These are also the ones you'll need most. Because I do fabrication on a daily full-time basis I sometimes use uncommon power tools, such as a 1/4-in. 90° electric drill or a 9-in.-long Bosch grinder with a 3-in. grinding disc. I find these uncommon tools helpful, but I encourage you to stick with the basic power hand tools.

ELECTRIC HAND TOOLS

Disc Grinder—The 7-in. *disc sander*—or *disc grinder*—is my favorite because it is not too heavy. It's a real workhorse and has many functions. Consequently I use mine a lot. This grinder has many uses: sanding a fender for painting; V-grooving pieces to be welded; or deburring parts. After holding a disc grinder in a raised position for several minutes, you begin to appreciate its light weight. It makes the tool easier to use.

A nice feature to have on a disc grinder is a *spindle lock*—a spring-loaded pin which locks the spindle to allow easy removal and installation of grinding discs. Grinders without a spindle lock take more time and energy to change discs. A separate wrench must be used—which always seems to be misplaced just when you need it.

> **CARE OF POWER TOOLS**
>
> An electric power tool used in a shop is extremely vulnerable, particularly its electrical cord. The cord is often walked on, driven over, tripped over, and rubbed or dragged across rough surfaces. This will fray, split and generally raise havoc with the insulation. Inspect your power-tool cords often because a damaged one is a hazard. Repair or replace a damaged cord so you won't get a heavy-duty shock.

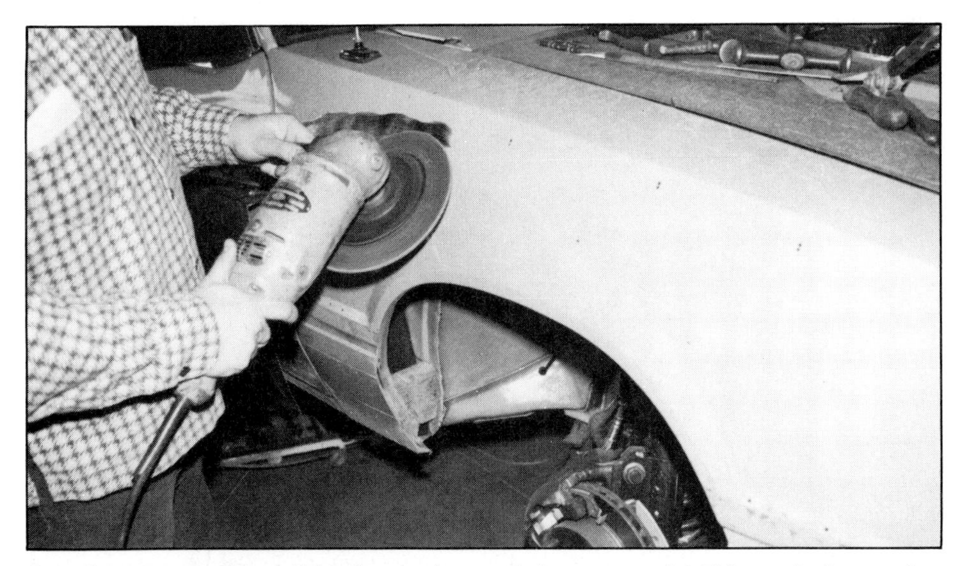

One of many operations a 7-in. disc sander can do is remove paint. This sander has a spindle lock, allowing fast disc changes.

Recommendations—Black and Decker, Craftsman, Makita, and Ingersoll Rand, to name a few, have good disc sanders. The Craftsman 7-in. grinder is a good value. It is also the lightest one available.

SAFETY EQUIPMENT

You must take certain precautions when grinding. A grinder throws off hot metal sparks or paint fragments at a very high velocity. These fragments can cause injuries. To avoid injuries, safety equipment *must* be used with a grinder.

Leather gloves worn while grinding will protect hands from burns. A long-sleeve shirt will protect your arms. Safety glasses and face shields are necessary companions to many kinds of shop equipment, particularly grinders. **Grinding requires the use of eye protection, not only for the person doing the grinding, but for anyone nearby as well.** A full face shield is excellent because metal fragments thrown from a power tool sometimes bounce behind safety glasses and into an eye. I prefer a full face shield instead of safety glasses.

Face Shields—A full face shield has some advantages over safety glasses. It not only protects your eyes the same as safety glasses, it also protects your entire face and part of your hair. An added benefit is a face shield can be worn over glasses. A double advantage. It saves the cost of prescription safety glasses and your glasses are protected. If you've ever seen what grinding spatter does to glass, you'll know what I mean.

Face shields have an adjustable headband and a replaceable face

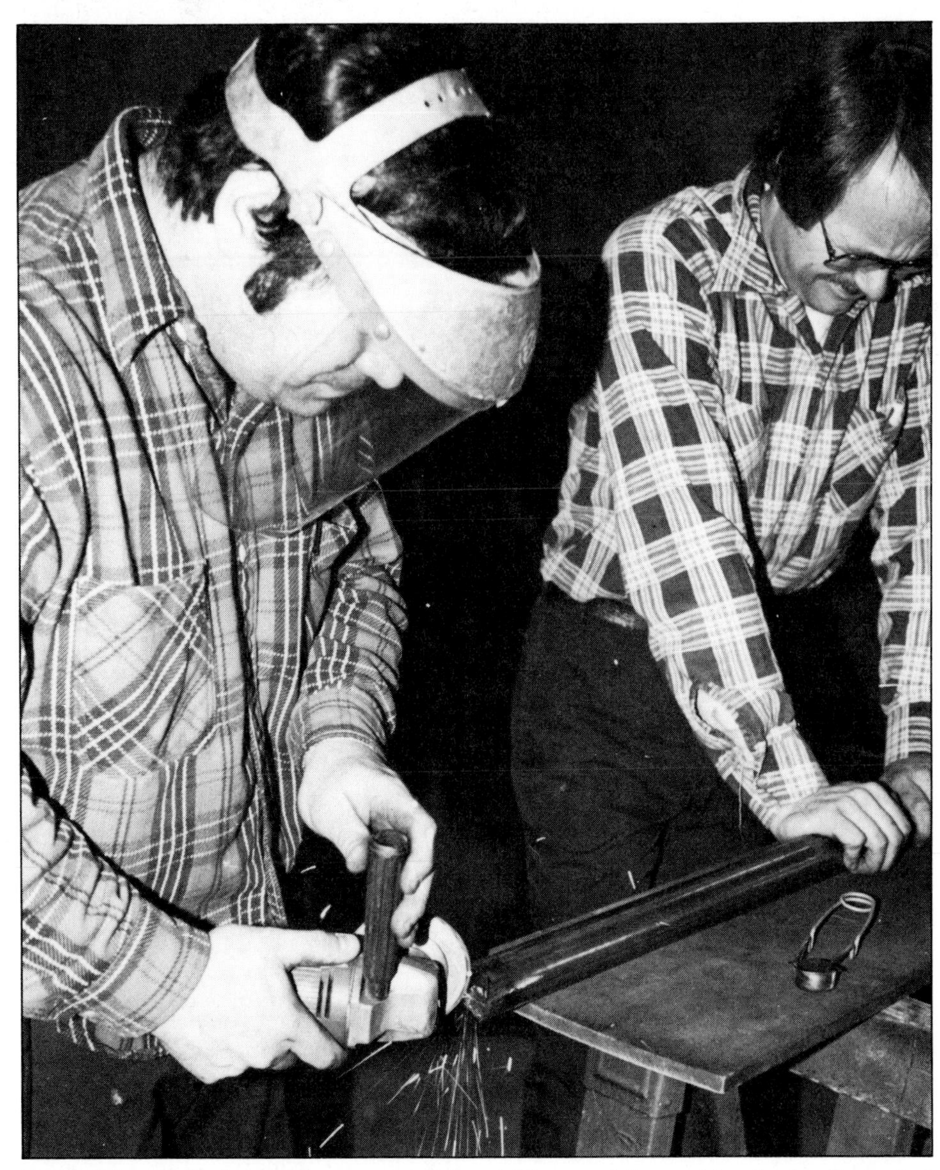

Bosch 9-in. grinder is a small, very powerful tool. It's especially good for shaping the end of a thick-wall tube. And it is easy to maneuver.

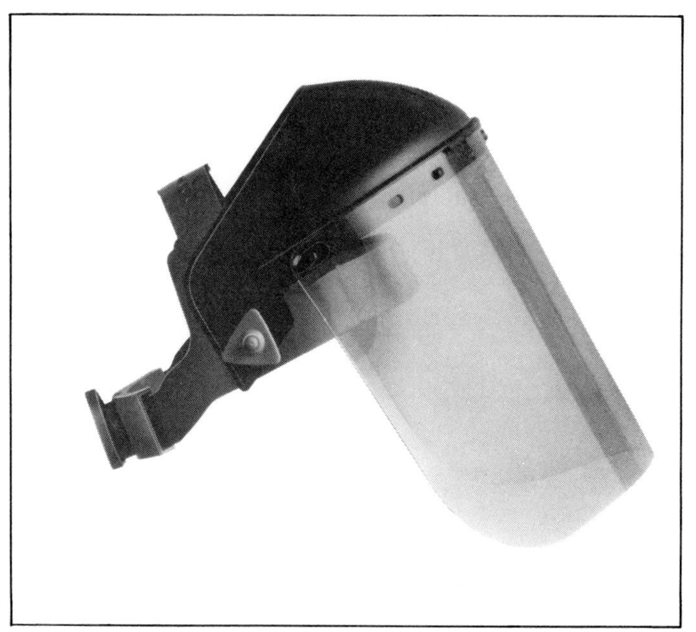

I like this type of shield. It protects your whole face. Shield is replaceable and inexpensive. Photo courtesy of Norton Co.

Chipper's goggles can be worn when you're using a die grinder. The goggles shield your eyes from metal chips flying off the cutter. Venting in the sides keep them from fogging—and make them more comfortable to wear. Photo courtesy of Norton Co.

shield. Replacing the face shield is necessary because the clear plastic gets scratched through normal use, gradually obscuring vision. So it must be replaced periodically. I urge you to replace a shield when it gets scratched. Not only must your eyes and face be protected, but you've got to see what you're doing. **A clear, safe look at your work is a must.**

Safety Glasses—There are a number of good safety glasses on the market. And there are some that will fit over glasses to save the expense of buying prescription safety glasses. The better safety glasses have plastic side-wing panels, or side shields, to protect your eyes from bouncing fragments. This extra protection is worth the extra cost or time to ensure that you get this feature.

Some safety glasses have tempered high-impact lenses. These lenses break less easily and cost more, but are well worth the price. Look for tempered lenses when shopping for safety glasses. As with the face shield, safety glasses should be replaced when damaged so you'll have a clear view of your work.

SABER SAWS

A saber saw is excellent for many kinds of cutting. This hacksaw is powered at high speed by an electric motor. Its short blade oscillates vertically against the work rather than horizontally as with a hacksaw.

I find a saber saw most useful when I must cut a hole in the center of a large part. For instance, a deck, or trunk lid that must have a large round hole in the middle for a fuel-fill neck cannot be put in easily with a bandsaw. A hole saw isn't big enough for the necessary hole. Similarly, a quarter panel may need such a hole. The saber saw is the answer.

To use a saber saw, first mark a cut-out line where you want the hole. Drill a small access hole inside the cut-out line—large enough for the saber-saw blade to go through. Then cut on the line with the saber saw.

Recommendations—Several different types and brands of saber saws are available. When shopping for a saber saw, look for one with multiple-speed or variable-speed control. Also, a long-stroke saber saw is best because it cuts smoother and uses blades more efficiently. With these features, your saber saw will be capable of cutting many kinds of metal in various shapes and gages. I prefer the Bosch saber saw because it's very maneuverable, the blades are easy to get, and they are easy to change. The blades are made in a wide range of sizes and pitch.

Saber-Saw Blades—Always use the correct blade. Generally speaking, a fine-tooth, *fine-pitch*—more teeth per inch—blade is for cutting hard materials, such as stainless steel. Also, the thinner the material, the finer the tooth pitch must be. Conversely, softer and thicker material requires a coarser saw-tooth—fewer teeth per inch.

For most metals, another general rule is that there should always be *at least two teeth against the material being cut.* For example, for cutting 1/8-in. material, the blade should have a *minimum pitch* equal to 2 teeth divided by 1/8 in., or 16 teeth per inch.

ELECTRIC DRILLS

3/8-in. Electric Drill—The *3/8-in.* prefix refers to the *maximum drill-shank diameter* the chuck will accept. The shank of a drill is the end that fits into the drill chuck. The shank diameter of a standard straight-shank drill bit is equal to the drill diameter. For example, a 3/16-in. drill bit has a 3/16-in.-diameter shank.

When compared to a 1/4-in. electric drill, a 3/8-in. electric drill is much more versatile. The 1/4-in. drill is not entirely limited to a 1/4-in. drill-bit. There are larger drill bits with smaller, 1/4-in. *shanks.* Regardless, there are other reasons for choosing the 3/8-in. electric drill over the 1/4-in. drill. In most cases a 3/8-in. drill is not physically bigger than a 1/4-in. drill, yet it is heavier duty, has more power, and will do everything a 1/4-in. drill can do—and more.

If a 3/8-in. electric drill isn't already in your collection of tools, choose one with *variable speed* rather than *constant speed.* Variable speed lets you alter drill speed, or rpm, to suit the drilling need exactly. It simply gives you more control.

The variable-speed feature is especially nice when starting to drill a hole. You start slowly, and when the drill starts cutting, you can increase rpm to the normal drilling speed.

1/2-in. Electric Drill—When you need high torque for drilling a large-diameter hole, the 1/2-in. electric drill is the one to use. In my shop, the 1/2-in. drill is a necessity. It's my "portable" drill press. I use it to drill large holes in assembled or installed pieces. These holes can't be done in a drill press.

Because of the high torque of the 1/2-in drill, an extra handle that extends straight out from the drill is essential. If the drill isn't equipped with a handle, there will be a tapped hole in the side of the drill to accept 1/2-in. threaded pipe.

The 1/2-in. drill is well suited to hole saws. The drill cuts at a relatively low speed, which preserves the hole-saw teeth. You'll find 1/2-in. drills with different speed and horsepower ratings. A common rating is 600 rpm and 1/4 horsepower.

Motors used in 1/2-in. drills are very powerful. **Take special care not to let the drill bit snag or hang up. And do not use trigger locks.** Its motor will keep running, causing the electric drill to rotate rather than the drill bit or saw. It can catch you by surprise and injure you! So be cautious when using a 1/2-in. electric drill.

Recommendations—The best drills I've used through the years have been made by Craftsman or Black & Decker. There are many types of drills made by other manufacturers, all readily available from tool stores, automotive supply houses, department stores and discount outlets. A stand-out are drills made by Makita. I've found them to be very durable. When shopping for a drill, check to make sure the chuck is the diameter you need before you buy it.

DIE GRINDER

Early in my metal-fabricating career, I learned to use an electric die grinder. I immediately found that this tool can be very useful. For instance, during the construction of a chassis, I frequently find a bracket doesn't quite fit because of excess material. The die grinder handles this situation in short order. It will remove metal very quick-

Steel and aluminum are easily cut with a Bosch multi-speed saber saw. This is one tough, reliable tool—and it's expensive. Roller supporting blade (arrow) is a key to this saber saw's durability and cool-running blade.

Black and Decker variable-speed 3/8-in. drill is nice for many drilling jobs. I use mine constantly.

Electric die grinders are commonly used for deburring exhaust system parts. A carbide cutting tool is being used in this Milwaukee die grinder.

I have relied heavily on my Black and Decker electric shears for years. They will waltz right through 16-gage steel. Aluminum sheet is no problem.

ly and is less tiring than filing, saving energy and time.

Other uses for the die grinder are *deburring*—removing rough fragments at an edge. Cleanup work on exhaust systems often involves the removal of *slag*—scaly buildup at welds. The die grinder effectively and quickly removes the scaly material.

Time saved without sacrificing quality construction is a bonus—the die grinder will do it.

Recommendations—Buy an electric die grinder. Air-powered die grinders are more compact but require an air compressor and an inline oiler. An electric grinder only needs an electrical outlet. If you have an air compressor, get an air grinder.

Get a die grinder with a 1/4-in. *collet*—a cone-shaped sleeve which holds the cutting or grinding tool. A collet smaller than 1/4 in. cannot use tools intended for continuous heavy-duty use. The 1/4-in. electric die grinder is built for serious craftsmen. It can take the punishment of continuous, hard use.

Care and Safety—Dozens of different types of cutting tools—both carbide and abrasive—are available for the die grinder. This makes it useful for working various metals.

Electric die grinders are high-speed tools. 30,000 rpm is not unusual, so **die grinders are dangerous** if proper care is not used. **Do not allow a cutting or grinding tool to hang up or chatter.** This can bend or break the cutting-tool shank and cause the tool to be thrown out of the collet. Obviously a low-flying 30,000-rpm cutting tool can do some damage.

The cutting tool must be kept tightened in the collet. Check it often to make sure. Also, keep a carbide cutting tool well lubricated with wax to preserve it and so it will cut more efficiently. The wax keeps the cutting tool from loading up with metal chips. Metal fragments fill the voids between the cutting teeth, quickly reducing the cutting ability of the tool. Soft metal, such as aluminum, is especially susceptible to this.

A good source of wax is your local hardware store. Wax sealing rings for toilets do the job and are very inexpensive. A bar of soap can be used when no wax is available.

Eye protection is absolutely essential to prevent eye injury when using a die grinder. The high speed causes an incredible amount of sharp, hot fragments to be thrown off at high velocity. So it's best to protect all of your face by wearing a full face shield rather than safety glasses. This will give you additional protection, particularly if a cutting tool is thrown from the grinder.

Recommendations—Milwaukee and Black & Decker manufacture quality die grinders. I am partial to the Milwaukee brand die grinder because it is durable and fairly priced.

ELECTRIC SHEARS

Portable electric shears are very handy for *blanking*—cutting out shapes from large sheets of metal. Some shears are very heavy-duty and can cut up to 12-gage (0.1046-in.) sheet metal.

Electric shears enable you to cut freely in any direction on full sheets of metal. A bandsaw can also cut thick sheet metal quickly, but it is limited by its throat depth. Electric shears do not have this limitation. Light-duty electric shears are used in a similar way, on a smaller scale—smaller parts and thinner metal.

I find electric shears limited in their ability to cut small radii. With this in mind, I plan ahead and don't try to cut tight curves with them. I restrict the use of electric shears to straight-line cutting. Used this way, they are very fast and easily controlled. It's easy to stay on a relatively straight cutting line. Naturally, power shears save time and effort.

Recommendations—Kett, Black & Decker and Bosch all offer a line of electric shears. The electric shear I use most often is Kett's Model K-100.

PNEUMATIC HAND TOOLS

Compressed-air, or pneumatic, power tools require additional investment in a compressor. The efficiency and durability of the tools merit that investment in many cases. I mentioned that I prefer an electric die

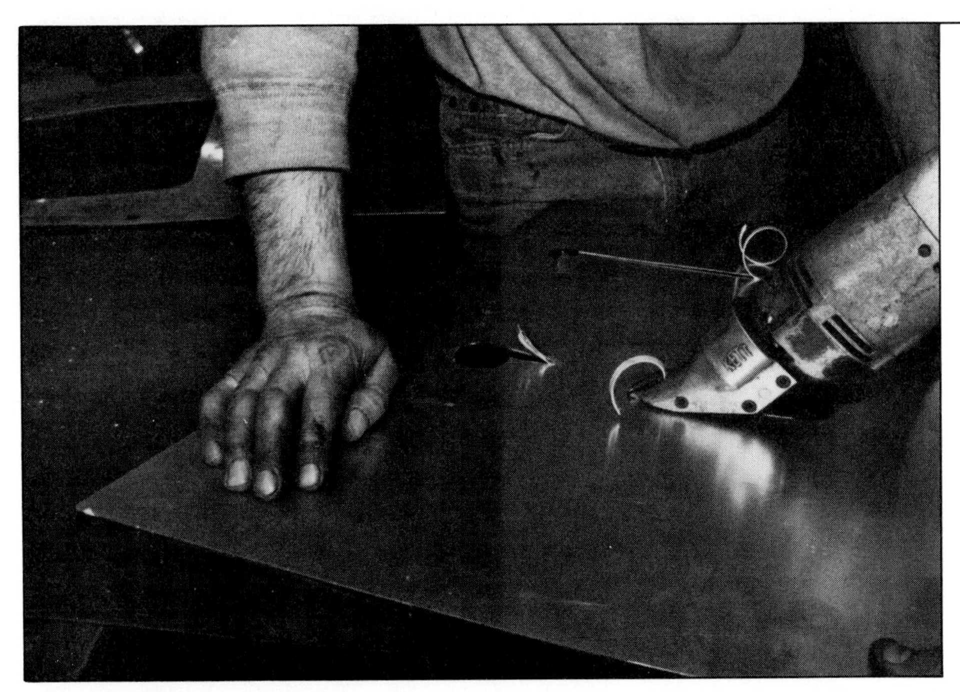

Kett's light-duty cutting head adapts to a common 3/8-in. electric drill. This tool has saved me many hours over the years.

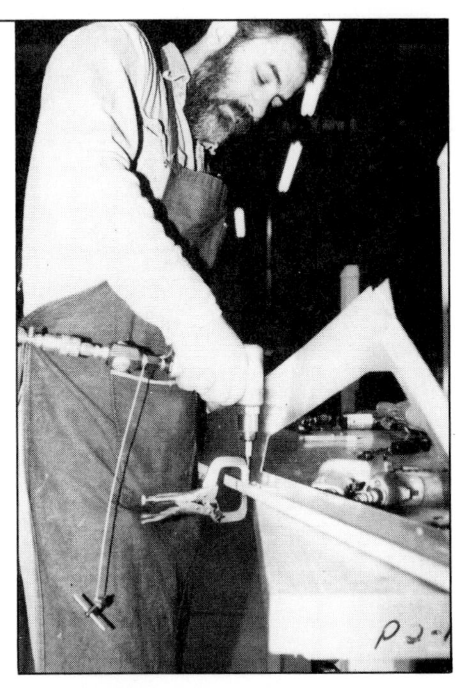

Don Maxwell, expert sprint-car builder, likes his 1/4-in. pneumatic drill. It's compact and reaches easily into tight spots. It needs an oiler in the air line to work at its best. Vise-Grips hold the metal in position for drilling.

grinder to an air-powered one. However, if you have a large-capacity air compressor, then a pneumatic die grinder deserves your consideration. Ideally, a 3—5-horsepower compressor is most useful.

Pneumatic 1/4-in. Drill—A compressed-air-powered 1/4-in. drill is great for drilling many small holes in thin metal. The big advantages with this tool are: it has plenty of power and it is small. It gets into tight spots easily. This may be impossible with an electric drill. Its high speed is excellent for drilling soft metals. An air drill also stays cool and the exhaust blows the chips away so you can see the work.

I often use mine for drilling small holes in aluminum. Its high speed and compact size make it perfect for drilling rivet holes when assembling aluminum *monocoque structures*—such as an Indy-car tub or wing.

Care and Use of a Pneumatic Drill—A 1/4-in. air drill—or any tool powered by a pneumatic motor—should always be used with an *inline oiler.* An inline oiler provides constant light lubrication to air tools while they are in operation. The drill should also be used only with the air pressure regulated to that recommended by the drill's manufacturer. This will make the drill work properly and last a long time.

Recommendations—Regardless of the drill you buy, stick to name brands: Sioux and Black & Decker for instance. Industrial-tool suppliers are the best place to start. Mac Tools and Snap-on also carry pneumatic drills.

PNEUMATIC RIVETERS

Pop-rivet Gun—An air-powered Pop-rivet gun is a great tool for large projects requiring many rivets. However I don't recommend buying a pneumatic Pop-riveter unless you often do projects that require a lot of rivets—like fabricating wings. A hand-powered rivet gun is adequate for normal occasional use.

The pneumatic Pop-rivet gun can pull up to 3/16-in.-diameter Pop rivets. Unlike the hand riveter, little muscle is required. As with any air tool, an inline oiler and regulated air pressure should always be used.

Recommendations—There are many air-powered Pop riveters available; most of them are good. Some are more durable and compact than others. Choose one carefully, making sure it is the right size you need. Consider how well the tool will fit into the spaces you most often rivet in. These are some of the facts you must consider before buying.

PNEUMATIC SOLID-RIVET GUN

The pneumatic solid-rivet gun—also called a *riveting hammer—*is used to install *solid rivets—*solid, as opposed to *hollow-core,* or Pop rivets. Like the power Pop-rivet gun, the pneumatic solid-riveter is useful when installing many rivets.

A solid rivet doesn't have the *blind* feature of the Pop rivet. There must be access to both ends of a solid rivet to install it. One end of the rivet is *bucked,* or backed up. The opposite end is *driven,* or *upset* by impacts from the rivet gun. This "mushrooms" the rivet, giving the rivet another head. It also expands the rivet to fit snugly in its hole. This locks the joined parts

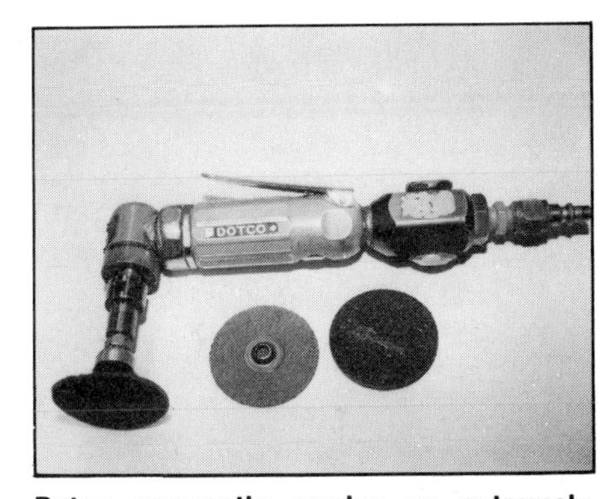

Dotco pneumatic sander, an extremely rugged and reliable tool. Dotco also manufacturers die grinders of equal quality.

DEBURRING TOOL

Make sure that you include a deburring tool in with your drill set. Nasty burrs can result from drilling a hole—burrs that cut. Remove them immediately with this tool.

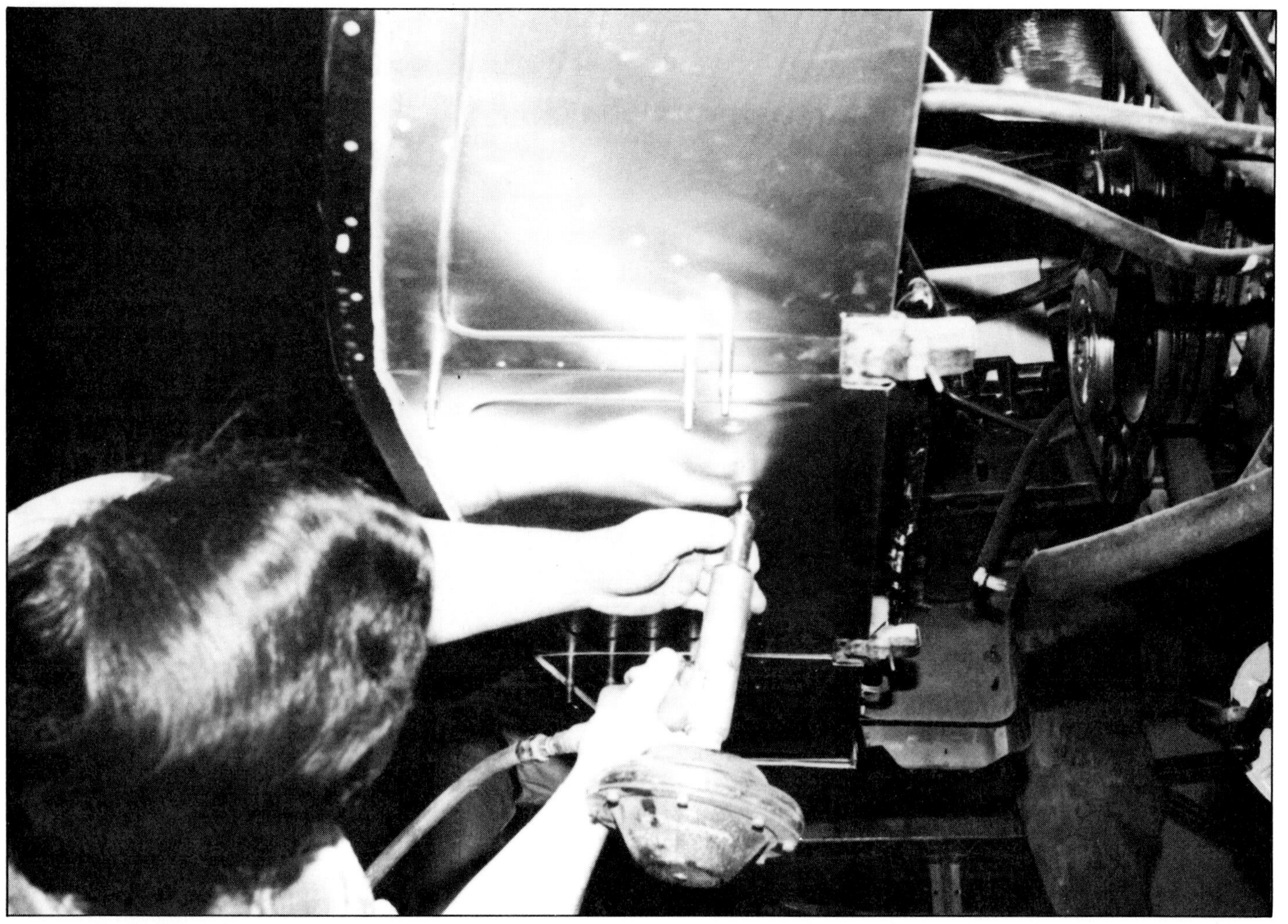

Unless you want to develop your arm muscles, a pneumatic Pop-rivet gun is best for installing many rivets fast. It requires minimum effort and saves as much time as sweat.

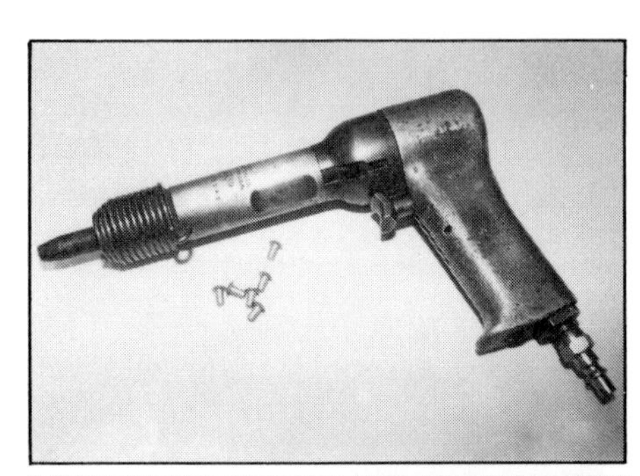

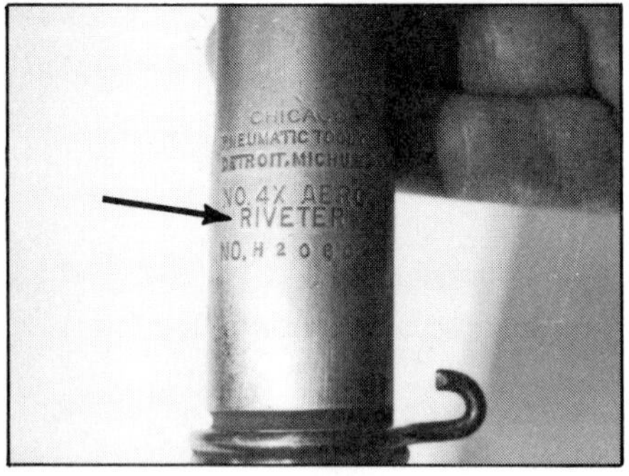

Rivet gun and solid rivets. Although they look similar, a rivet gun is not an air chisel. It says so on the nose of the riveter (arrow).

tightly together.

The solid rivet gives a much stronger joint than the Pop rivet. That's why it's used for assembling all types of monocoque structures, especially those for aircraft, Indy cars and formula cars.

Recommendations—When shopping for a pneumatic solid-rivet gun, make sure you have the right tool. An air chisel looks identical to the solid-rivet gun. On the nose behind the tool-retaining spring you will find the word RIVETER if it is a rivet gun. Don't be fooled.

There's an important difference between the riveter and the air chisel. The air riveter strikes a rivet at a certain speed and with a certain force. The air chisel cannot be used as a riveter because it hits too hard and too frequently. It's designed for cutting muffler and body panels. Using an air chisel on a rivet can drive the rivet straight through the hole in the metal and out the other side.

Rivet Buck—A pneumatic solid-rivet gun is used with a *rivet buck*. A rivet buck is simply a heavy flat-faced steel object held against the rivet end. The buck provides resistance at the end of the rivet. The hammering force from the gun, through the rivet and against the buck causes the end of the rivet opposite its head to mushroom against the panel.

Recommendation—An old heel dolly makes a fine riveting buck. It has a flat surface, sufficient weight, and is small and easy to handle.

OTHER POWER TOOLS

As your skills grow you may want to add some other power tools to your collection. The tools I've mentioned will enable you to handle most fabrication. In general, skill and patience will be the most valuable "power tools" you will ever have. Don't expect any tool to substitute for them. Use the power tools to ease your way. Take care of them.

3

LARGE EQUIPMENT

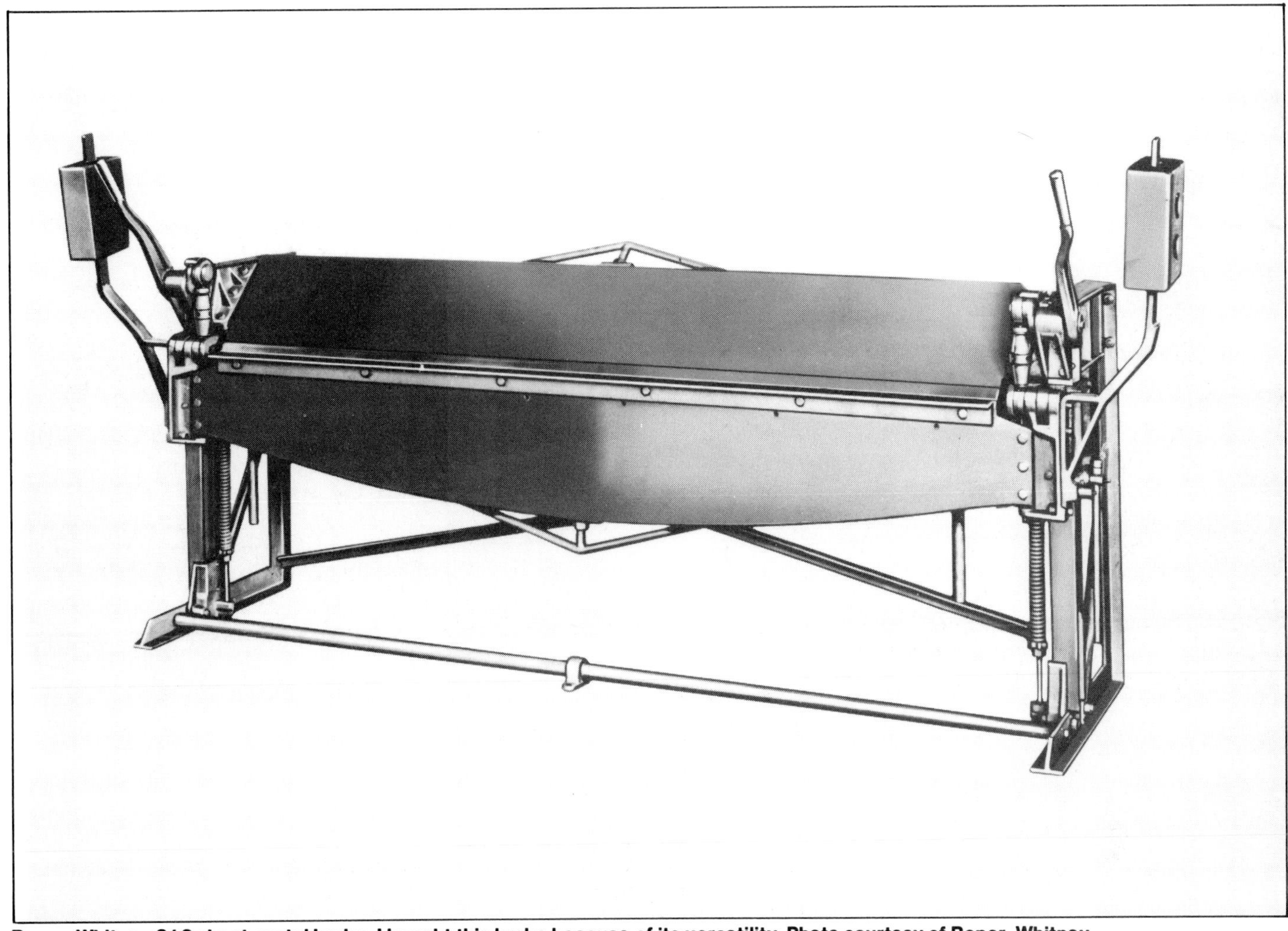

Roper-Whitney 816 sheet-metal brake. I bought this brake because of its versatility. Photo courtesy of Roper-Whitney.

Here's where I discuss more expensive tools. The first items I cover are within a hobbyist's budget. The others are more expensive, so you'll have to do some careful budgeting or find a "friend" in the fabricating business who will let you use his equipment. Some large sheet-metal equipment can be rented. Regardless, what you'll need depends on how deeply involved in metalwork you want to get. If all you want to do is build one simple item, you can either borrow the needed equipment, rent it, or have someone else do a given phase of the work for you.

Large sheet-metal equipment is used for doing jobs that hand tools either cannot do, or cannot do as well. I'll fill you in on those tools I used to complete some of the metalwork described in this book. I'll pretend you're setting up a moderate-size automotive sheet-metal shop. I'll discuss the models and brands I've been most satisfied using.

VISE

One of the most useful pieces of metalworking equipment is not very big or complicated. A *vise* will hold metal securely. Without a good vise you will have great difficulty doing certain jobs. For instance, to use a hacksaw or file you must hold the metal firmly. A properly mounted vise will will do this.

Never overtighten a vise. Don't hammer on the handle to tighten a vise. Even a quality vise will not stand up to this kind of abuse.

Recommendations—Don't buy a vise that's too small. Vises are rated by the width of the jaw face. Pick one with at least a 4-in.-wide jaw face. For my money, a Sears vise is a good piece of equipment. Others may be better, but their cost is usually very high. Be prepared to pay around $100 for a good vise. Bargain vises are usually not even worth their "bargain" prices, so don't make the mistake of choosing a vise that probably isn't durable.

Jaw Covers—Vises have sharp serrations or teeth on their jaw faces. These teeth are intended to grip the work. This they do, but the jaw serra-

A vise should be bolted solidly to a large heavy workbench.

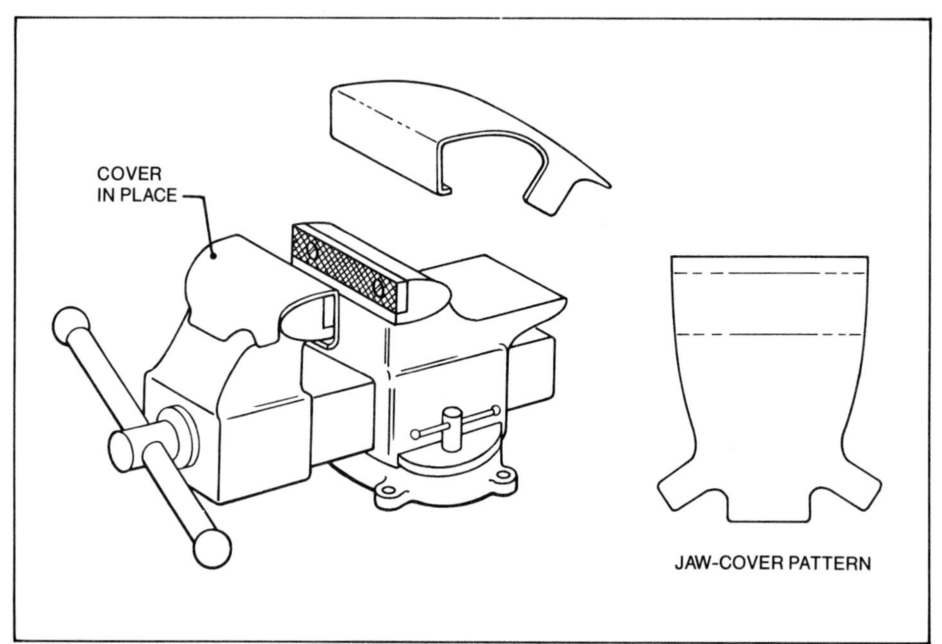

Soft vise-jaw covers will prevent the vise from marring work it's holding—3003-H14 aluminum, 0.063-in. thick works well. Make jaw covers to fit your vise using this pattern.

After years of hard service this Roper-Whitney 816 is still in top shape—no sign of wear. Homemade radius dies are underneath the brake. Box of bolt-in fingers—also under the brake—increase the versatility of the brake even further.

tions also put ugly marks in the surface of the metal. To eliminate this damage, make and install a simple pair of vise-jaw *covers* from soft metal such as aluminum or brass. Then you can secure the metal workpiece in the vise without damaging its surface. Using the accompanying drawing for reference, make a pair of vise jaws from 0.063—0.125-in. *soft* aluminum or brass sheet. When you have to hold a piece more securely, but don't care about surface damage, the jaw covers can be removed temporarily.

SHEET-METAL BRAKE

Probably the two most-common large sheet-metal tools found in a metal shop are the *sheet-metal bending brake* and a *squaring shear.* These two pieces of equipment are used to bend and cut large sheets of metal.

The sheet-metal bending brake, or *leaf brake,* is basically a very simple machine. It has few moving parts. Due to its leverage it can easily and accurately bend large, awkward-to-handle pieces of metal. The bending brake enables the worker to perform tremendously varied bending jobs with great accuracy.

Although the bending brake's basic function is to make simple bends, there is no end to the possible combinations of those bends. The final product is only limited by the imagination and planning of the worker. It's like a sculptor using a simple material such as clay. How he uses and combines it makes the difference.

Like any piece of equipment, each sheet-metal bending brake has its limitations. Some brakes have the capacity to bend heavier-gage metal than others. Some can bend *pans*—a rectangular or square shape with folded sides. There are brakes that can bend various radii with the use of special dies.

Brake length determines the total length of bend possible. For instance, a 4-ft brake can bend no more than 4 ft of sheet metal. If cost isn't a problem, buy the longer brake. It does everything a shorter brake can do, and more. The point is: *Brake length determines bend length.*

Recommendations—Choose a brake that's about 8-ft long. This length will

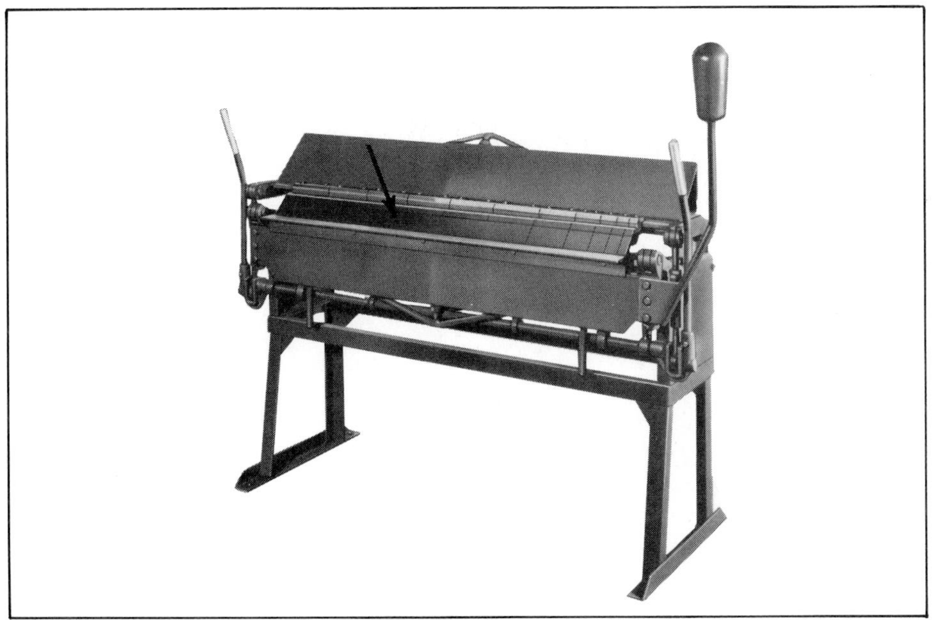

Combination brake with fingers (arrow) like this Roper-Whitney 420 would be a good addition to any sheet-metal shop. Photo courtesy of Roper-Whitney.

Williams Low-Buck Tools makes these two inexpensive brakes. Short brake will bend up to 1/8-in.-thick steel—fine for small brackets. Longer brake will bend 48-in.-wide, 16-gage (0.0598-in.) sheet steel.

Bending a transmission cover with the Williams Low Buck brake: an easy-to-use machine.

handle up to full 8-ft sheets of metal. Remember, don't limit your flexibility with too short a brake.

Combination Bending Brake—A very important feature to look for in a brake is its ability to bend boxes and pans. A bending brake with this feature is called a *combination* bending brake.

With a combination brake, you can install metal pieces called *fingers*—see photo on this page. Fingers greatly expand the use of the brake. Because the fingers protrude, you can make the second, third or fourth bend of a pan or box—at an angle to the other bends. A simple leaf brake cannot do this, because as the second bend is being made, the first bend will strike the *leaf.* Consequently a full bend cannot be made.

Radius dies—The combination brake can also be fitted with *radius dies*—special dies to form radiused bends—at the edge of the leaf. This feature means you can do fancy work such as dashboards and interior work with smooth curved bends rather than a series of sharp *brakes* or bends.

Recommendations—For all of these reasons, I'd choose an 8-ft combination box pan brake with fingers and radius dies. This brake is the most useful and versatile. I use a Roper-Whitney Model 816 combination bending brake every day.

Care of a Brake—Caring for a brake is simple. Don't overwork it by bending metals too thick for it. Each brake has a specific thickness capacity, or limit. This limit is stamped on a metal plate attached to the front of the brake. Don't disregard it. Bending metal which is too thick will spring the upper or lower beam. Your brake will then no longer be able to produce straight, crisp bends. Be careful never to bend metal with burrs or welding scale on the bend line. These irregular or hard surfaces can damage—nick or gouge—the upper beam bending edge. Bending wire rods will do similar damage.

Why is it important to avoid damaging the upper and lower brake edges? It's important because these edges are meant to clamp tightly against the bend line. Nicks or scratches on the brake edges will be transferred directly into your work. The bends will not only look bad, they'll be flawed. **Such flaws could cause structural failure of a fabricated part.**

SHEARS

Foot Squaring Shear—The second most-common piece of sheet-metal working equipment is a *foot squaring shear*—a large machine for cutting metal on a straight line. Easy to operate, the squaring shear is powered by a foot-operated pedal. It makes clean and accurate cuts very quickly. My shop is equipped with a 52-in. foot squaring shear, Pexto Model 152K. This particular shear has an automatic hold-down clamp that holds the sheet metal firmly in place while it is being cut. This is important because if the work should slip from its original aligned position, the cut would not be accurate.

A good safety feature of this shear is a *finger guard*—a device to keep your fingers clear of the cutting blade. Another safety feature is the *treadle stop.* This device protects your foot from being crunched under the treadle on its downstroke.

The 52-in. Pexto power shear is a workhorse. It's also very durable. I recommend attaching a small flourescent light above the shear to make it easier to see the cut line. Read all warning signs before using. Photo by Michael Lutfy.

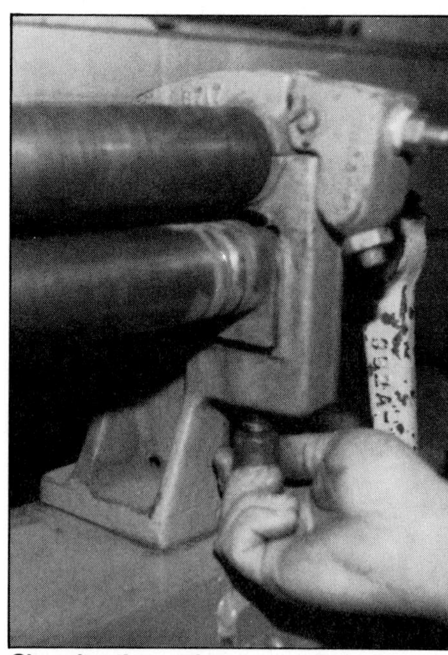
Changing the setting on one of the four adjusting knobs on this Pexto roller. Knobs move the bottom rollers up and down.

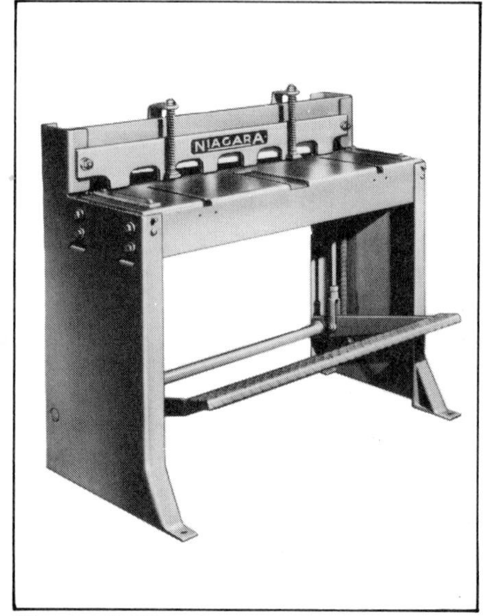

Niagara P-302 shear is tough. Metal hold-down secures the work while it's being cut. Photo courtesy of Niagara Machine and Tool Works.

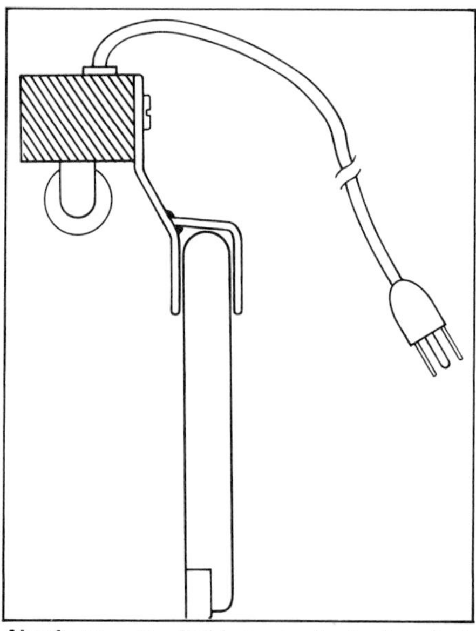
Aluminum or light-gage-steel bracket mounts a small fluorescent light directly above the shear cutting line. Light slides easily across the width of the shear.

Positioned 90° to the cutting edge, the *side gage* on a foot squaring shear allows you to measure the length of the cuts to be made as you feed metal through the shear. This is not only convenient, it saves time. The shear I mentioned has a nice feature: two side gages, one at each side, for measuring lengths of cuts. Side gages also help locate metal for 90° cuts.

Recommendations—Several sizes, or widths, of shears are available. I find a mid-size shear most useful for my purposes: over 48-in. wide, preferably 52 in. The reason is simple. Sheet metal comes in standard 48-in. widths. I know at least two guys who were *very* disappointed after they bought 36-in. shears. They were unable to cut standard-width sheet metal.

See the Line?—Side gages are great for gaging certain lengths of cuts, however it is frequently easier or necessary to cut on a blued-and-scribed or otherwise marked line. The problem is: It can be very difficult to see the line, particularly in the low light of the typical shop. You may want to duplicate a little trick I used to remedy this situation. I installed a small fluorescent light directly above the shear blade. As you can see from the photo and sketch, the U-type support bracket allows free movement of the light from side to side. This lets me move the light along the shear edge to the exact spot I need light most.

Care of Your Shear—A sheet-metal shear needs little care. Just keep it lubricated and clean. Don't cut across scaly surfaces or welds, or cut wire or rod. This will knick or dull the blade. Don't try to cut material that's thicker than your shear was intended to cut.

Like a bending brake, a shear has a small plate riveted to it that lists its *cutting capacity*—maximum thickness of different metals it can shear.

ROLLERS

Slip Roller—A *slip roller,* or *roll-forming machine,* is a tool used to roll flat sheet metal into a tube or cone. This machine consists of three long rolls, or rollers: one top roller centered over two bottom rollers.

Except for the additional roller on the bottom, a slip roller looks and operates similar to a wringer on an old washing machine. While the rollers

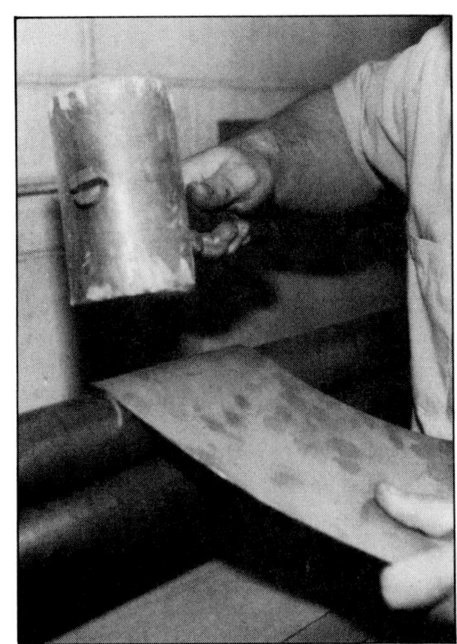

First inch or so at each end of the blank cannot be bent by the roller. Make these bends with a mallet over a pipe in a vise. Don't hammer on a roller like this. Rolling the blank produces a continuous curve.

are being turned, material to be rolled is fed between the top roller and the two bottom rollers. The material bridges the bottom rolls. The top roller exerts pressure against the material midway between the bottom rollers. This causes the material to bend—how much the material bends depends on the amount of pressure. Adjusting the rollers closer increases the pressure against the material, increasing the bend or decreasing the bend radius. This adjustment is accomplished with a knob at each end of the lower rolls.

Recommendations—Rollers are available in several sizes, or lengths. I recommend the 36-in. slip roller because I've found it will do most auto-fabricating jobs. I have had only one occasion during the last few years to roll anything longer than 36-in. The roller I prefer is Pexto's hand-operated Model 381D. With it you can roll metal up to 36-in. wide, and down to tubes as small as 2-in. diameter.

Cones—Cones are a basic and often used shape. They are used for tank bottoms, exhaust collectors and megaphones, wheel-house tops or simple body-panel shapes such as hoods or belly pans.

The ability to roll cones is an important feature of the slip roller. This is possible because the rollers are adjustable at both ends, allowing the pressure to be varied from one end of the material to the other. The greater the pressure difference, the greater the difference between the diameter of the big end and the small end of the cone.

P-401 Niagara roller, top, is handy for large jobs on heavy-gage metal. P-402, bottom, is for general shop use. Inset shows roller release for removing work. Photo courtesy of Niagara Machine and Tool Works.

The Pattern—To form a cone, the first and most important step is to *develop* an accurate pattern. To do this you'll first need the basic dimensions of the cone: diameter of the big and small ends, and the length of the cone, measured at the surface of the cone. Refer to the illustrations for details on developing a pattern for a cone.

After you've developed the pattern for the cone on cardboard and have it cut out, transfer the pattern to sheet metal. After cutting on the transferred line, you will have the *blank*—the flat piece of metal that will be bent and joined to make the part. The end of the blank with the large arc *must be at the loose end* of the slip roller—the end with least pressure. Because the large-diameter end of a cone has a larger circumference, or distance to travel, it will *slip* as the cone is rolled.

Tubes—When using the slip roller for rolling tubes, the rollers must be adjusted for equal pressure against the full length of the blank. This results in a smooth, constant-diameter tube.

Start bend in the middle so the blank can be rolled evenly in both directions from the center. Photo by Michael Lutfy.

To roll a cone shape, one end of the roller is tightened more than the other. As you roll, the larger end of the cone *slips* and curves less than the smaller end. That's where the term *slip roller* was derived.

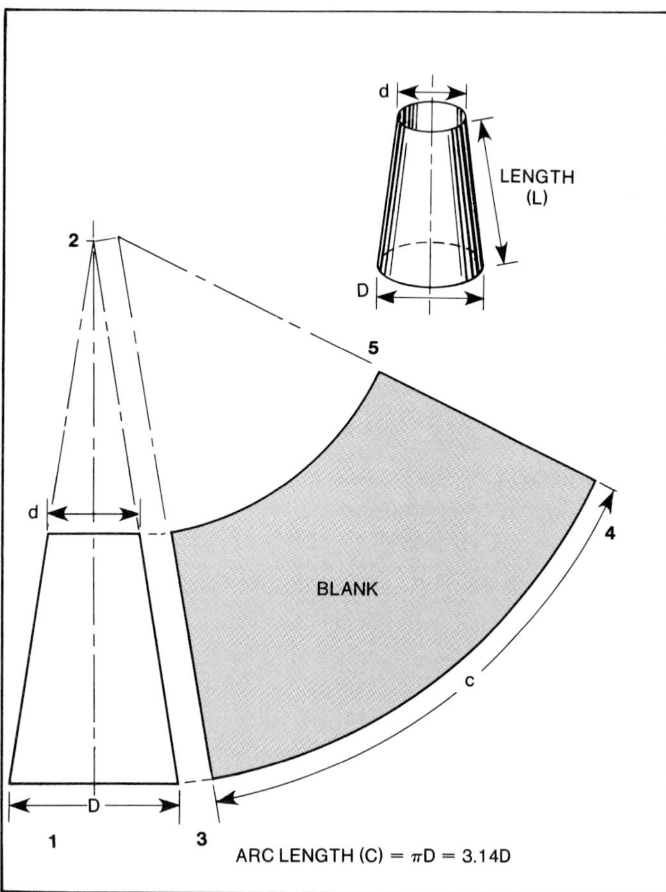

Cone development requires some knowns: diameters of the big and small ends, cone length, and circumference of the big end. Going by the numbers: 1. Draw a side view of the cone: 2. Project the lines at the sides to find the peak of the cone: 3. Draw a line parallel to the cone side. Project lines from the cone peak and and top and bottom of the cone, 90° to the cone side: 4. Swing an arc whose radius is equal to the distance between the cone peak and cone corner. Length of this arc should equal the circumference (C) of the big end: 5. At the end of the arc, draw a line from the end of the arc to the center of the arc. Draw a second arc between the two lines and it will automatically be the correct length to form the cone. You now have an outline for your cone blank.

Remove blank from the roller by lifting the upper roll using the cam lever.

Good-looking, functional beads can be put in aluminum and steel with the Pexto 622-E beader. I'm beading a fuel-cell-container end.

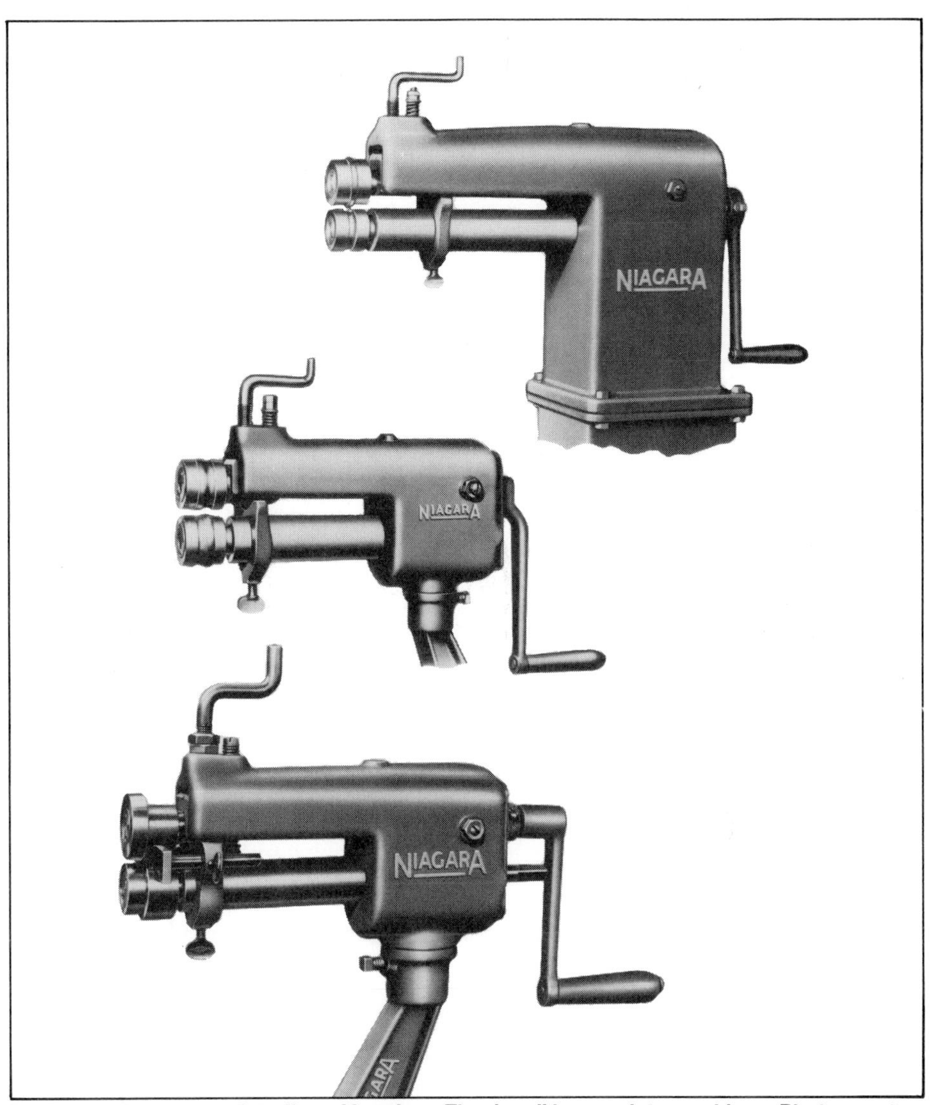

Niagara makes a complete line of beaders. They're all heavy-duty machines. Photo courtesy of Niagara Machine and Tool Works.

Equal end pressure causes smooth uniform bending of the metal over its full length.

There's a trick of the trade for rolling tubes and cones. The slip roller cannot bend approximately the first and last inch of metal. You can hand form these edges *before* rolling the piece with a mallet over the top roller as shown in the photo, page 35. To do this, lay each straight edge of the blank over a round section such as a tube with the desired curve. Then gently use a wooden mallet or slapper to curve the leading and trailing edges of the piece. After hand-forming, use the roller to complete the cone or tube.

When you finish rolling a tube or cone, it will be curved up around the top roll with the ends butting. To remove your work push the cam lever. This allows the upper roll to raise at one end so you can slip the workpiece off.

Care of a Roller—As with most metalworking machines, the slip roller must be kept clean and lubricated. Likewise, slip rollers also have a plate that lists its limitations. Respect these limits to preserve the machine.

BEADERS

Beaders are hand-operated metalworking machines which form *beads*—a rounded groove or depression in the metal. The main intent of a bead is to strengthen sheet-metal panels. Beads are also decorative.

How it works—A beader works by turning a hand crank. This drives an upper and lower die. The dies pinch the metal piece placed between them, forcing the metal into the die shape. A crank handle above the upper die adjusts the pressure to vary the depth of the bead.

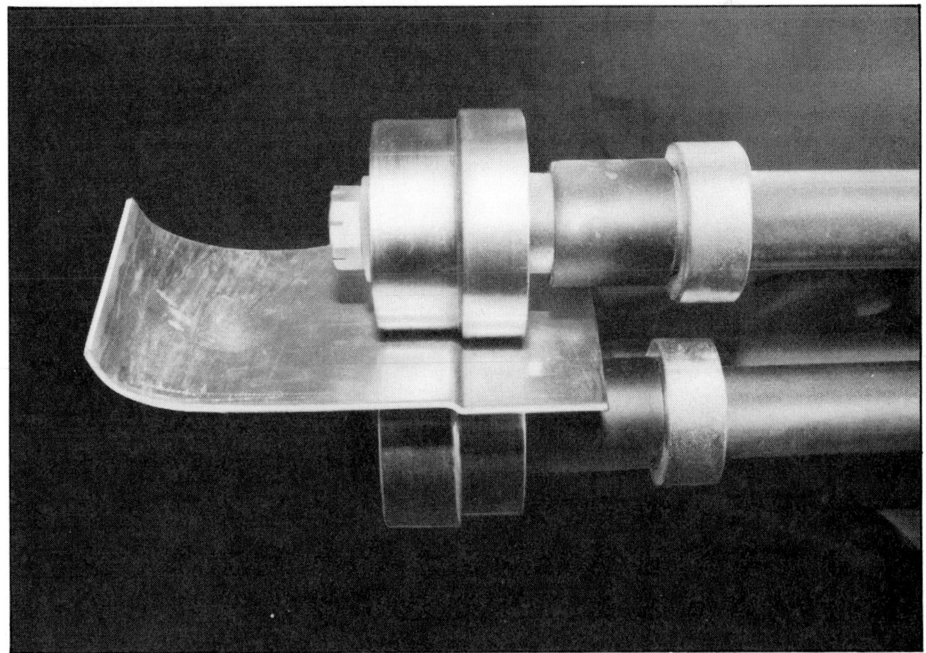

Stepping dies for your beader are easy to make on a lathe. Williams Low-Buck Tools supplies die blanks for their beaders.

Machining stepping dies in a lathe. Turn down the diameters to get an offset equal to one metal thickness. In this case, the offset was 0.063-in. Offsets on the dies should be opposite one another. Also, the dies must be *very smooth.* Polish them if necessary.

Another use for a beader: indenting, or tipping to form a curved flange. You can easily follow a line on the metal piece with a flat die underneath and a die with a narrow, radiused edge on top.

Beading dies came with this Williams Low-Buck beader. I bolted the beader to a good strong stand. Note the beader's deep throat.

Recommendations—I have two beaders. One is a Pexto Model 622-E with a 7-in. throat and a large selection of beading *dies.* See the photos. It can form many different beads in metal up to 7-in. from the metal's edges. I use this beader most often and recommend it, or other brands with a 7-in. throat, as best for the general craftsman. It can handle up to 1/16-in.-thick *soft* aluminum or up to 0.040-in.-thick mild steel.

I also have a heavy-duty Barth beader. With this beader I can bead 13-in. from the edge of metal. It is also capable of forming heavier metal, such as 0.063-in.-thick mild steel. This heavier-duty beader is very handy. Unfortunately, the heavy-duty Barth beaders are no longer available. You may be able to find a used one. If you don't have any luck, Niagara offers a comparable beader.

Stepping—A beader's most common function is to roll beads. However, the usefulness of the machine can be greatly expanded. I use mine to *step* or *joggle* metal—to offset an edge by one metal thickness so another panel of the same thickness can be overlaid evenly. This process requires special dies. These are easily made on a lathe as shown.

Indenting—An extra important use for a beading machine is to form a small indentation, or crease, in a piece that's to be bent. This indentation *on the bend line* serves as a guide for making the bend. As you can see from the photo, this operation requires relatively simple, but special dies.

The indentation enables you to execute more accurate, sharper bends, particularly bends that aren't straight, like a fender wheel-opening flange. About the first 15—20° of the bend is formed by the beader. The metal will then accept the remaining bend more accurately and easily.

To complete the bend, hold a dolly against the backside of the metal at the base of the bend while you tap with a hammer from the other side. This will stretch the metal to the desired angle. Using this process to make a bend avoids drifting off of the bend line.

POWER EQUIPMENT

So far all of the large metalworking equipment I've discussed has been hand-operated. Let's now take a look at the three basic metalworking machines which are powered by electric motors: the bandsaw, drill press and combination finishing machine. These pieces of equipment are essential to a well-equipped metal-fabrication shop. In addition, a lathe and mill are great to have. If you can afford them, great. Not many amateur craftsmen can.

BANDSAW

Cutting metal can be done in a number of ways, depending on whether the cut is straight or follows a curved line. Metal can be cut with a hacksaw, hand shears, power shears or an acetylene torch. It can also be cut on a foot squaring shear. And it can be cut on a *bandsaw*—a continuous saw blade that is powered by an electric motor.

A bandsaw cuts quickly, with great accuracy and little effort, regardless of whether the cut is straight or intricately curved. When used properly, a bandsaw can be very accurate. The edges of cut pieces require little cleanup work, which saves time.

A bandsaw can easily cut metal of nearly any thickness with the proper blade and cutting speed. I have found my bandsaw especially useful in cut-

Get a bandsaw that has its own blade welder, and one that is multispeed to cut various materials. Also, always wear safety goggles and use a push block. Never wear gloves, either. Photo by Michael Lutfy.

ting heavy-gage metal for brackets. A bandsaw is equally as handy for cutting light-gage metal quickly and efficiently. As an extra benefit, I also use my bandsaw for cutting out slappers and *hammerforms* from wood. Hammerforms are covered in Chapter 8.

A bandsaw has various speed settings and a choice of interchangeable blades. Each blade and speed setting is suited to a particular material to be cut. Wood, aluminum and steel each require a different blade and speed combination. Blade type and speed are usually indicated on a chart attached to the bandsaw, as shown above.

The worktable of a large bandsaw can be tilted for angle cuts. The blade guide can be raised and lowered to accommodate larger or smaller pieces. A light mounted above the worktable offers a clear view of the work. Some bandsaws even have a *chip blower*—a flexible-metal air hose for blowing away chips and keeping the cutting line clearly visable. Look for these features when shopping for a bandsaw.

Recommendations—I would buy a 20-in. bandsaw if I were shopping for a new bandsaw tomorrow. This 20-in. *width*—distance from the saw blade to the saw column, or back of the machine—would allow me to cut off 20 in. from my workpiece. Another popular bandsaw size is 14 in. I prefer the larger one because it can do more.

Many good bandsaws are available.

Blade type and speed are usually specified on a chart that accompanies or is attached to the bandsaw. You must use the correct blade and speed to cut any material efficiently. Generally, hard material requires a fine-tooth blade at a low cutting speed: Soft material requires a coarser blade at a higher speed.

Rockwell-Delta, DoAll and Powermatic make great 20-in. bandsaws. Their smaller 14-in. and 12-in. bandsaws are equally as good. The 12-in. Craftsman bandsaw, although small, is another good one. The Rockwell-Delta 14-in. bandsaw is OK for the general workman.

Blade Welder—Some bandsaws have a blade welder and grinder. This is important because you can buy saw blades in bulk, then make blades to the length your machine needs. This feature is a real money-saver. For example, I can buy one 12-ft welded blade for about $15.00 for my mane—but I don't.

I buy my bandsaw-blade rolled up in 100-ft lengths for about $48.00. I can make about 8-1/2 blades from the 100-ft roll. The difference: $1.25/ft compared to 48¢/ft—a 61.6% savings! As for the 1/2-blade, I use it when I get my next 100-ft roll, so there's no waste. If you use your bandsaw much, blade cost is significant. So take advantage of this savings. Although the dollar amounts change with time, the percentage savings should remain about the same.

Bandsaw Safety—I consider the bandsaw the number-1 finger-eater of all of the equipment mentioned in this book. Accidents are frequent because the bandsaw cuts at high speed

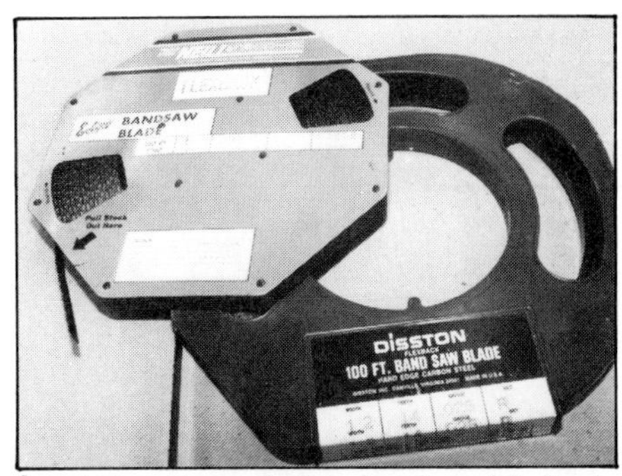

Lowest cost way to buy bandsaw blades is in 100-ft lengths. Cut and weld the blades as needed.

Possibly the most-often-used piece of equipment in a metal fabrication shop is the 12-in. disc sander. With so much cutting and welding going on, there's always a rough edge to sand.

and makes sharp, hot chips. I advise that you use two safety measures: **Wear a full face shield and use a wooden push block to guide and push your work into the blade.**

The face shield protects your face and eyes. The wooden block will keep your fingers at a distance from the saw blade. It's also easier to push against a wood block than the edge of a piece of sheet metal.

Care of a Bandsaw—To get the best use from a bandsaw, adjust the blade guide as close as possible to the piece being cut. This will increase cutting accuracy and prolong blade life. Blade tension should be kept adjusted so the blade won't slip on its drive wheel while cutting. Cutting speed is very important to blade life and efficiency. Refer to your bandsaw's instruction manual for recommended speeds for each material, and *stick to them.*

Clausing drill press is high quality. This one has variable speed in high and low ranges. Some even come with reverse. Note drill-press vise. A vise or C-clamps should be used to hold work that's being drilled.

Combination sander. Dust collector mounted to the rear of the machine makes for cleaner air in the shop. Photo courtesy of Rockwell-Delta.

To maintain a bandsaw, follow two rules: Keep it clean, and keep it oiled. The machine produces lots of chips, grit and dust. Cleaning a bandsaw often prevents those contaminants from damaging the machine itself. Oiling it often lets it operate smoother and with less wear. Stick lubricant is available for lubricating the blade.

FINISHING MACHINES

Disc Sander—A disc sander—also called a *disc finishing machine*—is often used in conjunction with a bandsaw. A disc sander is a large electrically powered machine used to sand metal or wood. It has a 12-in. paper-backed coated-abrasive disc positioned 90° to an adjustable work table. Some disc sanders even have a vacuum dust collector.

The disc sander is great for cleaning up or smoothing metal or wood pieces quickly. Different *grit* (coarseness) sanding discs are available. The selection of the abrasive depends on: finish desired, amount of material to be removed, and the material you're working with. After choosing the right abrasive, simply peel the back off the coated-abrasive paper and apply it to the disc.

Combination Sanders—A combination sander does what the name implies. It combines the disc sander with a *belt sander.* The belt, also a coated abrasive, but on cloth rather than paper, is a 6-in. wide band or belt mounted on two rollers.

Used in most metal shops, the combination belt sander is a compact unit that saves time, space and cost. The belt sander is best for the larger pieces, while the disc is handier for smaller work.

Recommendations—There are several good disc and combination sanders available. I've used a Powermatic combination sander daily for 15 years and it has never given me any major problems. Therefore, I can recommend it highly.

Use the same precautions with a sander as you would with a bandsaw: *Face protection and a wooden push block are necessary to avoid having chips fly in your eyes, or sanding your knuckles.* It's not always possible to use a push block, but use one when you can. I've suffered eye and hand injuries, and recommend neither.

Care of Finishing Machines—The same rules aimed at prolonging the life of a bandsaw apply to a sanding machine. Keep it clean and lubricated. Frequently clear away any grit, dust or chips. This will prevent the sander from fouling. Oil it often. Do these two things and your sander will stay in top working order.

DRILL PRESS

Like the hand drills discussed in Chapter 2, a drill press is used for drilling holes. There is one big difference: Instead of bringing the drill to the work, the work has to be brought to the drill. A drill press allows drilling many holes very quickly and precisely.

A 15-in. drill press with variable speed and a 1/2-in. chuck is the type most often found in a metal-fabrication shop. The 15-in. dimension is measured horizontally from the spindle center to the column.

The 15-in. drill press can do most jobs. Important features are the variable speed and a movable worktable. The variable-speed *spindle* in which the chuck is mounted allows for drilling different materials. The movable worktable adjusts vertically and swings horizontally to accommodate large pieces. A Rockwell-Delta 15-in. floor model has served me very well for many years.

Assuming you can get the work to the machine, a drill press is excellent for drilling with hole saws. There are two good reasons for this: Cutting speed can be adjusted low enough to preserve the hole saw, and it is very easy to secure the workpiece. The result will be a very clean and accurate hole, exactly where you want it.

Drill-Press Vise—There are several ways of securing work in a drill press. You can usually hold the piece in a

Rotex turret punch press saves time punching a wide range of holes, but it's also very expensive. It's best for punching holes in thin sheet metal. Any metal fabricator would love to have this machine, but only high-dollar, high-volume shops generally have them. Photo by Michael Lutfy.

drill-press vise—a vise meant especially for use with a drill press. When the piece is too big or awkward for the vise, it can be clamped to the worktable. If necessary, you can even bolt the piece to the worktable, in the name of safety.

Care of a Drill Press—A drill press gets dirty easily from metal cuttings and oil. Clean it by removing debris often. Wipe the worktable clean before and after each use to keep foreign objects from getting under the workpiece. Such objects can cause inaccurate drilling. So keep a drill-press table clean at all times. Keeping it lightly oiled will keep it unmarred longer.

Pay close attention to how far down a drill bit travels so you can keep the drill bit from touching the worktable. Constantly running a drill into the worktable will gradually ruin the work surface.

TURRET PUNCH PRESS

A Rotex *turret punch press* is a large hand-operated punch. It will punch holes repeatedly and accurately in mild steel or aluminum. It can handle up to 10-gage sheet steel (0.1345-in. thick). The greatest advantage of this machine is how quickly and easily it can make an accurate hole. Also, punch sizes can be changed very quickly.

Recommendation—Rotex Model 18-A turret punch press—see above photo—has 18 different punches ranging from 1/8- to 2-in. diameter. The punch throat is 18-in. deep. This offers great flexibility in using the machine. It is a *very* expensive piece of equipment, costing thousands of dollars—up to $5000! However, it is a terrific time and trouble saver. If you can buy one, great!

LATHE AND MILL

If you can afford them, or if you have access to a lathe and mill, you are fortunate. Although they are not absolutely essential to a metal fabrication shop, they sure are handy.

When a job requiring either a lathe or mill comes along, you can farm the work out. For instance, I've built several Indy-car tubs and their suspension components. Most of the job was sheet-metal fabrication, but it also required some special machined components to be used in the suspension and tub *bulkheads*. I simply took the blueprints detailing those parts to a professional machine shop for machining.

Recommendation—Rather than recommending a specific lathe or mill, I suggest that you get one that suits your needs or budget. Do this after doing a lot of looking and talking. A small machine shop or machinery dealer are good places to go for advice. Cost: It will be several thousands of dollars.

BENEFITS OF LARGE EQUIPMENT

The major benefit of having these pieces of equipment is making special parts and tools necessary for a particular fabrication job. In Chapter 7 on metal-shaping, you'll see how a small lathe is useful.

Large sheet-metal equipment helps to form flat metal into the sophisticated shaped parts that can be admired. This equipment is strong, reliable and sometimes indispensable. Respect its use, and learn to apply it well. You will find machinery is like a musical instrument—it produces results in direct relation to the practice and talent of the user. The instrument doesn't make the music, the musician does.

PROJECTS: GETTING STARTED

A total project like this clean-sheet-of-paper frame and suspension requires considerable planning. It also requires a working knowledge of materials and and how to apply them. Photo by Tom Monroe.

Almost everyone can *envision* a project. Almost as many people *talk* about a project. Many *begin* a project. Some people even *work* on a project. But few people ever *finish* a project. Why is that?

I think some people quit in the middle of a project due to a lack of interest or funds. But many more unfinished projects are the result of jumping into them not knowing what's necessary to carry a project through. You need to understand how to shepherd or guide your basic idea through to a pleasing final product. In this chapter I describe how to get from your idea—through the project—to the final product.

PLANNING A PROJECT

You must think a project through completely *before beginning it.* The time and energy used for tossing an idea around in your head is the first step toward ensuring a quality end product. The restless turning over of ideas and possibilities is definitely not time wasted. However, some people just don't know how to think over a project. They may be preoccupied with a paint color and a paint scheme instead of estimating material costs. Get your priorities in order first.

There are some basic considerations every project requires. Let's talk about them.

Need—Any project begins with a need. What is it that makes you want to do a particular metalwork project? Is it a functional addition to an existing vehicle? Is it a special vehicle that's unavailable unless you build it? Is it a project intended to improve appearance or offer a custom look? Will it combine function and styling? Is it a repair? You need to know your goal very clearly.

Determine exactly what it is you want to do. If there's a trade-off on style or function, decide which is more important to you? Write down your exact project requirements. Which feature/s can or cannot be compromised? Many of your choices will be determined by ease of construction, enviromental considerations, and which features are most important.

Example: For instance, let's assume you are going to build a fuel tank. Before starting the actual fabrication, you must decide how many gallons it must hold and where it has to fit. With these in mind you must also decide which is more important: fuel-tank capacity or location.

Location could affect how big you can make the tank and therefore how much fuel the tank will hold. Each gallon requires 231 cubic inches of space. Capacity and location are interrelated. Which is more important, how much the tank holds, or where it goes? Assuming there's a problem, you must either compromise, or find a way to fill both requirements at the same time. I cover this problem in Chapter 11.

By defining what your project must achieve and placing the features in order of importance, you will automatically point yourself toward the correct design, materials and techniques. Keep your goal and priorities in mind every step of the way.

HOW TO DO IT

The next step in working out your project is to ask yourself some very pointed questions—then *answer them:* How big a project is it? What will it cost? How long will it take? Where can I do this? What materials can I use? Will it require welding, bending or forming? Can I do all phases of this work myself? Is special skilled help available?

Talk to experienced workmen. Write down their suggestions. Check with various metal suppliers. *Write down the information they give you. Take time to get estimates of any outside work so you can establish costs.* The yellow pages of your telephone directory is a good place to locate metal suppliers and custom workmen.

It is extremely useful to talk to someone who has done a similar project. Find out how long it took and how much it cost. Remember, your time has to be planned. *When* can you do the work, and *how long* can you work at a time? Total time for the project almost always is affected by how long you can work at each session. You must also have a specific place to work. Will it be large enough and

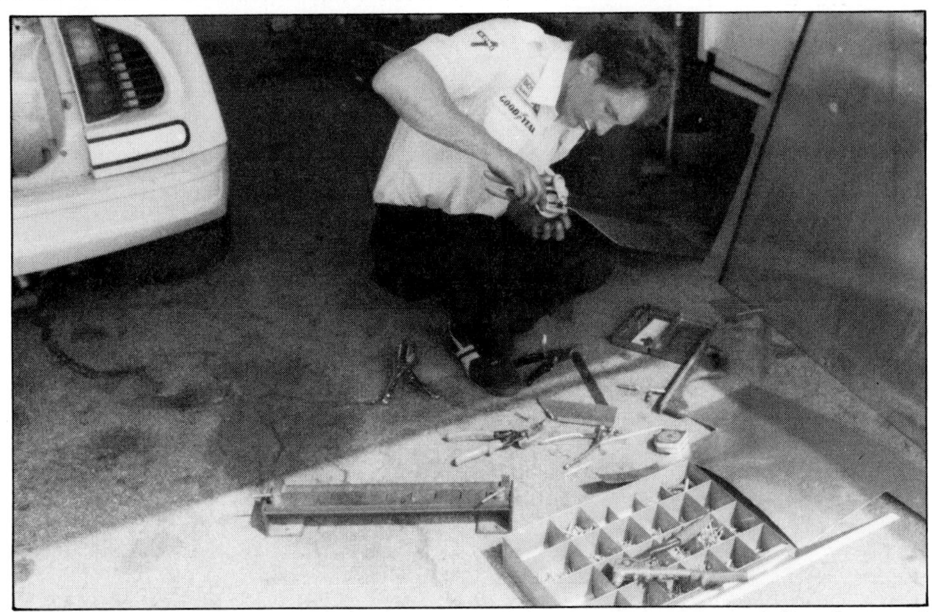

This fellow came to Road Atlanta well prepared. He's using many of the tools I've discussed and has a little trackside fabricating shop for his team. Mini sheet-metal brake is in foreground at left.

Every sheet-metal job can't be as attractive or simple as you'd like. Certain needs meant this transmission and drive-shaft tunnel had to be box-like rather than rounded. It worked fine, but I felt like I'd just gift-wrapped an elephant.

This is home to me: a clean, well-lit and organized shop. You can't see the fire extinguisher on the end wall, but it's there! I was anxious to get started on this drag-car frame.

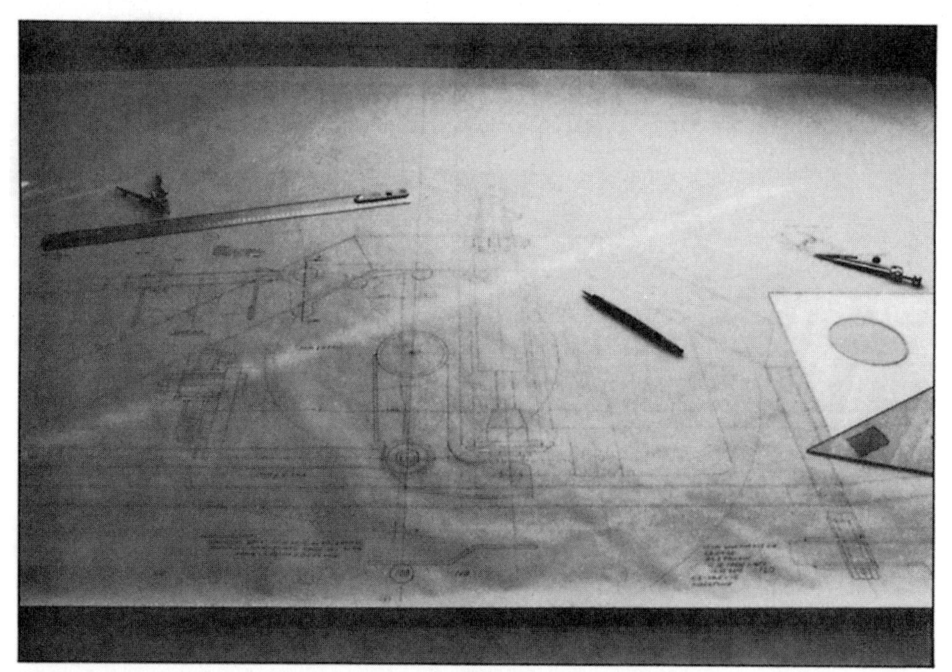

An engineered, full-size design is the best way to do a large fabrication project. Drawing is Tom Monroe's design for the car shown on pages 42 and 67.

Finished product based on the tub shown under construction on pages 48 and 49. Car was built for and raced under the direction of master Indy-Car mechanic, George Bignotti, in 1977. It appeared as shown in 1978 with Janet Guthrie driving. Photo courtesy of Indianapolis Motor Speedway.

available when you want to work? Do you have the necessary tools and equipment, or will you have to find them?

After all of these items are considered and resolved, you are ready to go on to a plan, a drawing or a blueprint—depending on the complexity of your project and your own experience. Correcting a mistake with an eraser is much easier than with a cutting torch or saw.

DESIGN

In most cases you can make a simple *full-scale drawing*—one inch on the drawing equals one inch in real life—of your project. For example, an air scoop or a fuel tank might be built from a full-scale drawing by a mildly experienced person without too much difficulty. For a complex project such as a race-car chassis, you must use the design expertise of an automotive engineer. The drawings or blueprints, if needed, must be detailed enough to serve as clear "road maps" for your work.

The amount of drawing detail needed for a specific project depends on both the complexity of the job and the extent of your skill. When in doubt, get more information and drawings from a professional designer. The cost of professional help usually saves trouble and money in the long run. It is well worth the cost.

A full-scale drawing is a good guide for a project for several reasons. It is clearer and easier to read than a reduced-size, reduced-scale drawing. I find using a full-scale drawing makes me less likely to make a mistake.

When I get to the stage in a project which requires fitting a component or a part into its intended place, the full-scale drawing enables me to do a trial-fitting. I can actually hold the given part against the drawing to make a quick check. This is impossible with a reduced-scale drawing. I must make many measurements and calculations during the fabrication and installation of a part. This increases the possiblity of an error. Each measurement or calculation is just another chance for a mistake.

Clarity—Keep a design as clear, clean and simple as possible. Any fool can make things complicated. Although it will take more thinking and rethinking, a simple design is easier to execute in metal, and costs less in time, sweat and money. And it generally looks better. It may take more time to develop a good design that suits your intended purpose, but it will be time well spent. Remember: **Keep it simple.**

CHOICE OF METALS

The choice of materials is crucial to a design. **An incorrect choice of materials could cause a failure.** The wrong *alloy*, or the wrong thickness of the right alloy, or the wrong *hardness* of the right alloy will cause serious problems. Read on for an explanation of these terms.

Serious problems mean different things: They could mean weld failure resulting in leaks or cracks. If the wrong alloy is used for an oil tank, a weld failure can cause leaks. A collapsed or broken control arm could be catastrophic. The wall thickness of a suspension control-arm tube must be adequate for the loads it must take. Using magnesium for a firewall could be disastrous because magnesium itself is an extreme fire hazard. **Serious mistakes in metal choices could be deadly.**

Less-serious problems are also caused by the incorrect use of materials, maybe only to the point of being troublesome. I know a pretty good fabricator who got tired of bending and breaking his snow shovel. So he made a *real* sturdy one. Only trouble was the thing ended up weighing about 14 lb. Snow shovelling with it was a job for Superman. He got a strong snow shovel; and he got a heavy one. It was good for a laugh between us, though.

Become familiar with the metals most frequently used in custom automobile fabrication. Three common steels and three common aluminums are used for this work. Although many other steel and aluminum alloys exist, you will probably only need to know about the six basic ones used in general car fabrication—and it will be to your advantage to know them well.

Metals are graded in several categories. First, a metal is coded by number according to how it is *alloyed.* An alloy is another element—manganese, silicon or chromium, for example—combined in small percentages with a metal during its manufacture. *Alloying* tailors a metal's properties: strength, hardness, and corrosion resistance. Other properties you'll be concerned with include how a material can be machined, heat treated and formed. Letter suffixes followed by an additional number/s with aluminum-alloy designations indicate mechanical or thermal treatment during manufacture. Heat treatment and mechanical working, or forming, of an alloy affect the properties—*condition, or temper*—of the aluminum or steel.

Second, metals are described according to the thickness of the sheet. Or, in the case of tubing, they are described by wall thickness. Thickness is expressed in *inches* in fraction or decimal form; or in a *gage. Gage* is a metal-industry term for metal thickness.

In addition to thickness, sheet metal is also described by its length and width. Round tubing is described by length, inside diameter (ID) and outside diameter (OD). For square or rectangular tubing, length, outside dimension/s and wall thickness are necessary. When ordering metal you will need to know the alloy number, condition—in the case of aluminum—and dimensions.

U. S. Standard Sheet-Metal Gages for Steel
For accuracy, always specify material in decimals of an inch when ordering material.

No. of Gage	Thickness (fraction of inch)
3	0.2391
4	0.2242
5	0.2092
6	0.1943
7	0.1793
8	0.1644
9	0.1495
10	0.1345
11	0.1196
12	0.1046
13	0.0897
14	0.0747
15	0.0673
16	0.0598
17	0.0538
18	0.0478
19	0.0418
20	0.0359
21	0.0329
22	0.0299
23	0.0269
24	0.0239
25	0.0209
26	0.0179
27	0.0164
28	0.0149
29	0.0135
30	0.0120
31	0.0105
32	0.0097
33	0.0090
34	0.0082
35	0.0075
36	0.0067
37	0.0064
38	0.0060

When ordering material, specify thickness in decimal form. Use this chart as reference when converting from gage to inch/decimal.

Material	Alloy	Temper	Thickness X Width X Length	No. of Sheets	Cost
Aluminum	3003	H-14	0.0625 in. X 4 ft X 8 ft	1	$XX.XX

Material Ordering—When dealing with any metal, you should be familiar with the language. It is a kind of shorthand language which describes the metal. It is not difficult, and you'll be able to order materials more efficiently if you know how to specify *exactly* what you want. How a metal supplier typically lists aluminum sheet is shown above.

ALUMINUM

Aluminum Alloy—Alloys used for aluminum sheet or tubing are combined with silicon, iron, copper and manganese. Many alloys also include magnesium, chromium, zinc and titanium.

The first digit in the four-digit code indicates the major alloy, or element, used to create a specific alloy. For in-

STRESS, YIELD POINT, TENSILE STRENGTH & ELONGATION: Terms you should be familiar with

To choose a material, be it aluminum, steel or whatever, you need an understanding of some basic terms used in metals catalogs. Even though these are engineering terms, you need a working knowledge of them.

Stress—Because one term builds on another, let's start at the beginning: *Stress,* usually expressed in pounds per square inch (psi), is the load in pounds for every square inch of *cross-sectional area.* For instance, if a 1-in. square bar is *pulled* with a 1000-lb load that tends to *stretch* the bar, the bar is stressed to 1000 lb ÷ 1 sq in. = 1000 psi. Changing the load changes the stress in *direct proportion.* Changing the cross-sectional area changes the stress *inversely*—halving the load halves the stress, but halving the area doubles stress, and vice versa. In simple formula form: $S = P \div A$; where S is in psi, P is expressed in pounds, and A is in sq in.

Although there are different types of stress, I purposely specified a pulling load because it creates a *tensile stress*—a stress that results when a material is being pulled. Why tensile stress? Stress ratings for metals are tensile stresses.

The two types of stress typically listed in metal-stock catalogs are *yield point* and *tensile strength.* First, understand that *any load* applied to *any metal* will cause it to deform—some will deform more than others. As load increases, so does deformation. When the metal is unloaded, it returns to its original shape—until it is stressed just beyond its yield point.

Yield Point—When a metal reaches its yield point, it will continue to deform, or yield, without any corresponding increase in load. Some metals will continue to yield, even though the load may reduce slightly! When this load is removed, the metal will not return to its original shape, but will remain permanently deformed.

Tensile Strength—Often called *ultimate strength,* tensile strength is the *maximum stress* a metal can withstand before it fails.

Percent Elongation—*Percent elongation* is the ratio of the deformation of a metal, immediately before it fails, to its original length. For instance, if a 2-in. length of tube is stretched to 2.40 in. before it fails, or breaks, its elongation is 20%, or (0.40 ÷ 2) X 100 = 20%. Percent elongation is important because it is an indication of a metal's *toughness.*

Toughness is the ability of a metal to absorb an impact load—very important in automotive applications. This is particularly true of many race-car components: roll bars, cages and frames for instance. Rather than breaking, the material gives, absorbing energy in the process. As a result, the stressed component will not be as highly loaded or stressed as a less-tough component with a higher tensile strength.

An indicator of toughness that's similar to percent elongation is *percent reduction of area.* When a metal fails in tension, it *necks down*—"pinches in." Pull a piece of putty apart and you'll see what I mean. The cross-sectional area through the break is smaller than it was originally. Generally, the more a metal necks down—has a higher precentage reduction of area—the tougher it is.

Yield Point vs. Tensile Strength—Now that you have an idea of what percent elongation and reduction of area are, you probably won't find it in any metal catalogs. However, another good toughness indicator is the proximity of a metal's yield point to its tensile strength. The farther away they are, the tougher the material.

For example: Mild steel has a yield strength of about 35,000 psi, and its tensile strength is 63,000 psi—1.80 times higher in tensile than in yield. At the other end of the spectrum, a popular high-alloy steel has a yield of 60,500 psi and a tensile strength of 95,000 psi—1.57 higher in tensile. These materials have an elongation of 38% versus 26%, and a reduction of area of 62% versus 52%, respectively. The high-allow steel is stronger, but the mild steel is tougher.

Tom Monroe, Registered Professional Engineer, Member SAE

stance, alloys with a 3 as the first digit like that in the above example are in the manganese group. Other alloy groups include: 2XXX, copper; 4XXX, silicon; 5XXX, magnesium; 6XXX, magnesium and silicon; and 7XXX, zinc. 3XXX or 6XXX—manganese or magnesium/silicon—aluminum alloys are the ones most often used in race- and custom-car metal work.

Temper—Temper, or condition, is indicated by the code-number suffix letter. This letter indicates how the aluminum was manufactured and tells something about its strength. Reading down the following list from F, the manufacturing processes become increasingly complex, but result in stronger alloys:

F—as fabricated—no additional processing.

O—annealed—softened by heating and cooling.

H—*strain hardened* by cold-working.

T—heat treated in a copper solution, sometimes cold-worked.

Hardness—Numbers to the right of the letter in the suffix indicate manufacturing process variations. Unique to the H and T groups, a higher number in the suffix indicates a stronger and harder alloy. The range of hardness available varies. In one group of alloys there may be many different tempers. In others, there may be fewer.

The harder a material is, the stronger it is. Example: 6061-T6 is nearly 100% stronger than 6061-T4 in *yield.* The yield strength for 6061-T4 and -T6 is 21,000 psi and 40,000 psi, respectively.

3003-H14—By far the most commonly used aluminum alloy in automotive fabrication is 3003-H14. To repeat, the basic number, 3003, is the alloy. The suffix, H14, is the condition and hardness.

H14 is *half-hard,* midway in the hardness range for this alloy. This hardness is ideal because it has good strength and is relatively easy to form.

Manufacturers refer to 3003-H14 aluminum in their literature as "very suitable for most forming operations." It is a *work-hardening* alloy with excellent welding characteristics. Work-hardening means a metal becomes harder and stronger as it is being *cold-worked*—formed while cold.

I use a lot of 0.050-in.- and 0.063-in.-thick 3003-H14 sheet aluminum

Car interior of 3003-H14 aluminum. It bends and welds easily. It's a natural for this curved dashboard.

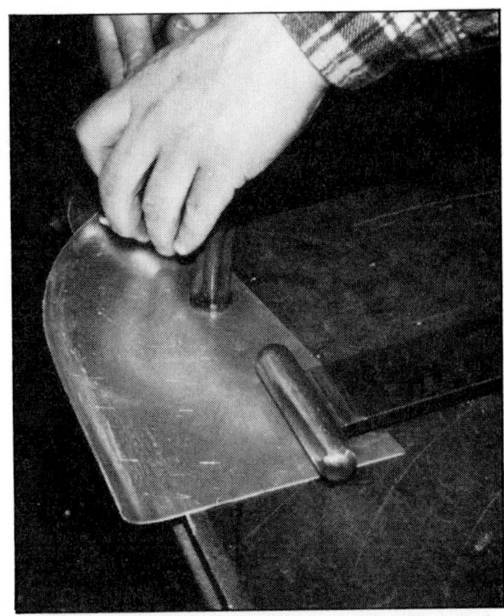

T-dolly and curved dolly were used to hand form a small interior part.

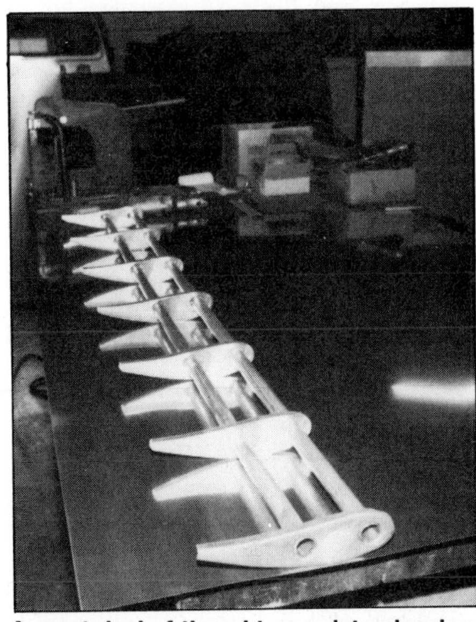

A great deal of thought goes into planning a project like a wing. An engineer designed the wing for good aerodynamics and strength. I helped choose the appropriate materials. Wing skin of 6061-T4 0.025-in. aluminum is light and strong. Bending it to fit the ribs wasn't difficult.

because it is suitable for a wide variety of applications. It's great for constructing oil tanks because it is easy to weld. I also use it for interiors because it anodizes well, and it is sufficiently strong for this and many other non-structural applications. It is also lighter than steel. It is the best choice when doing aluminum bodywork because it can be shaped easily, and the panels will retain their shape well after forming.

When using 3003-H14 aluminum you can take advantage of another one of its features—3003-H14 *work hardens.* This is good as it becomes stronger as it hardens. However, this makes it difficult to shape. The advantage is being able to correct the problem easily! Not all metals can be restored to workability once they have work hardened. You can *anneal,* or soften, the work-hardened area of this material by heating and quickly cooling it.

Using an oxyacetylene torch with its flame adjusted with excess acetylene, cover the area to be annealed with carbon. Then adjust the flame to neutral and heat the area evenly until the carbon burns off. Quickly dunk the metal in cool water. If the panel is too big for dunking, use a sponge and a bucket of water. Result: aluminum softened and ready for more shaping.

3003-H14 aluminum has another advantage. It can be easily welded with oxyacetylene. This means any job requiring welding is more easily made with this aluminum alloy, especially on parts with extensive welding.

6061—This is another sheet-aluminum alloy commonly used for car fabrication. This alloy is very strong, making it desirable for highly stressed components. The three most useful tempers or conditions of 6061 are: O, T4 and T6.

6061-T6 should be used rather than T4 when a part requires no bending or welding. Although it is the strongest of the three conditions, 6061-T6 is brittle and welding can be a problem. Welds tend to be brittle and crack easily. For the same reason 6061-T6 should not be bent at all, and 6061-T4 should not be bent on a tight radius. The high temper, or hardness, causes it to crack along the bend line. **Such cracks cause failures.** Highly stressed, complex parts with tight bends require 6061-O.

Because it's more easily formed, 6061-O or 6061-T4 is used in conjunction with 6061-T6 when a part being fabricated contains bends. The sections with the tight bends are fabricated from 6061-O or 6061-T4 and joined to the 6061-T6 panels. This is ideal for monocoque constructions.

T4 starts out quite strong, and when formed, it approaches the T6 condition *in the area of the bend.* When fabricating a complex shape

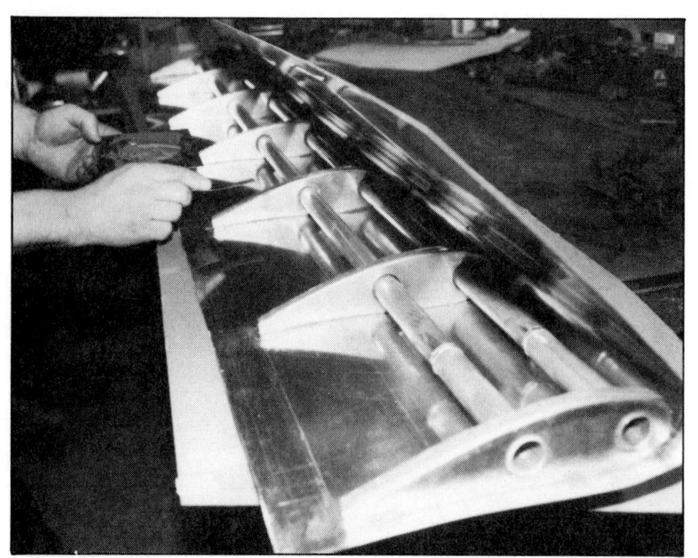

Wing ribs were hammerformed from 0.050-in.-thick, 3003-H14 aluminum. Spars —tubes running between the ribs—are 1-in. OD X 0.050-in.-wall, 6061-T6 aluminum. I'm using structural adhesive to attach the skin to the ribs.

Final wing assembly must be precise, straight and scratch-free. C-clamps and tube hold the skin tightly against the ribs for riveting and bonding. Cardboard under the wing and tape on the tube keep the skin from being scratched.

The only aluminum alloys used on this Indy-Car tub are 6061-T6 and 6061-T4. Panels requiring some forming were made from 6061-T4. Flat panels are higher-strength T6.

with tight bends and compound curves, such as a wing rib or a bulkhead for a tub, 6061-O is great. Although it's not at all strong, it also approaches the T6 condition when work-hardened. Unlike 6061-T4, 6061-O will accept tight, compound bends.

6061-T6 is joined to 6061-O or 6061-T4 by riveting *and* bonding, *never by welding*. This is to minimize the possibility of failure by cracking.

STEELS

Steel Alloys—Steel is also numbered or coded. The four-digit number describes the percent carbon content and the alloying elements in the steel. Sometimes letters are added to the basic number. These letters are generally initials for words describing the process used to manufacture the steel. A basic *carbon steel* such as 1020 is made with iron with about 0.20% carbon and traces of other elements. *Alloy steel*, such as 4130, is made of iron with 0.30% carbon and a complex mix of other elements, including its major alloying element, molybdenum.

Let's take a look at how the SAE (Society of Automotive Engineers) and AISI (American Iron and Steel Institute) classify and number the types of steel you're most likely to be concerned with. The first or first and second digits in the basic four-digit identification number indicate the basic type of steel: 1, carbon; 2, nickel; 3, nickel-chromium; 40 and 44, molybdenum; 41, chromium-molybdenum; and 43, nickel-chromium-molybdenum.

SAE-4130 steel is used for the bulkheads and roll bar. Rules require tub material to be a minimum 0.063-in. thick.

Panels and bulkheads are ready for joining by riveting and bonding. Completed tub weighed less than 100 lb.

When the second digit is not used to identify the major alloying element/s, it may be used to indicate the approximate percentage of the predominant alloying element. For instance, 25XX steel is made up of 5% nickel.

How about the last two or three digits? These usually indicate the approximate carbon content in hundredths of one percent. For example, 2520 is a nickel steel containing 5% nickel and approximately 0.20% carbon.

Luckily for you and me, we don't have to know all of the available steel alloys. We can stick to a few basic kinds, learn about them, and accomplish quite a lot with them. Let's consider the common steels used in car fabrication:

1020 CR—The most commonly used steel in car work is *cold-rolled* 1020, a *mild steel*. It is shaped when cold rather than when hot, as is *hot-rolled*. Cold-rolling produces a much stronger steel with tighter dimensional tolerances than does hot-rolling. Unlike hot-rolled (HR), cold-rolled (CR) steel has no *scale*—impurities on the surface. Cold-rolled steel has good strength, workability and it welds nicely. Mild steel is a carbon steel with a low-carbon content—0.20% in this case. Steel with 0.30—0.50% carbon is in the medium-carbon range: up to 0.95% content is considered high-carbon steel.

SAE-1020 mild steel is widely used in all kinds of race-car construction. Mild-steel 2 X 3-in. frame rails are used for this drag-car chassis.

Same drag car chassis, a little closer to completion. Lightening holes in crossmembers reduced frame weight, but not strength. Roll cage and braces made a strong mild-steel chassis.

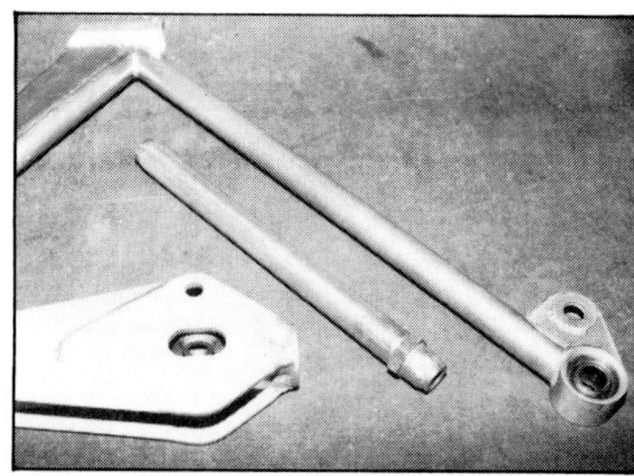

Front-suspension pieces are 4130 steel. After being preheated with an acetylene torch, they were heliarc-welded. Cadmium plating is applied to protect parts against corrosion.

I needed to form a small-diameter, deep dome to clear a water-pump pulley on this radiator-air duct. AK steel was a natural choice. Dome was easily formed with this special steel.

Cold-rolled 1020 steel is available in a number of different gages and/or thicknesses. I use quite a lot of 0.032-in.-thick 1020 CR because of its usefulness in making body panels. I frequently use 10-gage (0.1345 in.) 1020 CR also. This thickness is about right for most brackets. Thickness obviously depends upon the strength requirement of the finished piece.

SK or AK Steel—The second kind of steel you need to be familiar with is *SK*, or *AK*, steel. Although many people don't recognize it, low-carbon SK or AK steel is important to the metal fabricator simply because it forms so easily—an unusual quality for steel. It can be used when making body panels or formed brackets.

The letters SK or AK mean *silicon-killed* or *aluminum-killed,* respectively. Killing is the process of stopping molten steel from *effervescing,* or bubbling and combining with oxygen after it is poured into ingots. It is beyond the scope of this book to get into a detailed explanation of the process. Briefly though, SK or AK steel results through the use of silicon or aluminum as the deoxidizing agent, respectively. This results in a finer-grain steel. Generally, fine-grain steels are tougher and more ductile than coarse-grain steel. In addition, they do not work-harden as quickly as other steels. Therefore, SK or AK steel is very desirable for hand fabrication.

SK- or AK-steel sheets are generally available in the same thicknesses as mild steel.

Chrome Moly—The third steel you must be familiar with—one that is very popular in the aircraft and race-car industries—is 4130, popularly known as *chrome moly.* This name comes from the major alloying elements, chromium and molybdenum. Chrome-moly steel is very strong. Hot-rolled 4130 has a tensile strength of 86,000 psi compared to 69,000 psi for hot-rolled 1020—a 25% increase.

Because of its strength you can save weight by using a thinner-wall tubing. That's why 4130 is used mainly in suspension components. Again, it is available in standard thicknesses and sizes.

As you might guess, 4130 falls in the "too-good-to-be-true" category. First, it is extremely expensive. Chrome moly costs four or five times as much as low-carbon steel. In addition to the expense, it takes longer to make parts with chrome moly. It is hard, therefore it takes more time to cut. Not only that, bandsaw blades and other cutting tools dull or wear

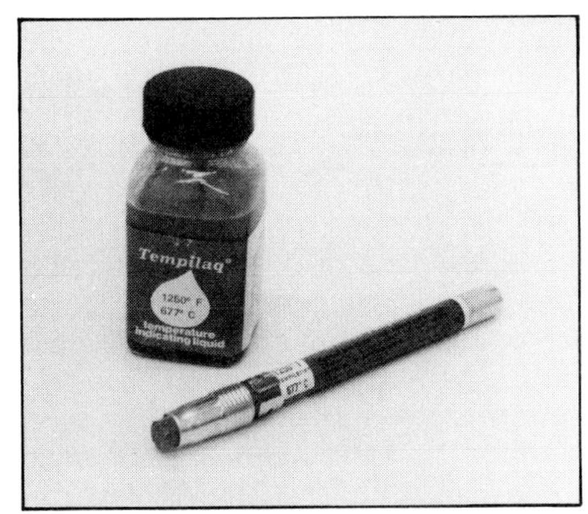

Apply Tempilaq liquid or Tempilstik crayon to area that's to be stress relieved. Tempilstik can be applied before or intermittently during heating. Tempilaq is painted on before. When metal is up to the rated temperature—1250F (677C) for 4130 steel—Tempilstick or Tempilaq melts. Let the heated area cool in still air. Do not water quench. Photo by Tom Monroe.

out much faster when cutting chrome moly. Besides being difficult to bend because of its high strength, chrome moly requires special treatment after welding or bending. Otherwise it may crack or break due to brittleness. This special treatment is discussed on the following page.

One final comment about using chrome moly—**DON'T**. There are few applications that warrant the use of chrome moly unless you're building an airplane, a Formula-1 car, or an Indy Car. Unfortunately, chrome-moly steel has become *trick*. It literally falls in the catagory of "a little bit of knowledge can be dangerous." Yes, it is stronger than mild steel. It is also *less tough* as described on page 46. Under impact, as in a crash, a chrome-moly-steel part is much more likely to break rather than bend, as would a mild-steel part. Not only does a part—such as a roll bar or cage—absorb energy as it bends, it remains intact and protects. So choose toughness when a life depends on it. However, if you have a legitimate application for 4130 steel, here are a few rules to follow.

Chrome-moly steel is frequently used for race car suspension components, such as those on Bobby Rahal's Galles-Kraco Indycar (top) and on Eddie Cheever's PC-18 Indycar (bottom). There are many different diameters of 4130 available to suit most any need. Photos by Michael Lutfy.

Which Chrome Moly To Use?—Chrome-moly steel is available in two conditions: normalized and annealed. When bending and welding is required—as for a fabricated suspension arm—use *annealed* chrome moly. Immediately before welding chrome moly, slowly preheat the area to be welded with an oxyacetylene torch. This will prevent thermal shock and cracking. Heat to about 500F (260C), or just too hot to touch. After bending and welding, the worked areas of the part become hard and brittle, so they must be *stress relieved* to prevent breakage.

Stress Relieving—Using an oxyacetylene torch, *gradually and evenly* heat bent or welded areas to 1250F (677C), or to a cherry red. Bright red is too hot. Don't do this in direct sunlight, or you won't be able to see the color. It is very important to bring the part up to temperature slowly and evenly to avoid *thermally shocking* the part—such as water on a hot light bulb—and doing more harm than good. With your torch about 6 in. from the work, *preheat* the area to be stress relieved. Once preheated, move the torch closer. *Evenly* increase the temperature of the material to 1250F by moving the torch back and forth. After the area is up to temperature and heated evenly, let it cool slowly in *still air.* **Do not water quench.**

To heat the part more accurately

Stainless-steel headers, for Bob Sharp's GT-1 turbocharged Datsun, are shown being constructed in Chapter 12. Stainless is being used more and more for exhaust systems. This is partially due to the increasing use of turbocharging, where high heat is a problem. The material to use is 321 stainless steel: expensive, but headers made from it will last forever.

use 1250F (677C) *Tempilstik*. A special type of crayon, Tempilstik melts the instant the surface it's applied to reaches its rated temperature—1250F in this case. *Tempilaq* can also be used. A liquid that's painted on, it also melts at its rated temperature. Tempilstik and Tempilaq can be obtained at most welding or heat-treating supply stores. If you don't have any success finding it locally, write Tempil Division, Big Three Industries Incorporated, 2901 Hamilton Blvd., South Plainfield, NJ 07080.

Normalized chrome moly can be used for large parts such as a race-car spindle or other suspension components. Normalized chrome moly should not be bent. It is brittle and might break.

OTHER METALS

Although stainless steel, magnesium and titanium are sometimes used in car fabrication, I think you should stay away from them at first. Each is highly specialized, requires special welding techniques and presents certain difficulties. Availability and cost are also problems.

PATTERN DEVELOPMENT

If you've finalized the design of your project, you're ready to translate the idea and drawing into metal pieces. Unless a part is *extremely* simple, the first step is to *develop* a pattern or patterns. Once you have the pattern/s, you can use it to make the metal blank.

Patterns are especially important during sheet-metal projects. Good patterns will speed up the fabricating process and help you produce a more accurate product. For instance, I often spend considerably more time developing a pattern than making the actual part. It is better to make mistakes in cardboard than in aluminum or steel—it's also considerably cheaper.

The pattern will also help with construction. By planning carefully, you can also develop a pattern to use fewer pieces. For instance, by adding a section to be bent up as a mounting flange, you won't have to weld on a separate one later. Also, you can include adjacent sides that fold into one larger piece. For example, a hood scoop can be made with folds rather than welded seams. This eliminates the extra work of welding the pieces together and smoothing the resulting warpage. The trick is to work hands-on and think about how to make things uncomplicated and orderly. Take time during pattern development. You will save time, trouble and money.

Pattern Materials—You'll need plenty of thin flexible cardboard—such as that used by cleaners for shirts—to make a pattern, or template. Called *chipboard*, it comes in big sheets and is available from printers, office-supply and art stores. You'll also need marking and measuring tools: T-square, protractor, pencil compass, HB lead pencils, assorted rulers up to 48-in. long, and felt-tipped pens and markers. Masking tape is also necessary. The sales clerk at the office-supply store will love seeing your shopping list.

Design—First choose a large, firm, flat surface on which to work. Make sure you have good lighting. Don't work in your own shadow.

For explanation's sake, let's assume your project is building a fuel tank. Fabricating tanks is fully described in Chapter 11. Knowing how many gallons it will hold and where it will be installed, determine the tank dimensions.

Sketch the tank full-scale, showing all the dimensions clearly. If necessary, draw more than one view of the tank. This will be the case if the tank is complex or a composition of various shapes.

Cut It Out—Using dimensions from your drawing, cut out cardboard pieces corresponding those of the final part. If you have a large piece of cardboard it is possible, and perhaps easier, to cut out the pattern in one big piece so it will include nearly all of the bends and flat surfaces. If the pieces for each flat surface are cut out separately, simply tape them together along the bend lines with wide masking tape. The result—a taped-together cardboard gas tank.

You can now use the cardboard gas tank, or whatever the part may be, to check it for fit in the car. If you find any problems, modifications are easily made with scissors and masking tape. *Don't hesitate to scrap a pattern and start over.* You will be time and

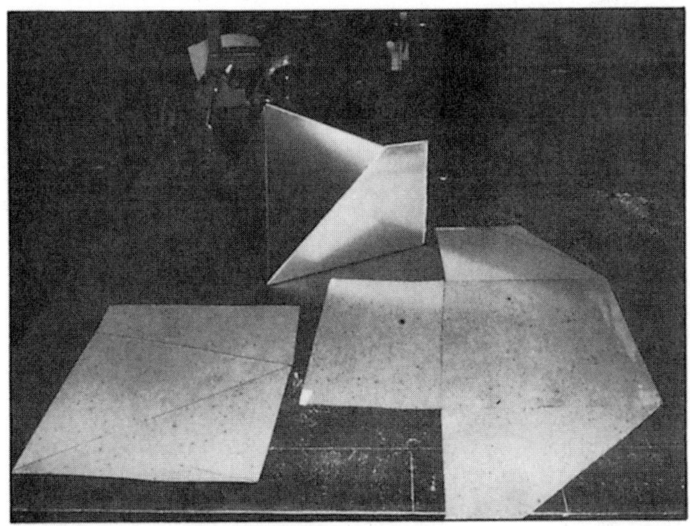

You'd be surprised how easy it is to develop an intricately shaped tank using a cardboard pattern. The cardboard mock-up can be used to check tank fit and volume.

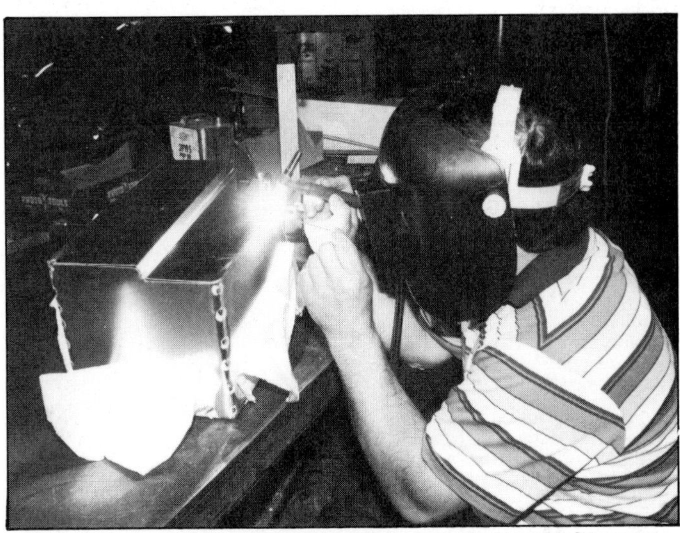

If you dislike scratches as much as I do, use aluminum with a protective covering. Peel back the covering where welding is to be done. Pat the cover back into place after the metal has cooled.

money ahead. Keep developing the pattern until you have one with the right fit and capacity. If you've modified your original pattern, remeasure the final pattern and recheck its volume. When it checks OK, you can then "go to" metal.

Developing a pattern in this manner gives an accurate representation of the finished part. It allows you to determine beforehand where to make the bends or welds. It also enables you to spot fit-problems very quickly, then correct them. It is called *pattern development* because the pattern develops—it changes as you cut and fit until it is right. It is rare to cut out the pieces, tape them together, and find them perfect, except on an extremely simple part. You must fit and refit, measure and remeasure, calculate and recheck calculations. Pattern making requires care and patience.

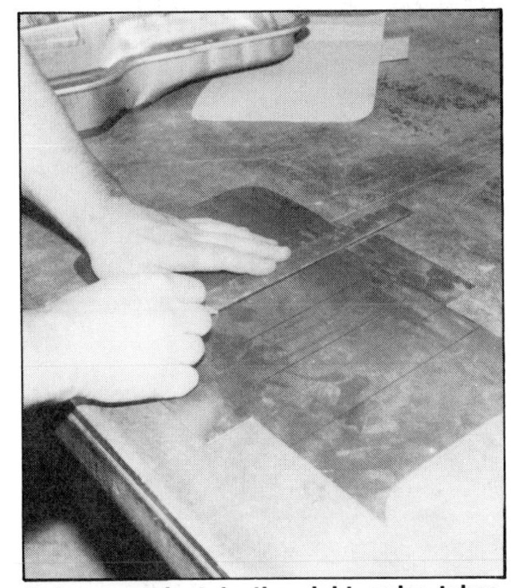

Doing a project in the right order takes planning. Custom oil pans require several different operations. First, a stock oil pan was used to develop a pattern. I then transferred the pattern to 0.040-in. mild steel.

After cutting out the blank, I clearly marked the beading and folding lines with a pencil. Don't use a scriber for this; tears may result.

METAL BLANK

Once the pattern is developed to your satisfaction, you can make the metal *blank*. Before putting the metal sheet on your work surface, make sure the surface is clean. There should be no way the surface can scratch or damage the metal. Steel is not easily scratched, but aluminum is, and scratches will ruin the appearance of the finished part.

Because aluminum is especially vulnerable to scratching, metal suppliers offer a special protective covering. Easy-to-peel-off paper or plastic comes applied to one or both sides of the aluminum sheet. Paper-coated aluminum is easy to draw on and the lines are easy to see afterwards. The metal can also be cut, formed and joined with the protective covering in place.

There is an additional cost for coated sheet aluminum, but the cost is nominal and well worth it considering the appearance of the finished part. Scratch-free aluminum is much more professional-looking.

Transfer the Pattern—Now that you have a pattern, the next step is to transfer it to the metal sheet. This should be done on the side of 1/8-in. or thicker sheet metal that will become the *inside* of the part. This allows you to bend the metal upwards, the same way you developed the pattern. For example, what will become a fuel-tank bottom should be down so the sides can be bent up.

Unfold the pattern and lay it flat against the metal. Trace the outline of the pattern onto the metal. Also transfer the bend lines from the pattern to the metal. Depending on the material, different tools can be used to do the marking.

Try marking with a soft black pencil first. I prefer using a pencil because it is very quick and just as accurate as *bluing and scribing*. I'll explain that method next. However, I've been doing metal marking for a long time. If after giving the pencil a try, you have difficulty "reading" the lines, try the next method.

To make lines on steel more visible, you can use *layout dye* and a *scriber.* Layout die is a highly flammable liquid, usually blue, which comes in a can. An applicator brush is usually built into the can lid. Or you can get layout die in a spray can.

After the layout dye is brushed or sprayed onto the steel, it dries quickly, leaving a deep-blue or -red color on the metal. Now, rather than using the pencil, use the *scriber*—a sharp-pointed marking tool—to scribe, or scratch, in the blued surface what will now be a highly visible line. **Do not scratch deeply into the metal. A deep scratch can cause the metal to break along the scribe line. Try to scratch only through the layout dye.**

No Scriber on Aluminum—When laying out work on aluminum, *never* use a scriber. Aluminum is soft and *any* scratch will encourage breakage or failure. Better yet, draw directly onto the paper covering of the aluminum sheet with a pencil or felt-tip pen. When marking directly on bare aluminum, use a soft black pencil—#2 lead—or a felt marker. This will make a very clear line because the aluminum itself is so bright.

Direct Layout—As you gain in experience, you'll find that you won't need a pattern when doing most simple sheet-metal projects. Called *direct-layout work,* mark bend and cut lines directly on the metal—just as you would a pattern. Again, use either a pencil, layout blue and a scriber, or a felt marker.

Center Line—A *center line* is a line which represents the center of a part or piece. Working from a center line makes direct-layout work easier, more accurate, and decreases the chances of a mistake. I consider it my "home base" for any lines drawn on either side. It serves as a handy reference point, allowing you to recheck side-to-side measurements quickly and to ensure symmetry. When working on a part involving angular sides, the center line is a reference from which you can check the angles.

You should now be able to take a project from the germ of an idea and into a metal blank. After making a sketch of a part, you can develop a pattern, or do direct layout work. Your toes are getting wet! Because you'll have to join those pieces, I discuss gas-welding next.

Beading is done before bending. **Otherwise, bends will interfere with the beader.**

Next I made bends in a pre-established order. Because of the unusual shape, some bends had to be made by hand.

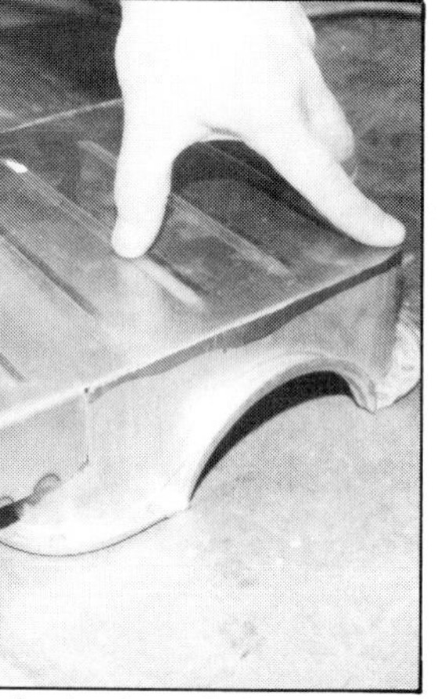

Patience is required when fitting a part. This will pay off in better welds. Note the "generous" radius bends on the add-on part.

Planning paid off in a good-looking, functional product. Oil pan is heliarc-welded. A drain will complete the project.

Layout dye on the floor of this Indy-Car tub (arrow) helped make center line visible. Normally I would use a pencil rather than the dye-and-scriber method because of the danger of creating a stress riser. However this is on a panel that was not to be formed and I was very careful to scribe only through the dye and not into the metal.

QUICK REFERENCE GUIDE TO COMMON METALS

Metal Alloy & Temper	Popular Sheet Sizes and Thicknesses	Welding, Bending and Shaping Characteristics	Common Uses In Car Fabrication	Comments
3003-H14 aluminum	4x8-ft sheets 4x10-ft sheets 0.040 in., 0.050 in., 0.063 in., 0.080 in.	Excellent for bending, good for shaping, very good for welding by either heliarc or oxyacetylene	Body panels, aluminum interiors, baffles, heat shields, surge tanks, air scoops. Many other applications.	Very attractive, anodizes well easily bent or shaped, polishes well, can be annealed to restore workability if it work-hardens
6061-T6 aluminum	4x8 ft sheets 4x10 ft sheets 0.040 in., 0.050 in., 0.063 in.	Bendable, but may crack if bent on tight radius. Weldable, but welds may tend to crack	Highly stressed structural applications like Indy-car tubs	Strong but brittle. Best joining method: riveting and bonding
6061-T4 aluminum	4x8 ft sheets 4x10 ft sheets 0.040 in., 0.050 in., 0.063 in..	Bendable, but cracks easily. Weldable, but welds may crack easily.	Ideal for monocoque construction. Good for wings	Good strength. Some brittleness. Riveted and bonded for joining or installing
1020 CR mild steel	4x8 ft sheets 4x10 ft sheets 0.0299 in., 0.0478 in., 0.1196 in.,	Excellent for bending. Good for shaping. Excellent for welding.	Roll bar mounts. Fuel-cell containers. Fuel tanks Hammer-formed brackets Body work. Floor pans. Many other applications.	Low cost. Excellent availability Strong. Cuts easily
1020 SK or AK steel	4x8 ft sheets 4x10 ft sheets 0.0299 in., 0.0359 in., 0.0478 in.,	Good for forming. Good for shaping. Good for bending. Excellent for welding.	General body work. Hammerformed brackets.	Very easily formed. Expensive.
4130 Chrome-moly steel	4x8 ft sheets 4x10 ft sheets 0.0299 in., 0.0359 in., 0.0478 in., 0.0598 in., 0.1196 in.,	Good for bending in *annealed* state. Not recommended for shaping components. Good for welding.	Fabricated suspension arms, other suspension roll bars, roll bar gussets, anti-sway bars, brackets, mounting plates for roll bars.	Stronger than 1020 CR mild steel of the same gage. Availability limited. Very expensive. Dulls cutting tools easily.

Although there are many steel and aluminum alloys available, these are the ones I use most. For a complete listing of materials and their properties, consult your metal supplier. He can supply you with a data book and stock list.

AREA FORMULAS

Square

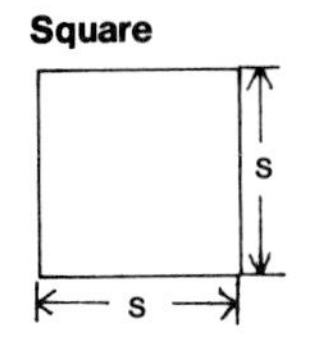

$A = s^2$

Rectangle

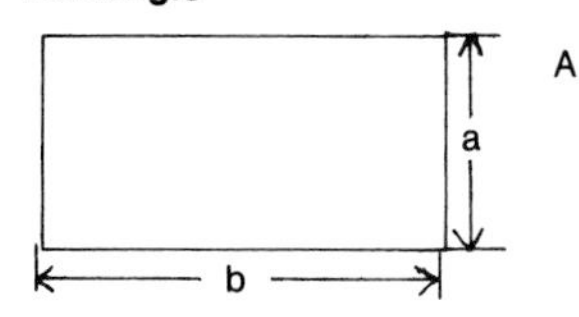

$A = ab$

Triangle

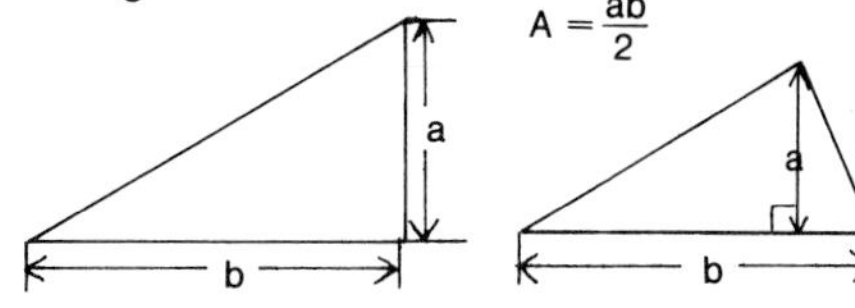

$A = \frac{ab}{2}$

Trapezoid

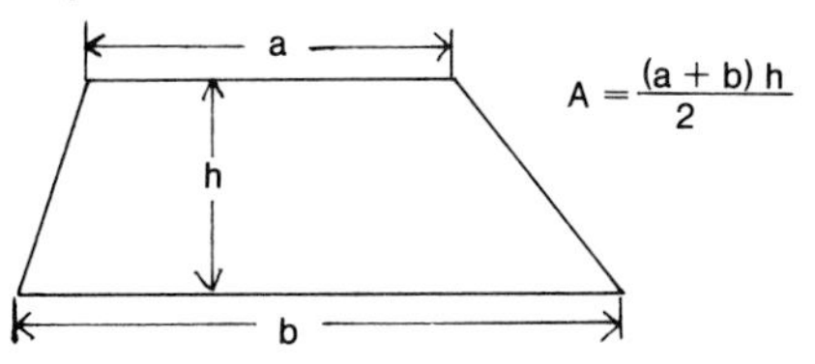

$A = \frac{(a + b)\,h}{2}$

Circle

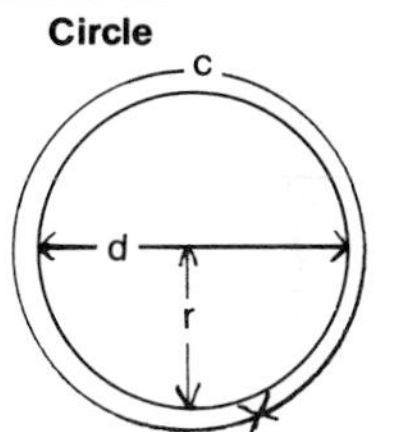

$A = \pi r^2$ or $A = \frac{\pi d^2}{4}$

$C = 2\pi r$ or $C = \pi d$

A—area
C—circumference
π—3.1416

Circular Section

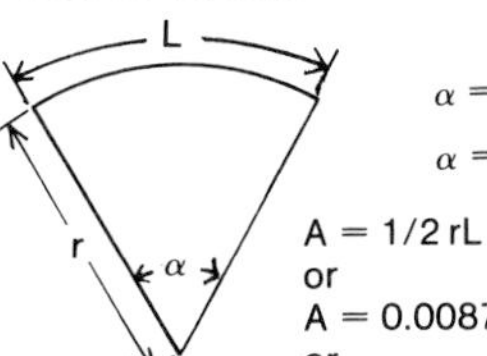

α = angle in degrees

$\alpha = \frac{57.3}{r}$

$A = 1/2\,rL$
or
$A = 0.008727\,\alpha r^2$
or
$A = \frac{\text{Area of circle x degrees of arc}}{360}$

Elipse

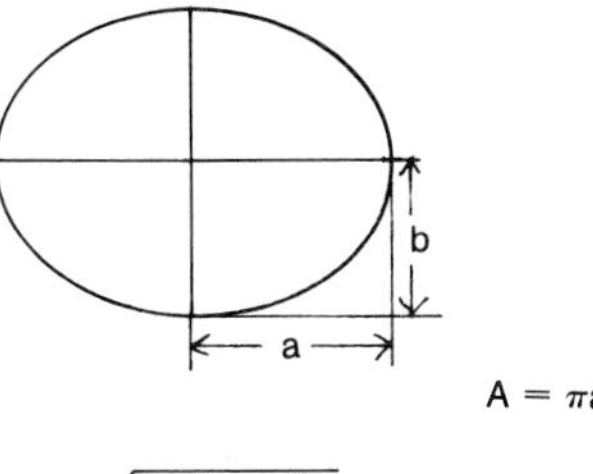

$A = \pi ab$

$C = \pi\sqrt{2\,(a^2 + b^2)}$ (approximate)

Parabolic segment

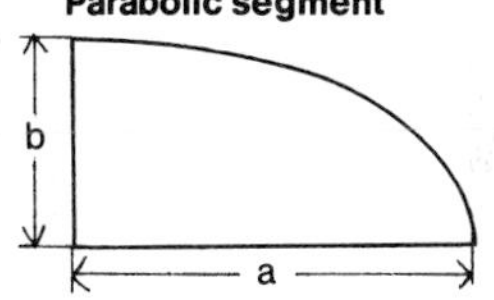

$A = 2/3\,ab$

VOLUME FORMULAS

Rectangular prism

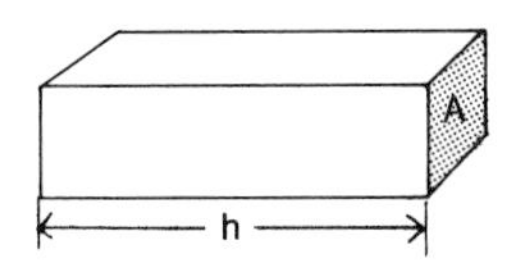

$V = Ah$

Cylinder

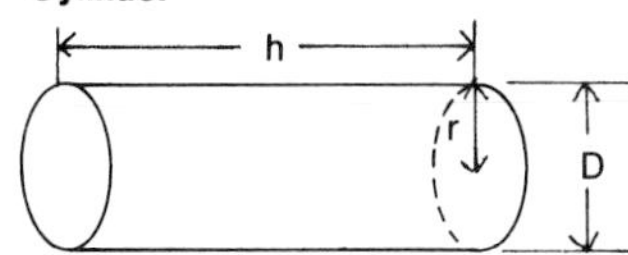

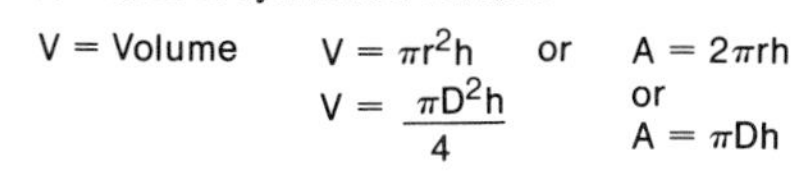

A = area of cylindrical surface

V = Volume $V = \pi r^2 h$ or $V = \frac{\pi D^2 h}{4}$

$A = 2\pi rh$ or $A = \pi Dh$

Pyramid

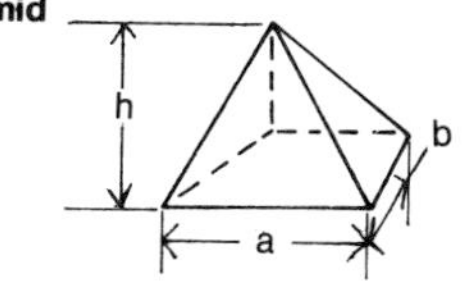

$V = 1/3\,abh$

GAS-WELDING

Gas-welding is a fine method and can be used effectively to fabricate headers. Unfortunately, gas-welding is frequently overlooked as a useful and efficient welding technique. Photo by Michael Lutfy.

Cut and shaped metal pieces must be permanently joined by the strongest method possible. This is where welding comes in. In this chapter I'll cover gas-welding: how, when, where and why it is used. I will deal with gas-welding steel, brazing, gas-welding aluminum and flame cutting. *Oxyacetylene* welding—commonly called *gas-welding*—is relatively inexpensive and one of the most frequently used methods of welding. If you're interested in fabricating high-quality parts, you must become proficient in gas- and electric-welding. Electric-welding will follow in the next chapter.

To distinguish between the welding equipment and the person doing the welding, I use the terms *welder* and *weldor. Welder* refers to the welding equipment. *Weldor* refers to the person using the welder.

WHAT IS WELDING?

Welding unites two pieces of metal through the use of heat. A burning gas or an electric arc creates the heat for the types of welding I cover. A *filler rod* is used to supply additional metal to form the weld *bead.*

When the two metal pieces being joined and the filler rod are heated to the correct temperature—a molten stage—they melt and *flow* together. They become permanently one material as they cool. A weld is as strong as the *base metal*—the material of the pieces being joined—*provided the welding is correctly done.*

The welding process intimidates many people. It shouldn't. Welding is a very simple and direct process, taking advantage of the effects of heat on metal. The "trick" of welding is to follow the rules while practicing until ample skill is developed. Skill and care ensure good-looking sound welds.

As with any developed skill, some people readily learn how to weld. Others have difficulty. For example, welding depends greatly on hand-and-eye coordination. If your coordination is good, you should have little trouble learning to weld. You'll also have to follow certain rules. Because certain welding techniques have strict rules, don't ignore them.

Although you can be assisted, the truth is: *no one* can teach you how to weld. In the end, you must teach yourself. It's like when you learned to ride a bicycle. Your father or mother may have run along beside you, giving you encouragement. Your friends may have yelled out advice and instructions, or maybe made fun of you. But in the end, when you finally went breezing down the sidewalk, you taught yourself how to ride. To carry this idea further, you probably fell a few times, getting some scrapes and bruises in the process. Likewise, you'll make mistakes while learning to weld. Don't worry about it and keep trying. You'll feel the same sense of achievement when you succeed.

To get started, you must have the right equipment. So let's take a look at it first.

GAS-WELDING EQUIPMENT

A very commonly used welding technique in car construction, *oxyacetylene welding* is named for the two gases used to make the flame—oxygen and acetylene. To become a really good gas weldor you must chose the right equipment and take the time to get familiar with it. You need to understand the capability of this equipment and how to use it safely.

Basic pieces of gas-welding equipment you'll need are: acetylene and oxygen tanks, pressure regulators, hoses and torches.

Tanks—Tanks, like soft drinks and ice-cream cones, come in three sizes: small, medium and large. The size tanks, or bottles, you choose depends on how much gas-welding and/or cutting you'll do. This is important as you'll soon see. Although sizes vary by manufacture, basic tank sizes by diameter and overall height measurements in inches are:

	Oxygen	Acetylene
Small	7 X 27	6 X 24
Medium	7 X 47	8 X 38
Large	9 X 56	12 X 46

Laws are different from one locale to another, but in reality you don't usually own a particular set of tanks. For instance, large tanks can only be leased from the welding supplier. You can buy small and medium tanks in most states, but when you have your shiny new tanks refilled, you may never see them again—you won't recognize them if you do. Although some suppliers will fill your tanks, most will only give you filled tanks of the same size. *Your* tanks will be filled and given to someone else.

If you lease oxygen and acetylene tanks, make sure you choose a size that matches your usage. Here's why. You'll pay for the gas, lay down a deposit for the tank, and after a certain time elapses—usually 30 days—you will be "hit" by a *demurrage* charge. This charge only amounts to a few cents per day. But at 8¢ a day you'll pay $2.40 a month and nearly $30 a year. As you can see, it's not economical to have leased tanks sitting around not being used.

Pressure Regulators—Because the pressure in the tanks will be considerably higher than the pressure needed at the torch, you'll need a way to regulate it. This is done with a regulator at each tank. There are two basic types of regulators: double-diaphragm and single-diaphragm.

For commercial and heavy use, the double-diaphragm regulators are the ones to have. They are more durable and have better pressure control. I use Victor double-diaphragm regulators. Airco and Linde also offer this type. If you're planning on doing occasional welding, single-stage regulators

TANK-PRESSURE GAGE
LINE-PRESSURE GAGE
VALVE
REGULATOR
PRESSURE GAGES
REGULATOR
TANK VALVE
LINE TO TORCH (RED)
LINE TO TORCH (GREEN OR BLACK)
OXYGEN TANK
ACETYLENE TANK

Acetylene and oxygen bottles and their pressure regulators, gages and lines. So you don't think you're losing your mind trying to thread them on: acetylene fittings (the red line) have left-hand threads.

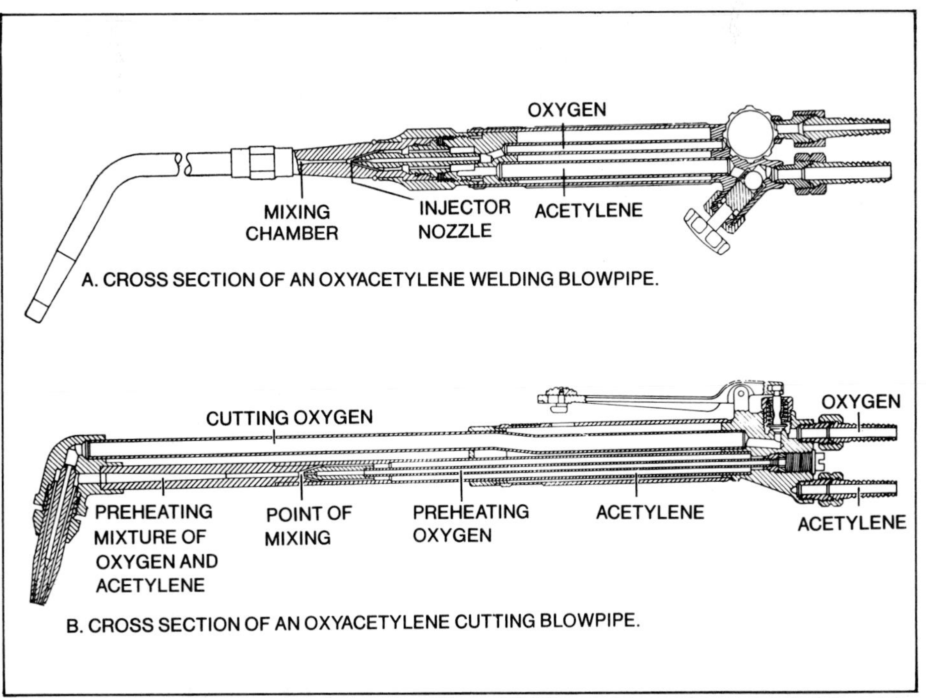

Welding-torch and cutting-torch cross sections. Welding torch can be converted to a cutting torch by installing a cutting tip. Drawing courtesy of Linde Air Products Co.

Good welding goggles are a must. Norton 5002-H goggles are the most common for gas-welding. However Norton 5024-H goggles give *more protection* and are more comfortable. Photo courtesy of Norton Co.

GAS-WELDING SAFETY RULES

Your shop should be tidy and uncluttered. Note absence of flammable material and the large fire extinguisher on the wall behind the frame and cage.

Handle gas cylinders with care: Because gas cylinders contain high-pressure gas they are potential bombs. A gas cylinder must be chained in place separately from other cylinders when being stored or used. When moving a cylinder be sure the *valve cap* is on. The cap will protect the valve from getting knocked off in the event the cylinder falls. Should this happen the cylinder would take off like a rocket, doing much damage, and possibly injuring or even killing someone in the process.

Always crack open oxygen or acetylene cylinder valves BEFORE installing the regulators: Opening a valve very slightly will blow out any dust or dirt that would otherwise be carried into the regulator and possibly damage it.

Never use oil or grease on any part of oxyacetylene-welding equipment: The combination of oxygen and oil or grease is volatile and will cause an explosion or fire.

After attaching regulators to a gas cylinder, back out the regulator adjusting screw BEFORE opening the cylinder valve.

Don't stand in front of the regulator when turning a cylinder on.

BEFORE using any oxyacetylene welding equipment be sure all of the connections are secure and leak free. You can check each connection for any leaks with oil-free soapy water. Shut the valves off at the torch, and open them at the tank to pressurize the lines. Wipe the soapy water around the connection and check for bubbles. A bubble indicates a leaky connection. Fix it.

will do. Craftsman or Purox single-stage regulators are two that I recommend.

Pressure regulators have two gages each, one for tank pressure and one for line pressure. The tank gage indicates when a refili is necessary so you won't get caught in the middle of a job and run out of oxygen or acetylene. The line-pressure gage lets you set and reset line pressure accurately. How much oxygen or acetylene pressure depends on the equipment, size of tip and whether you're welding or cutting. Regardless, pressures are set to give a sharp flame cone. Start by reading the instructions that came with your welding equipment.

Hose—Get at least 25 ft of hose. This will let you get in, under and around a car without continually having to move the tanks. The oxygen and acetylene lines are bonded together so they won't get tangled. Colors make the lines easily identifiable: green or black for oxygen, and red for acetylene. So you can't mix them up, acetylene-hose fittings have left-hand threads. Oxygen lines use conventional right-hand threads.

Torches—If you buy your welding equipment in a set, the torches and an assortment of tips will be included. You'll have all that's needed for practicing. You can determine what, if any, additional tips are needed. I discuss tips in more detail on page 60.

Cart—A welding cart is not an absolute necessity, but you'll soon think it is if you try doing your welding without one. A cart serves three purposes: It lets you move your welding equipment more easily, provides storage for this equipment and holds the tanks upright.

This last feature is very important. You don't want a tank to fall over and break off its neck. The tank will become an instant rocket. If you don't have a cart, secure the tanks so there's no chance of them falling over. This is easily done by standing them against a wall or sturdy bench, then hooking a light chain around them. Anchor the chain to the wall or bench at each end.

Goggles—Tinted goggles are a must when gas-welding. They not only protect your eyes from sparks and the bright weld, you'll also be able to see the weld bead, the parts you are joining and the filler rod. Don't try to weld without goggles.

Torch Lighter—Get a torch lighter, or *striker.* You have no options here. **Never light a torch with an open flame.** Turn to page 60 for more on lighters.

GAS-WELDING BASICS

Metal Table—You'll need a thick-metal-top table to weld on. Although the size of the table top depends on the size of the pieces you'll be welding, a 36 X 48-in. top will do for most situations. Rather than buy one, fabri-

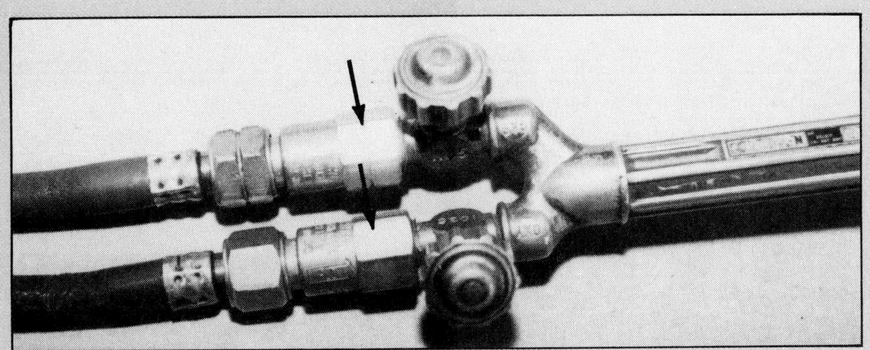

Safety item well worth the money. One-way valves between the torch and lines (arrows) prevent flame from traveling back through the lines to the cylinders, causing an explosion.

Never let regulated acetylene pressure exceed 15 psi. This is the maximum recommended operating pressure suggested by all regulator manufacturers.

Always have a fire extinguisher close by when welding or flame cutting. Check the effective dates for the extinguisher charge. An extinguisher with no charge is as useful as a square wheel, and a whole lot more dangerous.

Weld or cut in a well-lit area with adequate ventilation. Keep this area free of combustibles. Welding heat or sparks can ignite wood, cloth, fumes or gases. Welding near paint, alcohol compounds, resins, plastics or fuel containers is very dangerous. Don't forget that welding generates fumes and smoke too. Good lighting and fresh air are good safety measures that will make your work easier and more enjoyable.

Never use oxygen to blow dust off your work or clothes. This can turn a small, nearly invisible spark into a raging fire.

Never attempt to repair a regulator or welding torch yourself. In even slightly imperfect condition this equipment can be dangerous. Do not tinker with it. If a regulator or torch is not working, take it to an authorized repair shop. Your welding supply dealer can either provide this service or refer you to someone qualified to do the repair.

Always shut the valves at the bottles and open those at the torch when oxyacetylene equipment is not in use.

Always wear the right welding goggles. Anyone watching you weld must also have eye protection. Sparks and UV rays can cause serious eye injury.

Never use a match or cigarette lighter to light a torch. Always use a spark lighter.

Use the recommended working pressures and tip sizes for your welding equipment. Your equipment handbook was written by experts to help you to get best results from the equipment.

Wear suitable clothing. Long-sleeved work shirts, buttoned to the neck, protect your skin. Leather gloves protect the hands.

Always use oxyacetylene safety valves. Safety valves in the lines at the torch prevent a back-up of flame to the cylinder, and a resulting explosion.

cate the top from a 1/2-in.-thick steel plate. Use 3 X 3-in. angle iron of 1/4-in.-minimum thickness to fabricate the table base. Make the table about 30-in. high if you'll be sitting to do your welding. It should be about 34-in. high if you'll be standing. I strongly recommend that you sit down to weld whenever possible.

To save oxygen and acetylene, work should be elevated off of the work table with *fire bricks.* Heat from the work won't be lost into the work table if your metal is elevated this way. If fire bricks aren't readily available, use short sections of angle iron to support the work piece.

Practice—Sit down with your welding equipment and some scrap metal on which to practice. Get comfortable, and *practice, practice, practice* . . . It takes time and effort and repeated practice. Unless you have the patience to polish your skill, you are better off paying someone else to do your welding. You can't take short cuts while developing your welding skills. A good weldor is worth his fees. Even if you farm out your welding, you should learn enough about welding to know what to look for. You'll also be able to appreciate quality when you see it.

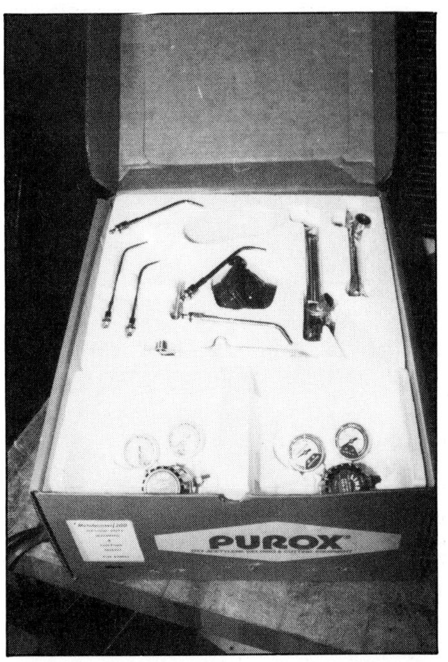

Purox makes several gas-welding kits. My kit came with two torch tips, regulators and gages, goggles, a lighter and an instruction book. I added three more tips to give the torch a wider range of applications.

Flame Types—An oxyacetylene torch will produce three basic types of flames: *oxidizing, carburizing* and *neutral.* Each flame is named according to its relative content of oxygen and acetylene. A neutral flame has a one-to-one mixture of oxygen and acetylene. An oxidizing flame has a high oxygen content. A carburizing flame— sometimes called a *reducing* flame—has a high acetylene content. Each of these three flames has a different appearance and has special uses. See the drawing of the three flames on the following page.

Of the three, the neutral flame has the widest application in welding. Of the two cones that make up a neutral flame, the inner, luminous cone is the hottest at over 6000F (3316C). Carburizing and oxidizing flames have limited welding applications and don't apply to the materials discussed in this book.

Flame control—A beginning weldor must first learn flame control. This is necessary so you can produce a proper

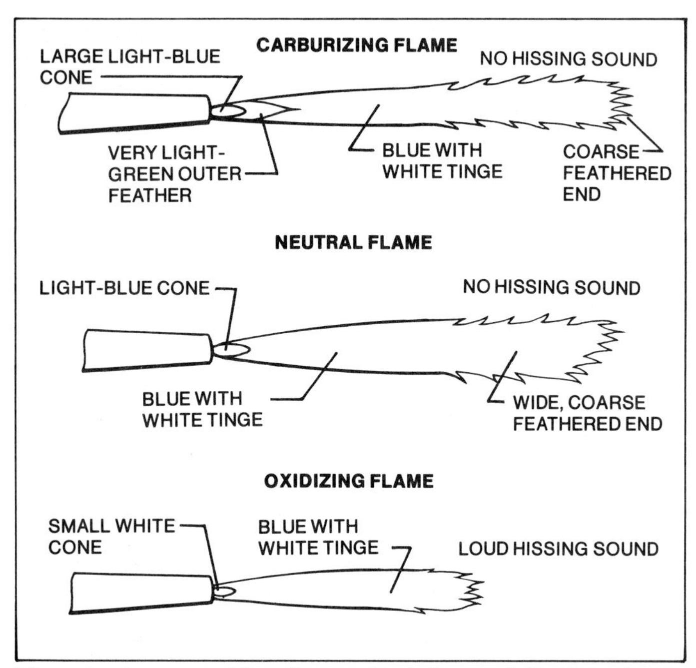

Three basic gas-welding flames. Each has a distinctive shape, color and sound. Neutral flame is the one you'll use most.

There's no magic involved in gas-welding. It simply takes a lot of practice. Practice on a piece of scrap metal. Keep the welding area clear. Seat yourself comfortably and wear eye protection. Light the torch with a flint-type torch lighter, never an open flame.

weld bead. You should be able to do this with or without a welding rod time after time. *You* must be in control of the torch and its flame, not vice-versa.

To learn flame control, practice on scrap steel. Don't start out practicing on a job. Get some scrap-steel pieces, your welding equipment, and sit down to practice undisturbed. This practice is the foundation for the skill you will develop. Begin with a single 1/16—1/8-in.-thick piece of clean mild steel.

Torch Tips—Choose an appropriate tip size for your torch. The instructions that came with your torch set will suggest the correct tip for each metal thickness. I can't specify the tip you should use because torch manufacturers have never come up with a size-code standard for tip orifices. So, although a #3 tip from one manufacturer might have a 0.060-in. orifice, another manufacturer's #3 tip might be significantly smaller or larger. Check the manufacturer's instructions carefully.

Choosing the correct torch tip is important because the wrong tip will cause plenty of trouble. For instance, a tip that's too large will weld the metal well enough at first, but before long the metal will get too hot and the bead will begin to pop holes. The flame looks very large. A common remark in a shop when a person is using an oversized tip is, "What are you doing? Trying to weld with a forest fire?"

A too-small tip will have other bad effects. It could cause the weld to look OK, but have very little *penetration* into the base metal. Penetration is the depth of a weld below the surface of the parent, or base, metal. Lack of penetration results in a weak weld. Too small a tip will also pop, and it will slow your welding. Check the manufacturer's instructions for a recommended tip size.

Prepare to Light—Having selected a piece of scrap steel and an appropriate tip, it's time to light the torch. With the valves closed at the torch, open the valves at the tanks. Adjust the regulators: oxygen at about 10 psi and acetylene at about 7 psi. Readjust the regulators once you have the torch lit.

There are two things to avoid when lighting a torch. A very loud explosion occurs when lighting a torch using equal amounts of oxygen and acetylene. This loud pop can startle bystanders. This may cause an accident. Another thing to avoid is lighting the torch using straight acetylene, although some old-fashioned manuals suggest it. Straight acetylene gives a quick, quiet light, but it produces great amounts of soot. I prefer to light with slightly more acetylene than oxygen. This gas mixture lights well. It causes only a soft popping, if any sound at all, and there's no soot to dirty up the shop.

Torch Lighters—A flint-type torch lighter is the only safe welding-torch lighting instrument! Several lighters of this type are available. None are expensive, so there is no excuse not to buy one. *Above all, never use matches or any kind of cigarette lighter to light a welding torch!* If you are foolish enough to try this, chances are you'll burn your hand severely. In addition, some weldors have learned the hard way that cigarette lighters, especially the butane-type disposable ones, act like a hand grenade when used to light a torch. Hands have been blown off by the explosion.

Adjust the Torch—Once you've lit the torch, adjust it to a neutral flame. With equal amounts of oxygen and acetylene, a neutral flame has no loud hissing sound. Check the flame chart on this page for color and shape. With experience you'll learn to recognize a neutral flame right away.

Begin To Weld—Use a comfortable bench or chair with good support. Have good lighting and good ventilation. Position your work in front of you, elevated off the work table with fire bricks or angle iron. Sit down and get comfortable. *Put on eye protection.*

Light the torch and adjust it. Wave the torch back and forth across the work piece to give it a slight preheat. Lower the torch tip close to the metal

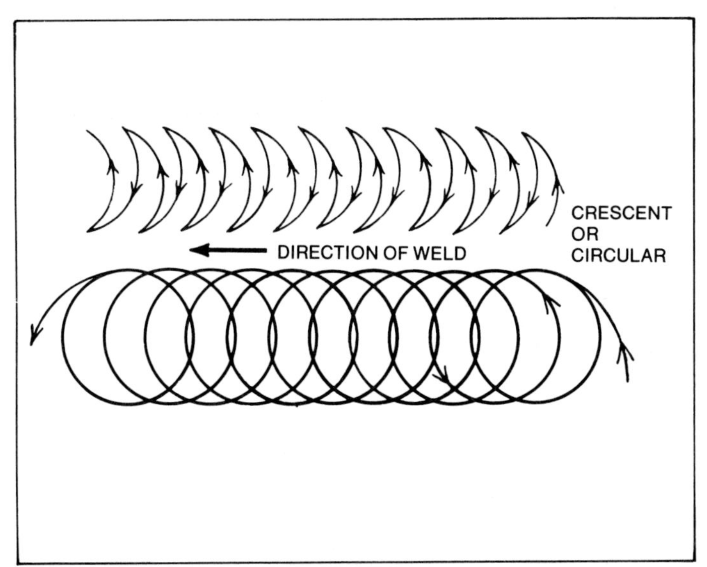

Move the torch right to left in an even crescent or circular motion. If you're a "lefty," go from left to right. The more consistent you can be, the better looking the bead.

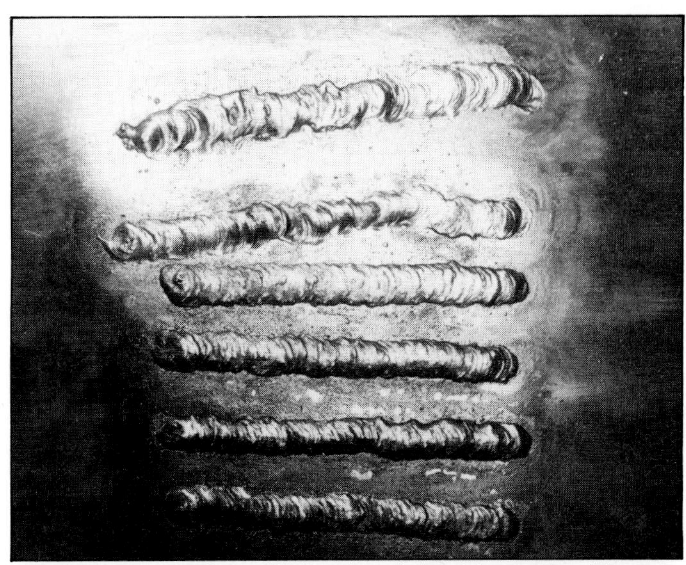

Welding with rod takes skill developed through practice. The top two beads were done too fast: They are lumpy and crooked. The following beads show improvement in regularity and rod deposit.

and hold it there until the heat melts a small pool of metal—about 1/4-in. diameter. Move the pool along by moving the torch from right to left if you are right-handed. Reverse the direction of motion if you're left-handed. As you move along the work, move the torch tip in small circular motions, or in movements like a half moon. I've used both motions with equal success, so use the one that suits you better. It's important to move the molten pool of metal smoothly and straight along the piece.

Keep in mind that steel becomes molten at about 2700F (1462C). You are not playing with a candle. Respect the temperature you're using and you'll avoid burns. Use leather gloves when handling the hot metal.

Tip Distance—The torch tip must be a correct distance from the metal regardless of whether it's the right size. If it is too close it will overheat the metal and possibly pop holes through the work piece. If the tip is too far away from the metal, the heat disperses too widely and is ineffective. If you're not getting a nice bead, check torch distance from the metal as well as torch tip size.

Bead Consistency—As you move the pool along as I described, the metal behind the tip will solidify and form the bead. You must repeat this control exercise over and over again until you can easily make beads that look the same time after time. Only when you've mastered this phase of flame control can you go on to practice the same exercise with filler, or welding

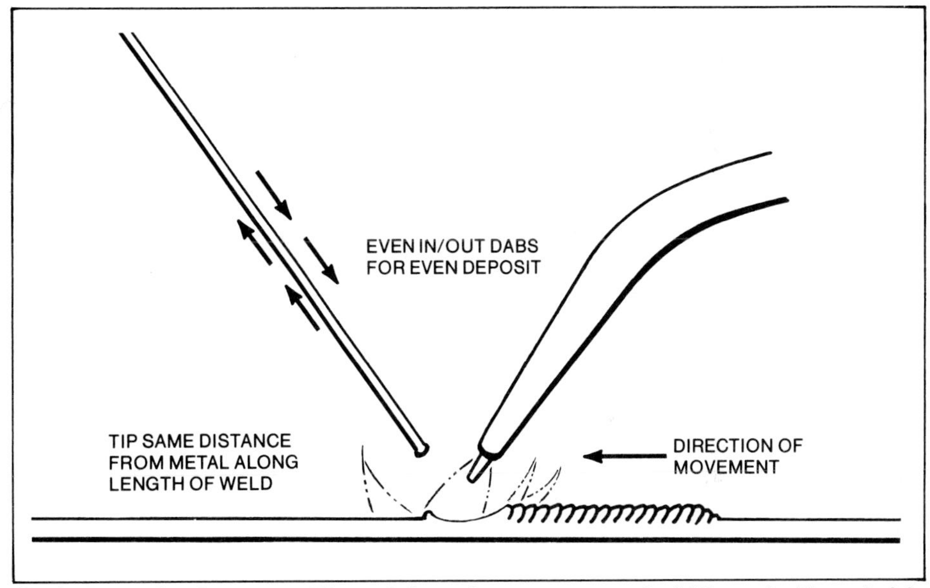

It takes coordination to use filler rod. Deposit an even amount of filler rod with each dab. Practice will develop rhythm.

rod. *Starting with a rod is not a short cut.* It just complicates things by requiring additional hand-and-eye coordination and movement as well as basic flame control. Using rod is a skill to *add* to the basic movement of flame control.

Adding Rod—When the smooth regular bead shows you are ready to learn the use of welding rod, use 1/16-in.-diameter rod. With the rod in your left hand—right if you're left-handed—start a pool of molten metal as before. Heat the end of the rod by inserting it into the flame. *Dab* the rod into the molten pool as you move. Dab and lift, dab and lift until the bead is smooth and regular. The idea is to add the right amount of filler to the pool at the right time.

The first beads you make will be lumpy and uneven. As you practice more and more they will get smoother and more regular. This is an hand-and-eye coordination exercise, involving rhythm. It will take time and patience to develop.

Keep practicing flame control until you really can control what the torch is doing to the metal. If you *are* in control, the beads will be straight and consistent time after time. They will be even in both height and width. Bead height becomes even as you learn to add the right amount of rod to the pool as weld. *Don't* move to the next

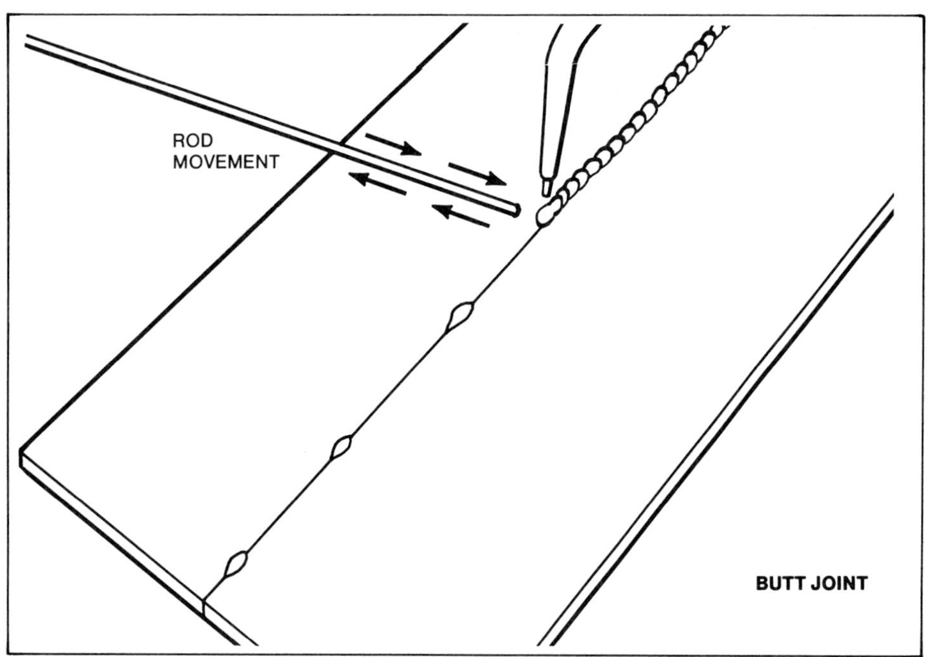

Butt-welds need plenty of tacks—one about every 1 in. Tacks help prevent warpage. Practice butt-welds until they are smooth and neat.

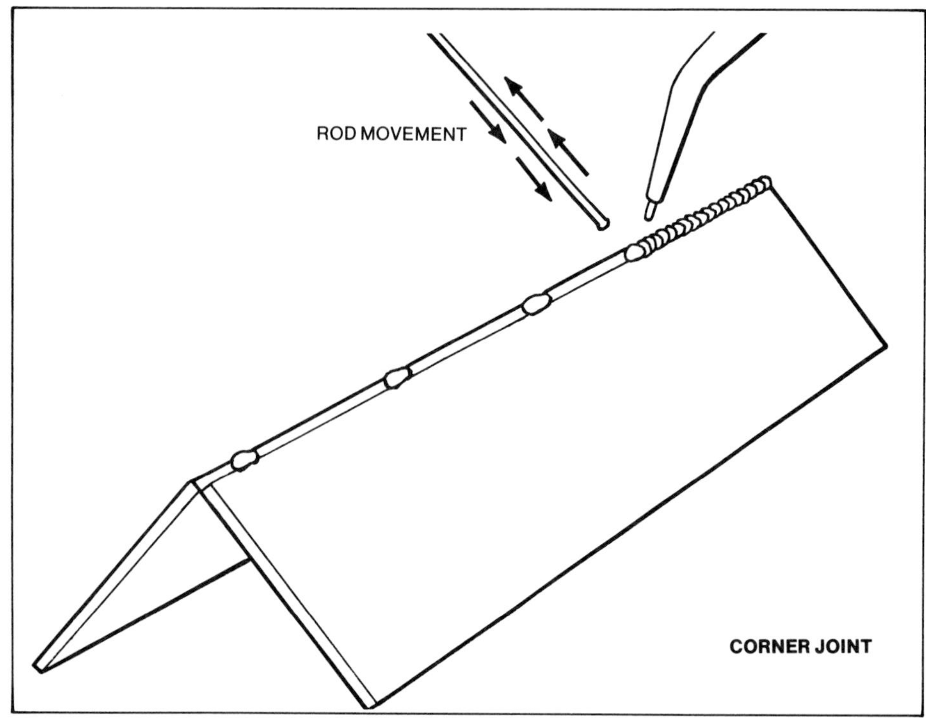

Outside fillet-weld. For maximum penetration the metal edges should *not* overlap. They should form a V. Space tacks no more than 1-in. apart.

phase of welding until you have mastered flame control.

TACK-WELDS

Tack-Welding is a series of very short welds or spots between two adjacent metal pieces. You'll quickly find yourself using tack-welds often. Tacking is a must to get a *reliable* final weld. The *tack* holds metal pieces securely for final welding. Tack-welding also preheats metal for a smoother, more accurate final weld.

Practice placing tack-welds at 1 to 1-1/2-in. intervals between two metal pieces butted edge to edge. Look for uniform size and good penetration. When your tack-welds are uniform and predictable you can go on to butt- or fillet-welding.

BUTT-WELDS

Sorry to repeat, but this is important: *Never use a project to practice your welding skill. Practice on scrap first.* Start out by *butt-welding* two pieces of metal. A butt-weld joins two pieces of metal lying edge to edge against one another. If there is a gap, it should not exceed 1/16 in. The gap is filled with welding rod during the weld.

To practice a butt-weld, select two clean pieces of mild steel—about 6-in. long. They should be the same thickness. Support the work pieces edge-to-edge on fire bricks or angle iron. Choose the correct welding tip for the metal thickness. Sit down and get comfortable. Light and adjust the torch to a neutral flame.

Begin by preheating the metal pieces. Next, tack-weld the pieces together in at least four equally spaced spots along the seam—about 1-1/2 in. apart.

Remember, a tack-weld is just a small spot of weld, well penetrated into the base metal. The tack-weld will temporarily secure the pieces and prevent them from spreading apart while you're making the final butt-weld. It also helps to distribute the heat evenly between the work pieces. Proper tack-welding pays off in sounder, higher-quality welds.

To finish the butt-weld, run a smooth bead from right to left along the seam edges. Weld in the opposite direction if you're left-handed. If theres's no gap or only a *small gap*—less than 1/64 in.—between pieces, little filler is needed for the bead, and you'll be welding faster. Larger gaps will require more welding rod and slower welding.

It makes sense to keep butt-weld gaps to an absolute minimum simply because welding pace slows with increasingly larger gaps. The reason for this is you must use more welding rod to fill the gap to get a sound weld. Again, master the butt-weld before moving on to other types of welds.

Test the Weld—Because the appearance of a weld can fool you, test your butt-welds to see if they are sound. "Pretty" welds can literally fall apart if there's little penetration. An ugly weld with good penetration is infinitely better. The type of test I describe can be used to determine whether any type of weld is sound, not just pretty. This is an important part of the learning process.

Solidly clamp the piece to be tested in a vise or to a steel table immediately below or beside the weld. Bend the piece over at the welded joint with a heavy steel hammer. If weld penetration is good the piece won't break at

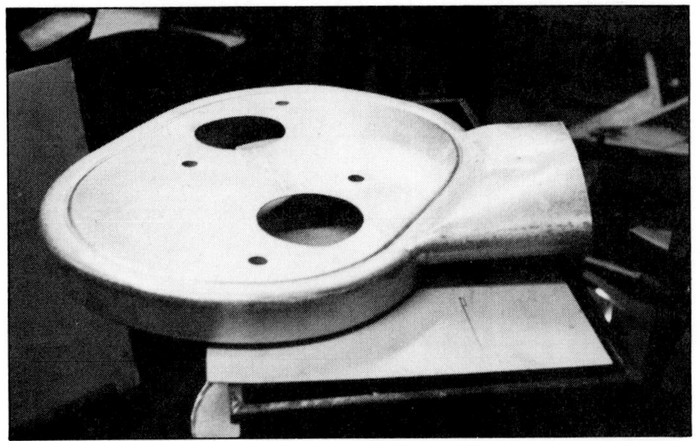

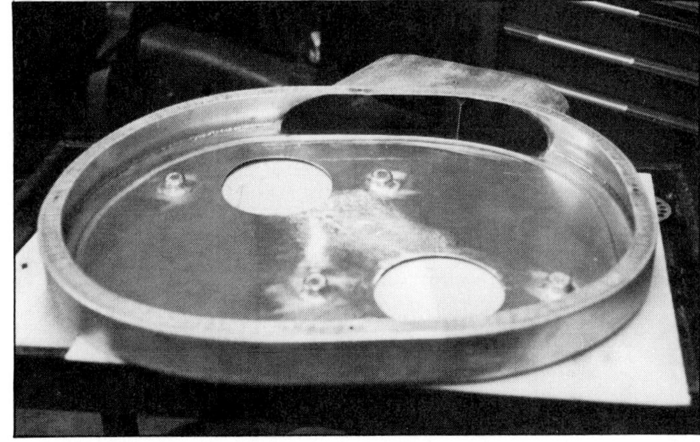

Aluminum air cleaner is a good project to use gas-welding. First, it's a non-structural part. And the hand forming on the air inlet was easier because gas-welding helped keep the aluminum soft and workable.

the weld. The metal should bend completely over, or break next to the weld, not through the weld.

There are other ways of testing a weld, but they are more difficult or sophisticated. This test is reliable, easy and proves the weld's quality. If your practice welds can withstand this test over and over again, you can make a sound weld. You'll be able to go on to a real project with confidence.

FILLET-WELD

A *fillet-weld* joins two metal pieces that overlap or are at an angle to one another. The weld *fills* an inside corner between the two pieces. Sometimes a fillet-weld joins metal at the outer corner of an angle. Sometimes it does both for maximum strength as illustrated at the right. As you might guess, fillet-welds are used extensively for chassis construction where many brackets are welded to the frame members. Fillet-welds are also commonly found on tank-mounting brackets.

Go through the same process to practice fillet-welding as you did to practice butt-welding. Stand one piece of metal edge-to-edge against and 90° to another piece. Tack-weld them together, then practice running an inner fillet-weld. Turn the piece around and run another weld on the outer edge of the corner. Check your weld using the bending test and keep on practicing until you're satisfied.

To get down to the nitty gritty: Suppose you have a project requiring oxyacetylene welding using cold-rolled mild steel. Simulate the actual welding conditions you'll encounter by making some test welds. Use the same material, thickness of metal and the same joint design. Chances are the final project will look and perform far better than if you had just jumped in "cold turkey" and attacked the project.

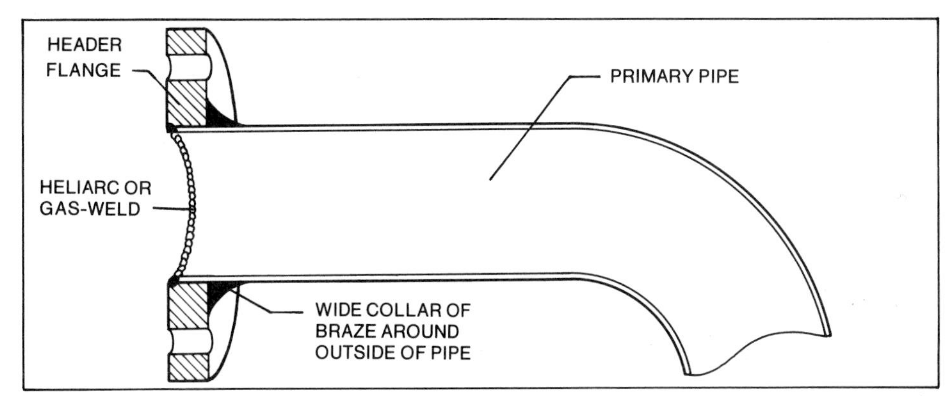

Brazing the primary header pipe around its periphery to the flange is a good way to gain extra pipe-to-flange strength. Pipe is heliarc or gas-welded to the flange ID.

BRAZING

An oxyacetylene torch set can be used for another important kind of metal joining. *Brazing*—akin to soldering—differs from oxyacetylene welding as it uses a non-ferrous-alloy rod—brass or bronze—with a lower melting point than that of the base or *parent* metal. The most common filler rod or wire you will use is 1/16 and 3/32 in.

Brazing involves preheating the base metal only to the point where it will accept the molten filler rod and *flux*. The base metal is not heated to the melting point as it would be for a conventional gas-weld.

Flux—A cleaning agent, *flux* removes impurities from the base-metal surface. It improves the flow of the molten filler material and floats impurities out of the joint and to the surface of the bead. A *fusible material*—it melts—flux usually comes in powder or paste form.

Without flux, the bronze would not adhere properly to the base metal and the joint would be weak. Powdered flux, available in cans, is applied to the rod. This is done by dipping the heated filler rod into the flux.

Brazing rod is also available with the flux coating already applied. Flux-coated brazing rod is more expensive than the combination of bare rod and a can of flux, but it is easier to use. You don't have to pause to dip the rod into the flux. It also enables you to run a long bead without stopping for more flux.

Brazing Advantages—Brazing is not as strong as a weld, but it has some advantages. It doesn't require as much heat because the base metal is not melted. This is a distinct advantage when joining small, thin pieces of sheet metal which might otherwise warp if an oxyacetylene weld were used. It's often used to repair small cracks or holes in exhaust systems. Brazing can also be used to repair metal castings that are not easily repaired by oxyacetylene or electric-arc welding.

Heat for Brazing—Using the correct heat for brazing is crucial. Start with a neutral flame. Too much heat will let the filler material flow too fast. Extreme heat will also cause scale to form on the base piece. Scale will prevent the base metal from accepting the molten brass. If scale forms, sand or grind it away. You can then try again on a fresh metal surface.

Too little heat when brazing prevents the base metal from accepting the brass. The first objective is to heat the base pieces only enough to allow the filler material to flow. No more. The base metal must glow *slightly*. This indicates it is ready for the brass. If the metal is too cold, the brass will form in lumps rather than flowing out smoothly.

Mistakes in Brazing—There are two common mistakes to avoid while brazing: excess heat and dirty base metal. Avoid excess heat by using the right size torch tip. And adjust the flame so it's neutral. Beautiful, strong joints can be made if you use the right tip. Refer to the torch manufacturer's recommendations for tip size.

Dirty base metal is the second common problem in brazing. *Anything* other than the base metal where it is to be brazed should be considered dirt. It could be paint, oil or corrosion. These impurities on the surface must be removed before brazing as they will keep the filler from adhering to the metal. If the metal you are going to braze is not new stock, expose new material by grinding the area to be brazed. Grind down to bare metal.

Brazing Exhaust Flanges—Brazing is useful in many ways, but I especially like it for reinforcing exhaust-header tubes at the cylinder-head flange. Although the tubes are heliarc-welded on the head-side of the flange, exhaust tubes need extra strength here because of high fatigue loads due to heat and engine vibration. Form a large wide bead of braze around the outside of the tubes at each tube-to-flange joint. This bead acts as a collar, spreading the loads at these joints over a broader area of the flange and the pipe. Because this distribution lessens stress on the exhaust tube, it makes a longer-lasting, more-durable joint.

To form a brazed joint at the exhaust-tube and flange, first weld the ID of the exhaust-header tube to the flange with a heliarc-welder as described in the following chapter. After you've finished welding all of the tube-to-flange joints, make sure the area to be brazed is clean. Preheat the flange and braze each joint around the outside of the tube. Sometimes bolt holes are so close to the exhaust pipe that you must allow for bolt-head clearance when you braze the flange and tube.

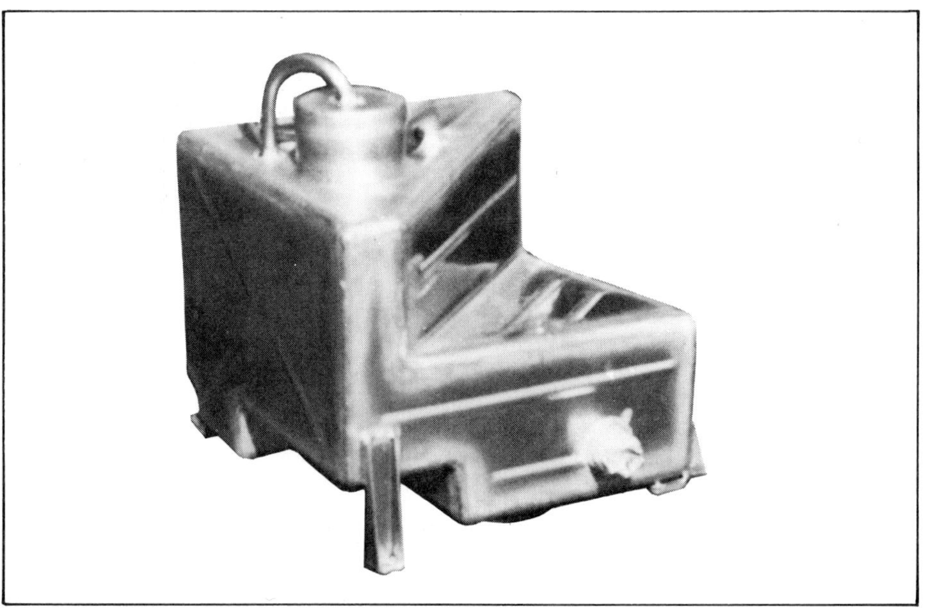

Oil tanks are another excellent aluminum gas-welding project. Joints remain strong and ductile. Fittings and bungs are heliarced.

GAS-WELDING ALUMINUM

Having learned to gas-weld steel, the next step is to learn how to gas-weld aluminum. Although many people are unaware of this welding method, I encourage you to try it. I find that it is very handy for welding *non-structural* components. It is different from gas-welding steel, but not any more difficult.

Yes, aluminum *can* be gas-welded. It was the first welding process to be successfully used on aluminum. And gas-welding continues to be used in custom metalwork. It is surprising to find that many people are not aware of this useful process. Many times when I am gas-welding aluminum, or even talking about it, people are amazed. They usually say, "I didn't know that was possible," or, "I thought you could only heliarc aluminum." I've even had people ask to see it done. They had to see it to believe it. Then there are the old-timers who say, "Oh, I saw that done years ago."

Gas-welding aluminum may not be common in the general welding trade, but it has a number of uses in custom-car fabrication. For instance, it is great for joining aluminum body parts. It doesn't harden the weld area, it leaves it soft and easy to work with. Because it doesn't become hard and brittle, a gas-welded aluminum joint resists cracking and is easier to metal-finish than a heliarced joint.

Uses and Advantages—Oil tanks and other liquid containers such as water surge tanks and overflow catch tanks are often gas-welded. Aluminum air and brake ducts, and even aluminum air-cleaner housings can be gas-welded. Gas-welding is good on any aluminum component which must hold a liquid, duct air, or be metal-finished easily at the weld area.

Gas-welding aluminum doesn't require expensive equipment. It can be done quickly and 100% penetration is easy to achieve. The advantages of low cost, very effective penetration, and the speed with which you can weld, make this process very appealing.

The oxyacetylene process is the most-common way to gas-weld aluminum. It can also be done with oxy-hydrogen. This oxy-hydrogen process is recommended when welding thin aluminum sheet because its cooler flame is easier to control. Regardless of whether acetylene or hydrogen is used, gas-welding is normally used on 0.040—0.063-in.-thick aluminum.

PRACTICE

Before you attempt gas-welding aluminum be sure you've mastered flame control as described earlier in this chapter. After you have this in hand, practice your welds on scrap aluminum. Remember, gas-welding alu-

There are several good aluminum fluxes on the market. Don't be sold aluminum *brazing* flux instead of aluminum *welding* flux; there's a big difference between the two. My favorite is Alcoa-22 aluminum welding flux. You can get it from Force Chemical Co., Industrial Blvd., Paoli, PA 19301. 215/647-3575.

minum is not necessarily harder than gas-welding steel, it's just different.

As with any type of welding, learning to gas-weld aluminum takes practice, time and patience. And don't be discouraged if your welds don't look great at first. Stick with it and you'll pick up another valuable skill.

Differences—There are three specific differences between gas-welding steel and gas-welding aluminum: Flux must be used—as with brazing. *Lower gas-line pressures* are used. You must use *special welding-goggle lenses.*

Aluminum Flux—The use of flux while welding aluminum is extremely important. Aluminum-welding flux removes oxides which naturally form on all aluminum alloys. If these oxides are not removed, you cannot gas-weld aluminum successfully.

Flux removes impurities by chemically combining with the oxides and absorbing them. The flux floats these impurities to the surface of the weld, leaving sound metal underneath. Impurities are left on the surface in the form of *scale.* This scale can be brushed away after the metal has cooled.

Aluminum flux comes in powder form in a plastic or glass container. In a glass or ceramic vessel, mix the powder with water, using about three parts flux to one part water to form a free-flowing paste. Copper or steel containers will contaminate the mixture and ruin it. Mix only a small amount at a time. A little flux goes a long way—and it's not cheap. Mix only what's needed for immediate use. You can always mix more if you need it. And don't save flux for later use once it has been mixed with water.

The welding rod must have flux on it. Cut the rod in half and dip one end into the flux paste. Then use a small *flux brush* to paint the flux directly onto the rod. Heating the rod a little afterward helps make the flux stick to the rod better.

I don't apply flux to the work piece until I have made all of my tack-welds with the flux-coated rod. Once all of the tack-welds are made, I apply a *thin layer* of flux to *both sides* of the area to be welded. This works best.

Tack-Welding—When you are tacking pieces together, make sure your rod has plenty of flux on it. Tack-weld your work about every 1—1-1/2 in. There shouldn't be any gaps. The more tack-welds you use, the less your work will distort while making the final weld. Tack-welds help to conduct heat evenly throughout the work, limiting warping. The bead also has less tendency to blow through.

Another good reason for using a lot of tack-welds: It preheats the metal so the flux will stick better to the area to be welded. The flux will not run down the work as easily and be wasted. *Remember flux is needed only at the weld.*

After the welding is done, completely remove the flux from the metal. Flux left on the aluminum becomes a starting place for corrosion. Also, flux residue will keep paint from adhering to the metal surface. Flux can be removed with hot water and a *soft* fiber brush or a clean rag. Soft clean material must be used so you won't scratch the aluminum. Remember, scratches are where cracks start.

Line Pressure—When gas-welding aluminum, regulate the line pressures lower than what you'd use for brazing or gas-welding steel. Aluminum melts at a much lower temperature than steel, so less heat is needed. The low line pressure produces a cooler flame with less tendency to blow holes through the material as you weld. For example, set both oxygen and acetylene or hydrogen at 2 psi when welding 0.040—0.063-in.-thick aluminum. The larger tip size you use, the higher the line pressure needed to produce the proper flame. However line pressure should never exceed about 4 psi.

For best results use these low line pressures with a neutral or a slight reducing flame to assure a sound weld. An oxidizing flame will cause the formation of aluminum oxides. This will result in a poor bead, poor fusion and a defective weld. Be sure the welding tip is clean. Care will pay off in sound welds.

Special Lenses—When gas-welding aluminum, the flux is melted by the heat of the flame. This melting gives off a yellow glow so intense that conventional gas-welding lenses cannot filter it completely, causing impaired vision. You must see the base-metal pool *clearly* when it forms and while you're making the weld. This means being able to see the tip of the flame clearly—the point of highest intensity of the bright yellow glow. Otherwise your ability to gas-weld aluminum is greatly handicapped. You will pop holes, deposit too much rod, or make an unreliable and unattractive weld.

Cobalt-blue lenses were once the most common ones used. However, they are no longer available due to federal regulations by the Occupational Safety and Health Administration.

Fortunately, T.M. Technologies, (P.O. Box 762, N. San Juan, CA 95960), offers the *T.M. 2000 lens*, specifically designed for oxyacetylene welding and brazing of flux-coated materials. It is a green glass that offers clarity and eye protection from the orange light associated with gas-welding aluminum, making aluminum gas-welding both safe and possible. The lens is designed to fit standard welding headgear. I recommend it.

Remember, with aluminum gas-welding: Use the proper flux; use lower gas pressures than you would with steel; and use special welding lenses.

Preheating—After the work has been cleaned, tacked and fluxed, you are ready to make the final weld. Preheat the work *before* starting to weld. This is always good practice with any type of weld or material being welded. This reduces the chances of cracking the metal.

Welding Motion—When gas-welding aluminum there is no need to move the torch in any direction other than forward. The welding flame should be held close to the work so the inner cone of the flame almost touches the pool of metal. This heats a very small area and greatly reduces the danger of burning through. It is very difficult to keep from burning through thin aluminum. Hold the torch and rod so some of the flame is on the rod. The molten pool of aluminum is not hot enough to melt the rod.

As you move along, *keep* the welding rod close to the weld area. This will keep the rod hot so it will melt when you dab it into the molten pool. If you are right-handed—the torch is in your right hand—weld from right to left so the flame points away from the completed weld and preheats the area to be welded. Work in the opposite direction if you're left-handed. When welding thin aluminum, the welding flame should be directed at an angle of 30° or less to the work surface.

Welding Rod—The most common rod diameters are 1/16 in. and 3/32 in. Use this rule: It is best to use the rod size closest to the thickness of the metal being welded. When welding body parts keep the rod deposit—filler material—to a minimum. The less filler you add, the less you will have to remove when metal-finishing the welds. Filler rod #1100 is best for gas-welding aluminum.

Using a cutting torch requires good hand and eye protection. A clean area means there's less chance of sparks causing a fire. A fire extinguisher and water bucket should be nearby. And don't do as I'm doing, do as I say: Wear long sleeves!

FLAME CUTTING

Another important function of oxyacetylene equipment is *flame cutting*—using oxyacetylene to melt metal and *blow it away* with a burst of oxygen to "cut" it. Flame cutting, if used properly, is very useful and saves a lot of time. It can be used to cut blanks of steel tubing which will be fitted more accurately for use in a roll cage or frame. Or use it to cut thick pieces of sheet steel which will be final-trimmed later. You can also flame-cut large irregular shapes that won't fit in a bandsaw. A stock car body can be lightened extensively by using flame cutting to gut the interior. Flame cutting is a great time saver if done carefully. It won't replace a bandsaw, but it has its advantages.

Careful use of a flame-cutting torch can produce clean results. Practice and proper tip size are crucial. Skill will show by the *clean edge* of the final cut—straight with little slag. How clean a cut is depends on the skill of the torch operator. For best results, flame-cut the bulk of the material, then go over the edge with a power grinder. The grinder will precisely define the shape of the cut edge and remove any slag.

Practice—Practice flame cutting 1/4-in.-thick scrap steel. Choose a torch tip recommended by your welding manual for flame cutting 1/4-in. steel. Set the oxygen and acetylene regulators at the pressures recommended by the manual for flame cutting. Draw some lines with soap chalk to follow on the steel—some straight, some curved. You're ready to light the cutting torch.

To avoid burns and eye injuries, wear heavy leather gloves and *conventional* welding goggles—not those for welding aluminum. Light the torch. Wave it over the steel to preheat it slightly. Hold the torch with both hands: the thumb on one hand operates the oxygen and the other hand guides the torch along the cut line.

Hold the torch with the flame directed at the edge of the metal. When it melts, press the oxygen lever and guide the flame along the line. Repeat this process all along the cut lines you drew. The cutting speed is largely determined by the metal thickness. If you move too fast, you'll get a shower of sparks back in your face as the cutting flame won't blow completely through the metal. Because it takes more time to melt thicker material, a slower cutting speed is required.

Here is a handy trick I often use when flame cutting. I clamp a piece of angle iron to the metal and use it as a guide for cutting straight lines. This makes a very straight cut that is easily cleaned up with a grinder.

Keep practicing flame cutting on scrap until your skill grows. Only when you are sure of your ability should you use flame cutting on a project. You will see progress as you learn to handle the cuts cleanly.

Caution—Flame cutting involves a great deal of heat. And molten steel and sparks spray off the metal. **Always have a fire extinguisher and a large bucket full of water handy when flame cutting. Always protect your clothing, skin and eyes.** Those sparks are dangerous. Think about what you're doing and pay attention to where the molten steel and sparks are falling.

After you're familiar with the various gas-welding methods, the next step is electric-welding. Knowing oxyacetylene welding will make the different types of electric-welding go smoother for you. In the next chapter I'll tell you how to do them, and where they are most useful.

6

ELECTRIC-WELDING

Frame and roll-cage required many feet of high-quality heliarc welds. Note heliarc foot control in foreground and torch hanging from front frame rail. Photo by Tom Monroe.

Rather than using burning gas as the heat source, electric-welding uses an electric arc. As I said before, whether you do your own welding or pay someone else to do it, you should be able to discern a quality weld from a poor weld. Although you may not become the best weldor on the face of the earth, you should have no problem learning enough to understand and test welds.

We'll look at three common types of electric-welding in this chapter: arc-welding; *heliarc*—also called *TIG-welding; wire-feed* welding—also called *MIG-welding.* They are all done with electricity. I will explain how each type of welding is done and how it is best used. You'll see examples of good sound welds and learn how to do them yourself. Remember what I said about gas-welding? You *must teach yourself* to weld. All I can do is point you in the right direction. It's then up to you to go in that direction. There are some safety tips I'll share with you, too.

ARC-WELDING

All types of electric-welding covered in this book are *arc-welding.* Heat is produced by an electric arc that crosses a gap between the welder's electrode and the work. One type of electric-arc welding receives the name *arc-welding* because it was the first to use an electric arc, thus the name. Consequently, the others got different and more descriptive names.

Arc-welding is very simple. A flux-coated consumable electrode, or rod, is mounted in an *electrode holder.* The holder is at the end of a cable which comes from the welder.

Let's look at the arc-welder in detail:

Welders—The arc-welding machine is the power source. It can be a d-c

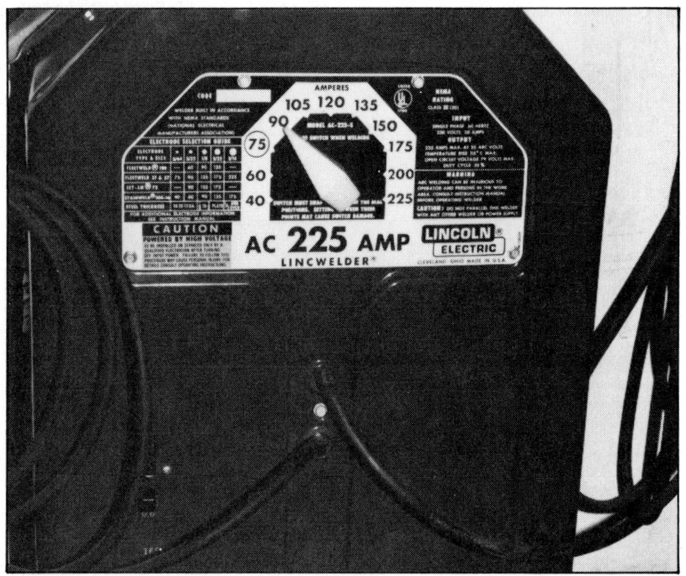

Lowest-cost electric welder—the a-c arc welder, or "buzz box." OK for welding mild steel, a-c welders are relatively inexpensive. Photo by Jeff Hibbard.

ELECTRODE
FLUX COATING
ARC
MOLTEN POOL
SLAG
GAS SHIELD
WELD BEAD
CRATER
PENETRATION
BASE METAL

Shielding gas is created as electrode coating melts. This coating combines with weld impurities to form slag on the weld bead, insulating it to slow the weld's cooling rate. Don't chip the slag off until the welded area has cooled a few minutes.

generator type, but is more often the less-expensive a-c/d-c transformer type that converts a-c current to d-c current for welding. Arc-welders come in various capacities, all rated according to their maximum *continuous current* in amps. When the welder is used at a setting over this rating, its *duty cycle*—minutes of continuous use—is limited.

It is important to choose a machine with enough capacity to handle all of your needs. I have found welders with a 90—220-amp range most useful. They are able to weld 0.090—0.500-in.-thick metal.

The arc-welder has a power control at the front or top of the machine. Amperage can be adjusted quickly to supply more or less welding heat. Two cables, as with all types of electric-welding, are needed. The first cable—the ground lead—goes from the machine to the bench or workpiece. The second cable—the electrode lead—goes from the machine to the electrode holder. Both cables are heavy-duty, flexible rubber-coated copper. Although the cables are designed to withstand rough use, check them periodically for cracks and loose connections.

Electrodes—Again, the electrode—or welding rod—supports the electric arc and produces filler metal for the weld joint. The high temperature generated by the arc forms the molten pool. The heat also melts the flux coating on the rod and releases gases to shield the molten pool. This protects the molten pool from contaminants—oxygen and nitrogen. The melted coating also produces slag. This *slag* insulates the fresh weld, slowing the rate of cooling. This helps to prevent brittle welds.

Sold by the pound, welding rod is available in several types and sizes for welding steel. Nickle rod is available for welding cast iron.

Don't buy more welding rod than you need. The coating on the rod can deteriorate from moisture. Consult your welding-supply salesman for your particular needs. An *experienced* salesman will recommend the rod you need for the welding you intend to do.

Uses—Arc-welding is most commonly used for welding thick metal. It works well on frames and heavy brackets. I used arc-welding on home-built hot rods a lot when I was still in high school. It got the job done and worked well.

Arc-welding is the one most used by amateurs—and with good results. The equipment is not expensive and is readily available. Sears, Lincoln Arc Welder Co., Miller and Hobart have good reliable arc-welders.

Practice—Learning to arc-weld isn't any different than other welding techniques. It takes practice and patience. A good way to learn basic arc-welding is to weld beads in the flat position. Weld the beads on clean 1/8—1/4-in.-thick steel. Experiment with different-size rods and heat settings. Practice doing flat-position welds until the beads are straight and consistent in height, width and in rod build-up. When you can do a consistently smooth and spatter-free bead, practice welding joints.

Testing—Just as you tested other types of welds with a hammer and vise, test your arc-welds. Secure the piece in a vise, and hammer it to see if you can break the weld. If the weld holds while the metal bends over or breaks elsewhere, your weld is sound. Don't be fooled by a pretty weld. Looking good isn't enough. A weld must be *strong* from proper penetration or it may break.

TIG-WELDING

Tungsten-inert-gas welding or TIG-welding—is also referred to as *heliarc-welding*—or just *heliarc*. It is probably the single most-useful welding technique in a custom-car shop. It is useful because it welds mild steel, stainless steel and aluminum easily.

Not only is the heliarc-welding process simple, heliarc welds are attractive and strong. The working end of the heliarc is a water-cooled *torch*. Unlike a gas-welding torch, an electric arc between a *tungsten electrode* and the base metal develops intense heat. This heat melts the base metal to form a molten pool. Filler rod—similar to that used in gas-welding—is inserted into the arc. The filler rod is melted into the pool to form the weld bead.

An inert atmosphere—argon or sometimes helium gas—is fed from the torch nozzle to envelop the electrode tip and the molten weld. This at-

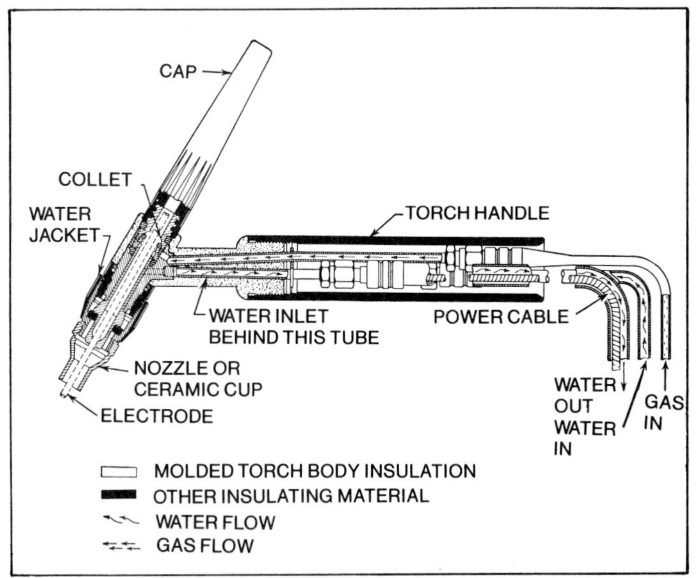

TIG-welding-torch cross section. As can seen by this drawing, the cable supplies electric power, gas shielding and cooling to the torch. Photo courtesy of Linde Air Products Co.

Miller Dialarc HF TIG-welder. It has a 10–310-amp range. A-c/d-c capability means it can be used to weld aluminum. Welding unit, water tank and gas bottles are mounted on wheels, making it mobile. Photo courtesy of Specialized Vehicles, Inc.

mosphere *shields,* or protects, the weld and the tungsten electrode against oxidation and contamination from the air, somewhat like the flux for gas-welding aluminum or brazing. If you haven't already guessed, the name *heliarc* comes from *helium* gas and the electric *arc.* As the weld is made, the molten metal cools and solidifies, joining the parts. If the base metal is as clean as it should be, there will be little or no spatter or smoke as you weld.

The heat produced by heliarc is not only intense, it is concentrated. As a result, heliarcing produces welds with good penetration on thick material. The welder has many heat settings. This wide range of heat settings is an advantage. You can set the heat very low, making it relatively easy to weld very thin material. You can also set power for higher heat for thicker material. And because its heat is concentrated, you can easily join thick material to thin material. This requires considerable skill with other welding techniques.

To do this the weldor directs the heat more toward the heavy metal so both thicknesses of metal flow together nicely. This gives the weldor great control. Many difficult welds are possible. Heliarc is easily done in all *positions*—horizontal, vertical and overhead—as the weldor's skill increases. You can make beautiful welds without the spatter or scale that are problems with other types of welding. It makes great clean welds on mild steel and stainless steel.

Welders—The choice of a welder is important. I use the very popular Miller Gold Star 300-amp a-c/d-c welder. I recommend this type of welder for a custom metal shop. By "type" I mean a welder with four specific features: wide heat range, a-c/d-c output, a foot control, and a small torch.

A welder must have a *wide heat range.* It is important to be able to weld a wide range of metal thicknesses. You must be able to adjust the welder so you can weld metal as thin as 0.035 in. Then there are times you may need to weld heavy-gage steel—such as brackets up to 1/2-in. thick. This requires much higher heat. So choose a welder with a wide heat range. Don't limit yourself.

A-c/d-c output is a great advantage. An a-c/d-c welder can weld steel *and* aluminum: d-c output is used for welding steel; aluminum requires a-c output. Because I frequently weld steel one minute and aluminum the next, I would be lost without the a-c/d-c feature.

A-c current is used for welding aluminum because of the type of arc it makes. It has a cleaning action that removes aluminum oxides which would otherwise contaminate the weld. The advantage of this is it eliminates the need for flux, which is essential for gas-welding aluminum. With no flux the welds do not have to be cleaned.

A *foot-operated* power control is not only helpful, I consider it essential. The foot control does more than just turn the arc on and off. A foot-operated *rheostat,* the foot control also lets you adjust power as needed *without interrupting the welding.* This flexibility helps you to make good, sound welds in both steel and aluminum. And time is not lost while you stop to make power adjustments.

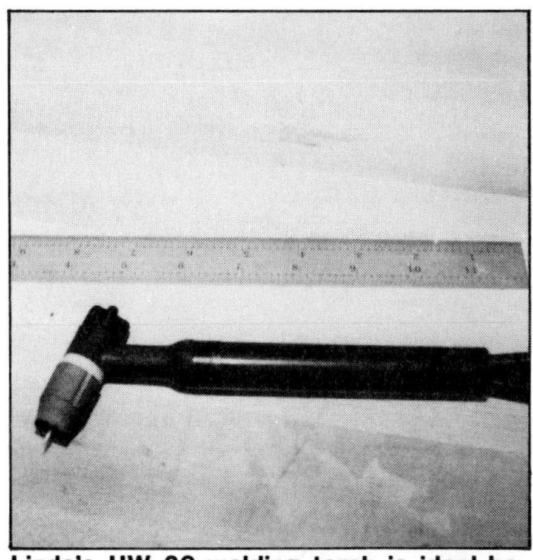

Linde's HW 20 welding torch is ideal because it will fit in tight spots and it is water-cooled.

To use the foot control, start welding by pushing the pedal part of the way down. Increase power by pushing the pedal farther until the heat is high enough to produce a molten pool. As the weld is nearing completion, let up on the pedal slowly. This gradually reduces the heat until the power shuts off. Turning the power off abruptly causes a pit, or crater, at the end of

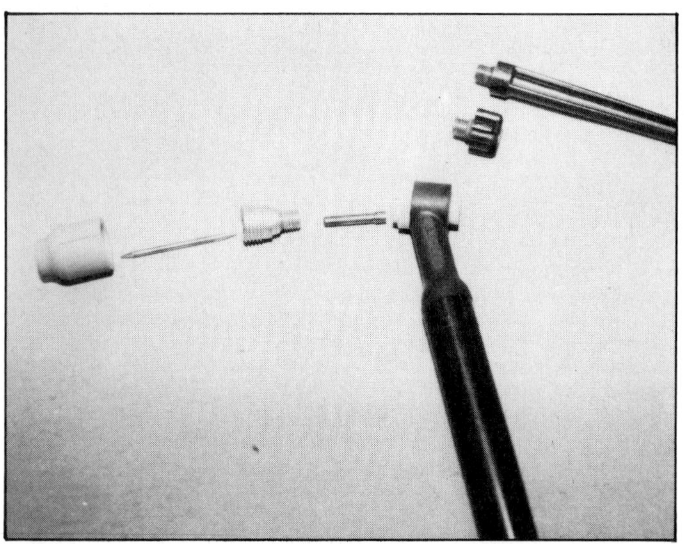

HW 20 torch is versatile. It accepts 0.040-, 1/16-, 3/32- and 1/8-in.-diameter electrodes. Two styles of electrode caps are shown above. Cup, electrode, collet (threaded), collet body are left of the torch.

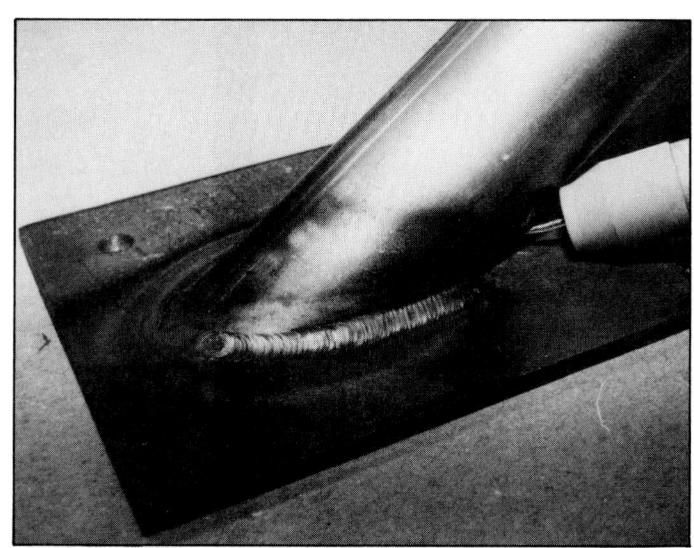

Welding in a hard-to-reach place calls for a gas lens, or long, small-diameter cup. It enables you to extend the electrode and move in closer.

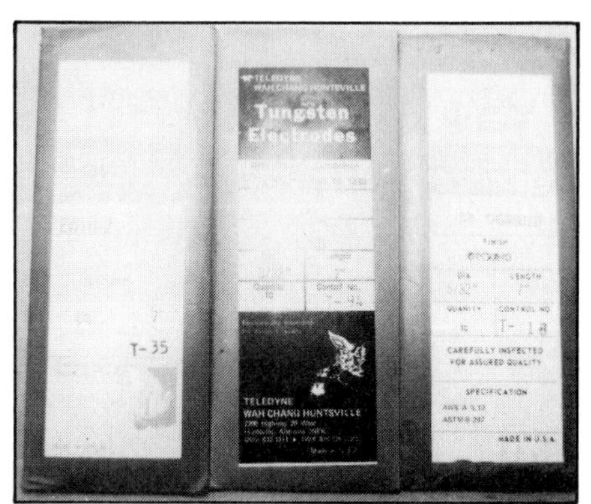

Check with your welding supplier for advice on electrodes. Many are available, and he'll suggest which one will best satisfy your welding needs. In most cases, the 2% thoriated electrode will do the job.

the weld bead. A crater looks bad and may cause a weld failure.

The fourth feature I look for on a heliarc welder is a *small torch.* I use a Linde HW 20 torch. It is very light and easy to use—it doesn't tire your hand. This is important because some welds will be long and your hand must remain steady. Another plus of a small torch is it is easier to use in tight places. And it allows you to do delicate welding.

Several good heliarc machines are available in different styles. Some have d-c only and no foot control. Each welder model has certain advantages and certain limitations. Miller makes a small welder with d-c only and an air-cooled torch. The power is turned on by a small switch in the torch handle. This type of welder has its limitations. Because this welder lacks a-c you can only use it for welding steel. If you need more or less heat you must stop welding, go to the welder and change the power setting. This machine is fine for a hobbyist or someone who doesn't intend to weld aluminum. The resetting is not difficult, but it slows you down. I know two talented amateur car builders who use and are satisfied with this welder.

Before buying a welder, think over your needs carefully and then list them. Using this list, choose the heliarc that is best suited to these needs—after shopping and comparing.

Torches—Let's take a closer look at torches. The *torch collet and collet body* hold the electrode firmly in place. This collet-holding system is very convenient. It allows quick replacement or adjustment of the electrode. Replacing or extending the electrode is done by loosening or tightening the *end cap*—see drawing. End caps come in different lengths. I like a short cap because it doesn't get in the way when you're trying to weld in small places.

Cups—Torch *nozzles* or *cups* are usually made of ceramic because of the high heat. The cup, available in many sizes and styles, directs the shielding gas—argon or helium—around the tungsten electrode and the arc. Cups vary in size and shape. There are extra-long, small-diameter cups for welding in small, deep places. Others are short with a larger diameter for broader shielding. Then there are standard cups between these two extremes.

Small torches are water-cooled, not air-cooled. Both the power cable and torch get too hot for air-cooling to be effective. And the longer you weld, the hotter they get. The cooling system is simple and effective. A water tank—usually about 30 gallons—is attached to the side or top of the welding machine. When the welder's main power is switched on, water is pumped from the tank, through a tube—which is routed parallel to the power cable and to the torch—and back to the tank through a return tube. Another tube carries the inert gas from a bottle at the machine to the torch's ceramic cup. *Take care not to burn or cut any of these tubes.* They are expensive and hard to replace or repair.

Electrodes—There are many varieties of tungsten electrodes. The most commonly used electrode is the *2% thoriated, centerless-ground*—boy, what a mouthful! I'm going to save you the pain by not attempting to explain this. But believe me, it can withstand more heat than other types of electrode.

Available in many diameters, the most frequently used electrodes are 0.040, 0.094 and 0.125 in. I always try to use an electrode diameter about the same as the thickness of the metal I'm welding. This helps to get a good consistent bead.

The heliarc-welder's tungsten electrode is *non-consumable.* It does not melt into and become a part of the bead. Rather, the electrode is the means of *forming the electric arc* which creates the heat to melt the base metal and the filler rod—much like the oxyacetylene flame of the gas-weld.

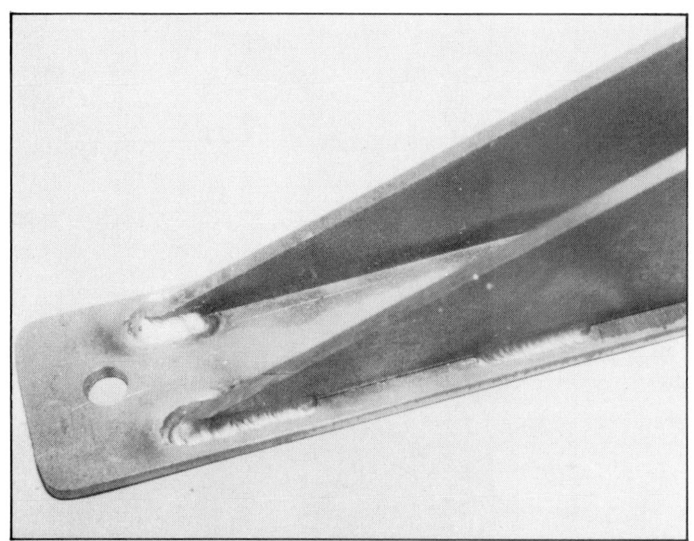
Bracket tips can be strengthened by running weld beads around the points. It's important to taper off the heat gradually when welding steel so you won't leave a pit in the bead.

Aluminum fillet weld will make a very strong joint on this fuel-filler neck. It's not uncommon to weld on both sides of the joint.

Also like the gas-welder, a separate filler rod is used to form the bead. This differs from arc-welding where the electrode forms the arc *and* is consumed as it melts to form the bead. Non-consumable doesn't mean a tungsten electrode doesn't get used up. It *does* get used up in two ways, but not by *melding,* or combining, with the weld.

Although they are originally ground for welding steel, the tungsten electrode is consumed as its tip is *dressed,* or reshaped, over and over by grinding. To weld steel, the electrode must be ground so its tip is pointed. To weld aluminum, the electrode must be formed into a round, ball-like tip. In forming these ends some of the electrode is used up. I'll explain how tips are dressed later in this chapter.

Second, a heliarc electrode can be used up after it gets *contaminated.* Contamination occurs when the electrode touches the molten pool or welding rod and is coated with the filler and/or base metal. When this happens, *you must stop welding.* If you continue welding, the foreign material on the electrode will gradually burn off, causing an unattractive weld that will be porous and faulty. The tip must be redressed to remove the contaminant. Redressing the tip gradually uses up the electrode.

When learning to heliarc you'll find yourself dipping the electrode into the molten pool or accidently touching it with the welding rod. This should be expected at first. However, as you gain experience it shouldn't happen as often. A weldor with many years of welding experience will rarely contaminate an electrode.

Although a little more complex, a person who has mastered the skill of oxyacetylene welding has a definite advantage when trying to learn TIG-welding. In both types of welding the weldor feeds the rod with one hand and holds the torch with the other. Fine control is needed in heliarc-welding because you have both hands working and a foot control to coordinate. You must be able to move your torch and rod along the work in an extremely steady and accurate way. At the same time, your foot increases or decreases the heat at the weld. This coordination requires a lot of practice and skill.

HELIARC-WELDING STEEL

Heliarc-welding on steel is very useful. There are a great many steel parts on a car. Frames, roll bars and all kinds of brackets can be welded with beautiful, high-penetration, scale-free welds. Stainless steel can also be welded with heliarc in the same manner as mild steel. Heliarc-welding on stainless steel requires stainless-steel filler rod. The tight, no-gap fit is even more important on stainless than on mild steel. Steel is heliarced with d-c current and the high-frequency setting on START.

If it's not already done, grind the electrode to a sharp point. Although smaller, it should look like the end of a pencil. The point should be clean and precise to make it easier to control the electric arc and the size of the bead. With a fine point on the electrode you may begin welding.

The steel must be clean and free of oil. Start out with thin steel—about 1/16-in. will be fine. Lay one piece down and practice making beads first. As your welds improve, move to thicker steel and continue practicing. After you've made some nice consistent beads—see the nearby photos — move on to join two pieces together.

After you've grown confident with welding steel and your welds have tested out OK, go through the same process with aluminum.

HELIARC-WELDING ALUMINUM

Heliarc-welding aluminum is used a great deal in custom metalwork. It requires a welder with a-c-welding capability. The welder is set with the high-frequency control on CONTINUOUS. Heliarc-welding aluminum can also be done easily in all positions. It's an advantage to be able to weld in any direction without metal dropping when you are working on a car.

Surface Cleaning—Aluminum must be very clean for heliarc. It must be free from all kinds of contaminants such as paint, dirt and especially oil. So before you try to weld any aluminum make sure it is completely clean and oil-free. Remember, another important tip when welding aluminum is a very close fit. The weld is intended

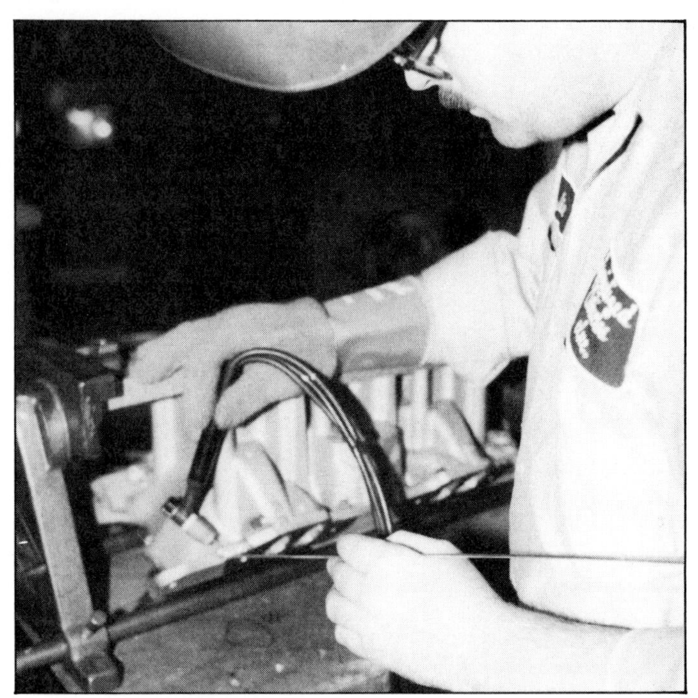

Sooner or later you'll need a special rod. Jim Reese is building up an aluminum casting using 4043 rod and a gas lens on the torch to get into tight corners. Photo courtesy of Specialized Vehicles, Inc.

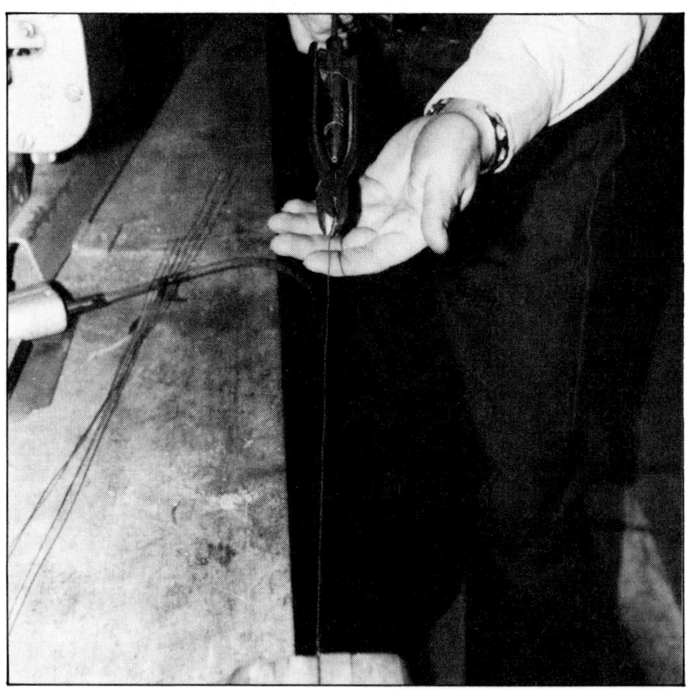

Straightening MIG-welding wire for TIG-welding using aircraft safety-wire pliers and a vise to pull it. Small-diameter filler rod allows deposits to be kept to a minimum when little filler is needed.

Best welding book I've found is published by the American Welding Society. Get it from a welding-supply store or the American Welding Society, P.O. Box 351040, Miami, FL. 33135.

to join pieces, not to putty up holes between them.

Gas Shield—To heliarc aluminum the metal must have a really good argon shield from the atmosphere. The argon flows down through the torch onto the weld. This argon shield protects the weld from gases in the air. If you try to run a bead along two pieces with an air gap between them, your chances for the ideal weld are slim. The argon shield can't work nearly as well when there's an air gap between the metal pieces.

Electrode Dressing—To heliarc aluminum you must *dress*—prepare—the tungsten electrode differently than you did for steel welding. The electrode must have a rounded, ball-like tip. A rounded end on the electrode keeps the electric arc steady. It stops the arc from weaving from side to side as you weld. Take the time to dress the electrode properly.

To make a ball-end electrode, set your machine on d-c+. Strike an arc on a piece of thick copper. The tungsten electrode will melt at the end, forming a ball. If it fails to melt, try using more heat. As soon as it melts, stop. You have formed a small half-ball at the end of the electrode. Try to make the ball no bigger than the diameter of the rod. A round end too large is as bad as no round end at all. It will make the electric arc waver during welding. Turn the welder back to a-c, and your round-tipped tungsten electrode is ready to weld aluminum.

When you have cleaned the aluminum thoroughly and have formed a ball end on the electrode, you are ready to start practicing beads. Start out making practice beads with rod on 1/8-in. aluminum.

Heat—Aluminum is a very good conductor of heat—it absorbs it very quickly. Consequently, you need more heat when starting a weld. The base metal gradually warms up as you continue welding and the weld pool may become too hot. When this occurs, ease up on the foot pedal to reduce heat. This feature of the foot-controlled heliarc makes it possible to make consistent beads, even though the work is getting a lot hotter as it is welded.

Rods—The correct use of filler rod is necessary to produce a sound aluminum weld. Common filler rods are 1100, 4043 and 5356. When welding tanks, or any project using 3003 aluminum, 1100 filler rod does a fine job. Rod 4043 is used on 5052 or 6061 aluminum. When welding 6061 aluminum to 3003 aluminum— example: a 6061 boss, fitting or bracket to a 3003 tank—use 5356 rod. Weld joints made using 5356 filler rod have excellent strength and good ductility. So choose your filler rod carefully.

Vent Hole—Here's a little hint that can save you some trouble. When heliarc-welding a tube frame, roll bar or cage, always *vent* the tube that's being welded if both ends are closed. You can do this by drilling a vent hole in the tube, page 131. Expanding air that would otherwise be trapped in the tube escapes, preventing the weld from being blown out. Air that is heated in a closed container expands very powerfully.

Practice—Learning to heliarc, like

any other welding, depends on practice. Start with a good metal work surface. It must be clean, steady and clear of obstructions. Unnecessary parts or tools on a bench may keep you from repositioning the workpiece quickly and easily. Don't weld on a wooden bench top. You'll burn the surface in a hurry. A large, thick steel work surface is ideal for heliarc-welding. It won't burn. And with the ground cable attached to the bench, the workpiece is automatically grounded when it's on the bench.

Get comfortable. Sit while welding whenever possible. This will help reduce body movement so your hands will be steadier—a must when heliarc-welding. Position the workpiece in front of you where it's within easy reach—near the edge of the work surface is ideal. This allows you to rest your elbows on the bench. Good arm support and back posture are needed. Don't lean over to your work. Get in close to the work so you can sit upright. You may be welding on the same piece for a long time. Leaning over will cause you to tire quickly. Tired weldors make mistakes.

For this reason I organize my work so I weld complex pieces in the morning. Being fresh and very steady allows me to make much nicer-looking welds. This is particularly important on a tank because the welds won't be metal-finished. They'll be left as is for the world to see. So they should show your best welding skill. Do it when you're fresh.

If you are right-handed, start from your right and move to the left. Reverse directions if you are left-handed. The arc is started and held steady until the metal melts, forming the weld pool. The metal filler rod is added at the front or leading edge of the pool. Keep the arc length short to get good penetration. Weld smoothly across the distance of the weld. Use the foot control to decrease heat *gradually* as you come to the end of the weld.

Shrinkage, cracks or craters may occur at the end of the weld when you stop the arc. Such shrinks, cracks or craters can cause a defective weld. This problem may be prevented by using the foot control to reduce heat *slowly* at the end of the weld.

Aluminum heliarc-welding requires a lot of practice in all of the different welding positions. Weld sideways. Weld up and down. Keep trying for

Miller Syncrowave 300 is the Cadillac of heliarc welders. If you can't weld it with this machine, give up!

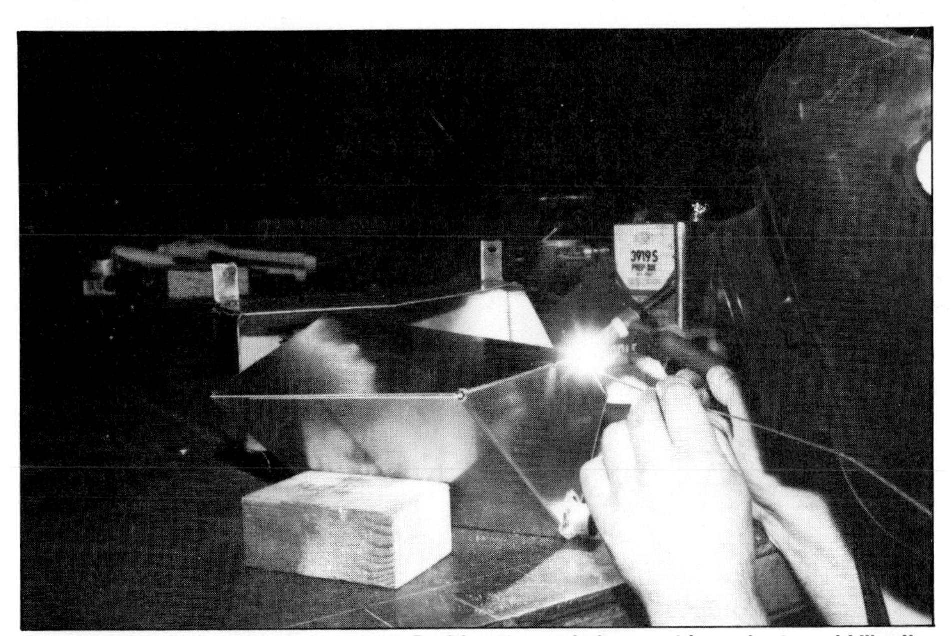

Rest your hands to get better control. Position the workpiece so it's easier to weld like I'm doing with this wood block. These add up to a better final weld.

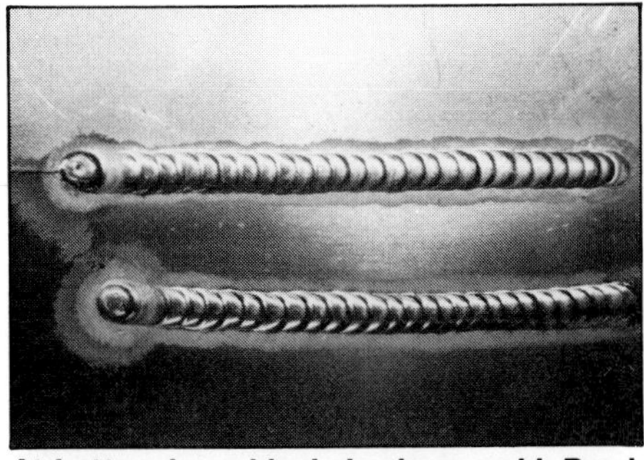

At bottom is an ideal aluminum weld. Bead is consistent, the deposit is even and the end has *no crater.* Crater is prevented by reducing heat slowly at the end of the weld. Top weld is good except for the small crater at the end. This is caused by cutting off the heat suddenly.

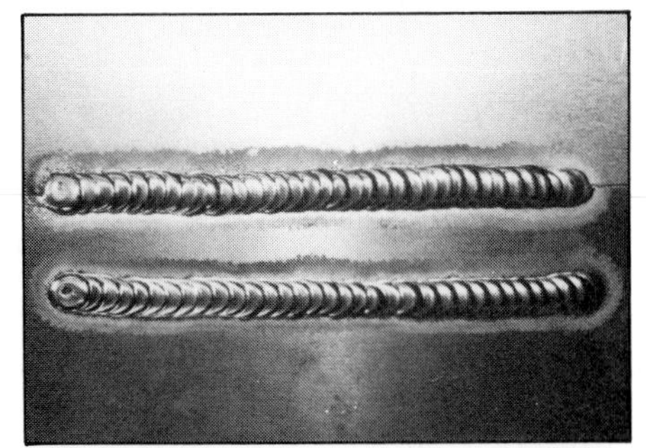

Both aluminum welds are very good. Shininess means correct amperage was used. Good fit allowed the top butt weld to be as even as the plain bead below it. I purposely released the foot pedal quickly at the end of the beads to show the crater effect.

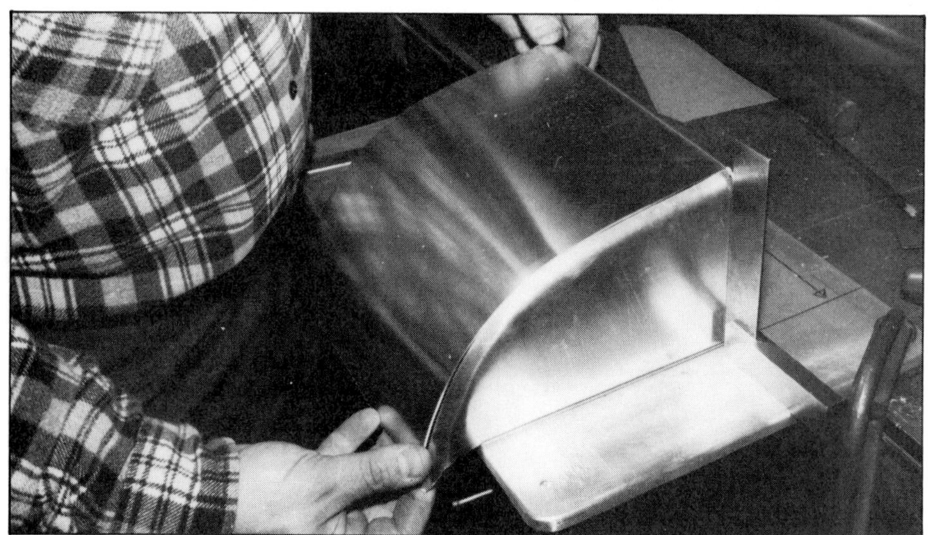

A good weld requires a tight-fitting joint. Large gaps are difficult to fill: The better the fit, the better the weld.

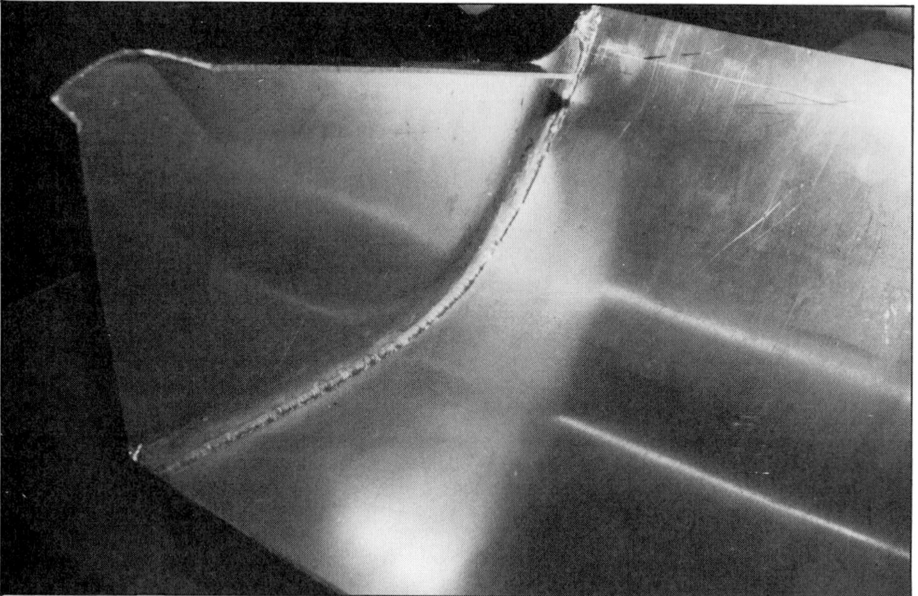

Underside of welded-on cap shows even penetration. Penetration is important for strength, and it prevents weakening of the weld if the part is metal-finished.

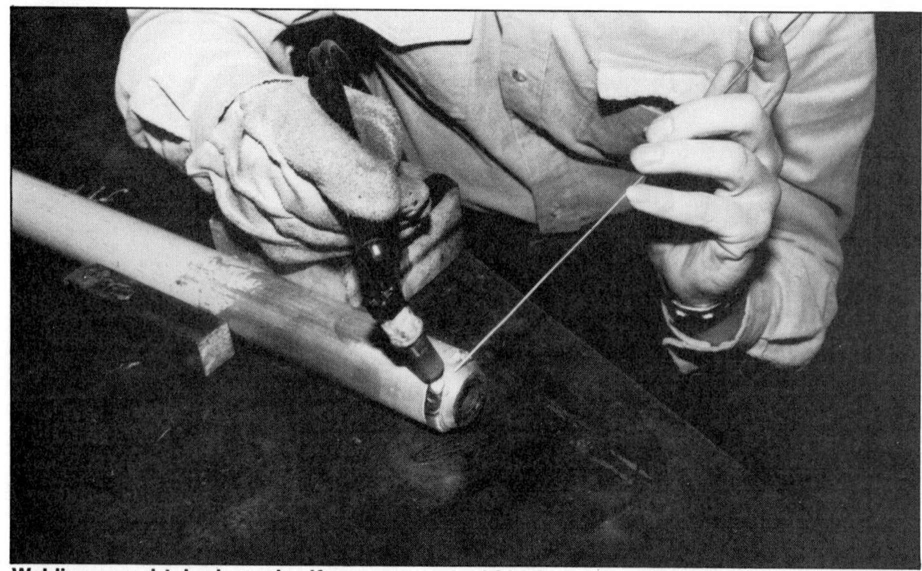

Welding round tube is easier if you use a steel block to keep the tube from rolling. After tack-welding, roll the tube as you weld.

consistently smooth beads in each position. Make several practice joints as well as flat beads. Keep making plain beads and joints until the beads look very good, time after time. The ideal weld on aluminum looks very shiny. Every ripple is very evenly spaced. A dull or hazy-colored weld indicates the weld was made too hot or too cold. It may fail in use.

Because it's easier than welding aluminum, learn to heliarc-weld mild steel first. You can familiarize yourself with the welding equipment and fine-tune your welding skills before tackling the problems associated with welding aluminum.

Testing—Test your welds for penetration as used for checking gas welds, page 62. I have seen beautiful heliarc-welds fail. This can happen easily if weld penetration is poor. A heliarc-weld bead may look nice, but it's no good if it has little penetration into the piece itself. The heat, speed and skill of the weldor are essential for maximum penetration.

On flat welds visually check for penetration on the underside of beads and workpiece. The underside will show penetration by the amount of, or absence of metal *drop-through.* If penetration is poor, slow down and increase the heat as you weld.

Tubing—When you are happy with your beads and penetration, try some tubing joints. Try welding different joints on both round and square tubing of different thicknesses. The fit should be as close as possible. Gaps in a joint lessen your chance for a nice-looking bead. Rather than concentrating on running a nice bead, you would be trying to fill the gap with rod. That makes it difficult to get a good bead.

Plug-weld—A *plug-weld*—also known as a *rosette weld*—joins two pieces by filling a hole. The weld should penetrate the sides of the hole in the upper piece of metal and the bottom of the hole formed by the lower piece.

Because rosette welds are so useful, they are worth practicing. Clamp two pieces of thin scrap aluminum or mild steel together, one with a 1/4-in.-diameter hole. Slowly and carefully heliarc the pieces together through the hole. Heat should be high enough to get good penetration, but without burn-through. Run the weld bead around the inside periphery of the hole, filling the hole completely to join the two pieces. Keep the plug-

Excellent plug-weld application. Plug-weld gives added strength to threaded plug in this thin-wall-tube suspension member. Photo by Tom Monroe.

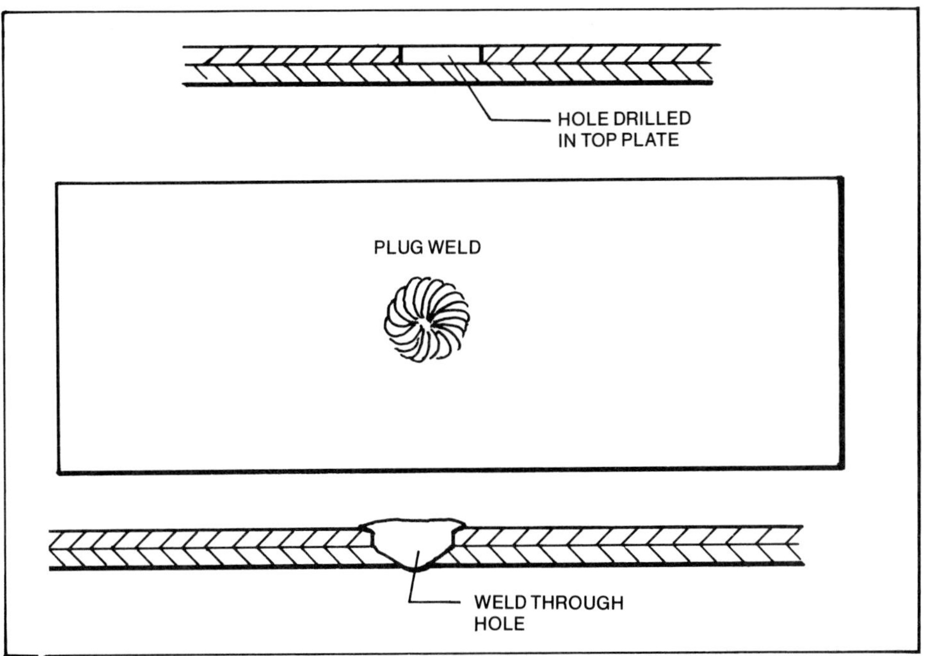

Plug-welds are a great way to hold two surfaces together. You may have a tank that requires an internal baffle—you could weld through a hole in the tank wall, into the baffle. The plug weld fills the hole and provides a strong joint.

weld reasonably smooth and neat.

Plug-welds are useful when attaching roll-cage mounting plates to frames or floorpans. Another use for a plug weld is even simpler. Sometimes you want to close up a small hole in sheet metal—such as Cleco holes used for temporary assembly during construction. To practice doing this type of plug-weld, drill or punch several small holes—about 1/8-in. diameter—in steel or aluminum sheet. Fill each hole completely and as smoothly as possible. You want a nice lump-free finish. Use a low heat when doing this type of plug-weld. A slow, gradual ending to the weld is especially important.

MIG-WELDING

Metallic Inert Gas welding, or MIG-welding, is another electric-welding process using inert-gas shielding. As with TIG-, or heliarc-welding, MIG-welding can can be used on steel and aluminum. The big difference between TIG- and MIG-welding is the MIG-welder uses a *consumable* electrode. This electrode is in wire form.

Unlike a TIG-welder's tungsten electrode, the MIG-welder's electrode is the filler material. The *continuous wire electrode/filler is fed* through the welding gun. This is where the term *wire-feed welder* comes from. The weldor adjusts the machine to set the amount of heat and the *rate*—speed in

Portable MIG welder carries its own shielding gas. This type of welder is often used in stock-car shops for good reasons: MIG-type welders can make many clean, strong welds very quickly.

MIG-welding produces lots of sparks and smoke. Protect yourself with helmet, elbow-length leather gloves and long-sleeved shirt, buttoned to the neck. Also, be sure the area is free of flammables and well-ventilated.

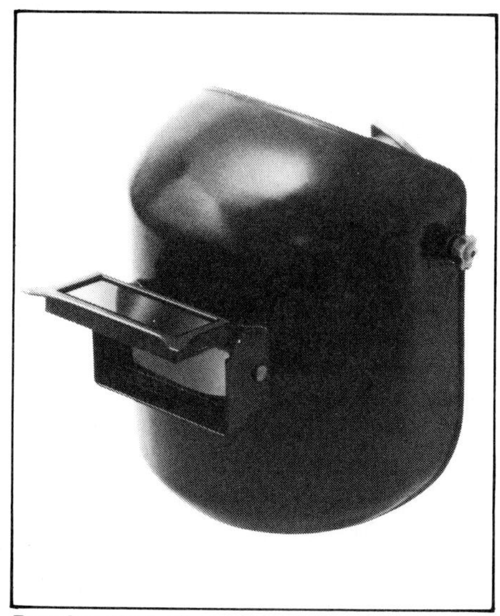

Popular type of welding helmet. Reasonably priced and durable, the 8260 Norton has a flip-up lens with a clear inner shield: a handy feature when starting welds and protecting your face while chipping slag. Photo courtesy of Norton Co.

in/min—of wire fed into the weld. The combination of gas shielding and electrode feeding makes clean strong welds at a relatively fast rate. The speed of MIG-welding and its good quality, make the MIG-welder very appealing and practical. Additionally, MIG-welding can be done in all positions with relative ease. Consequently, MIG-welders are gaining acceptance from custom-metal fabricators.

Stock-car constructors use MIG-welders for the same reasons. A stock car sheet-metal interior can be installed quickly and easily with a MIG-welder. MIG-welding is also frequently used to make roll bars, frames and kit cars.

Controls—Heat, wire-speed and gas-flow controls are all at the welding machine. They are simple to understand and easy to use. A single trigger on the welding gun starts and stops the welding process. The key to good MIG-welding is practice—learning to set the controls correctly and the actual welding.

You can weld all thicknesses of steel easily with a MIG welder. When welding thin steel you simply change to a small diameter wire and use less heat. Wire about 0.030-in. diameter is used for thin steel. The wire comes on spools and is reeled off then fed into a cable. This cable carries both the wire and the shielding gas to the welding gun. The shielding gas can be $C0_2$ with argon, or just argon. The $C0_2$/argon combination is more expensive, but it produces a better looking weld.

MIG-welding also creates a good bit of smoke. It is important to have adequate ventilation. Be sure to follow all of the general safety rules for welding when using a MIG-welder.

Chipping Hammer—When continuing a weld bead after stopping to change a rod or whatever, you must knock the slag off the end of the bead before continuing to weld. This is necessary because welding over slag will cause *inclusions*—weak spots—in the weld bead. You'll need a *chipping hammer* in order to knock off the slag.

A chipping hammer is basically a chisel and center punch with a handle. Turn the hammer in one direction to strike with the chisel end. Turn it in the other direction to knock the slag off with the center-punch end. After completing a weld, clean it by removing all of the slag.

CAUTION

The biggest hazard of all arc-welding is burns. The most common burn is *arc flash* to the eyes. Arc-welding produces ultraviolet rays that cannot be properly filtered by gas-welding lenses. **Be sure to wear a full welding helmet with a lens dark enough to protect your eyes.** Don't let anyone work near you or watch you work—even momentarily—unless they have the same eye protection. Otherwise their eyes will receive burn damage.

To protect yourself from other weld-related burns you'll need additional protection. Gloves, preferably leather gloves, are a must. They'll protect your hands from UV rays, sparks, radiated heat and the hot workpiece. The metal will be too hot to handle without them. You should also protect your arms from UV rays by wearing a long-sleeved shirt with the gloves overlapping the cuffs. And if you don't like the idea of dancing around trying to put out a fire at your ankle, wear cuffless pants. Cuffs do a great job of catching sparks that cause pant-leg fires.

BASIC METAL SHAPING

California Metal Shaping used nine pieces of 0.060-in.-thick, 3003-H14 aluminum for Chuck Looper's sprint-car fuel tank/tail section. Photo by Tom Monroe.

You are finally going to put your metalworking and welding equipment to use. Use the information in this chapter to do projects such as add-on spoilers, air dams and other body parts. Let's look at how to apply the techniques of *metal shaping* and how to use selected special tools and equipment.

As opposed to *metal bending,* metal shaping is the process of forming *compound curves.* As previously discussed, metal bending is a way to form a simple, or *straight,* curve from flat stock. A line drawn on the surface of the metal at and parallel to a simple bend will be straight. A line drawn on the surface of a compound-bend *in any direction* will be curved—it involves two or more curves.

You can use a sheet-metal brake with radius dies, or smaller hand tools such as glass pliers to make straight bends. When you know how to do this you have a head start on making compound bends and doing more sophisticated projects.

There are many ways of shaping metal. Some may seem crude. Others sound highly sophisticated. Either can be very effective in producing beautifully shaped parts.

Metal shaping requires considerably more skill than is required to do metal bending. Shaping metal was mentioned earlier when I discussed using slappers, dollies and special hammers. Additionally, you should become familiar with some special and more expensive sheet-metal-shaping equipment: *pneumatic planishing hammer, English wheel, Kraftformers* and *power hammers.* This equipment is discussed later. Let's examine what happens when metal is shaped.

Metal shaping involves changing the shape of metal—usually from a flat sheet—to a new shape with compound curves. Forming compound curves is achieved by *shrinking* or *stretching* the metal. Most of the time

C-clamps and Clecos hold aluminum pieces tightly to the station buck for checking and fitting. Bucks are discussed starting on page 81. Note simple curves in the side pieces. Center section is a compound curve.

Basic hand tools were used to shape this race-car nose. Leather-faced slapper, body hammers and file are at lower right. Nose is tack-welded and ready for final welding after removing from buck.

Metal-finished nose is ready for the painter. It fits precisely because the station buck was carefully planned and constructed.

both shrinking and stretching are necessary to achieve the desired shape.

STRETCHING

Sheet metal is easily stretched. Under pressure—as from hammering or rolling—the metal thins, elongates and curves. It stretches. If you can remember, it's similar to what your mother did when rolling dough for a pie. The dough spread and stretched under the rolling pin and got thinner. Similarly, metal spreads and thins under pressure. The amount of pressure determines how much the metal stretches. Keep this in mind: It isn't difficult to stretch sheet metal, but it is *easy* to overstretch it. Overstretching thins the metal too much, causing problems.

Excessive hammering or rolling can cause two problems: The metalwork hardens, making it brittle and difficult to shape. It can stretch the workpiece beyond the shape you want and "wahoo" it into an overblown curve. *Controlling stretch is the trick.*

SHRINKING

Shrinking metal is much more difficult than stretching. You must *gather* the metal, compressing or forcing it together to shrink it. The reverse of stretching, shrinking makes the metal thicker. Shrinking works in partnership with stretching to form compound curves.

Cold shrinking is done by using hand tools such as a mallet to pound the metal over a smooth, flat steel or hardwood surface. It is distinguished from *hot shrinking*—a process used to shrink steel by heating and hammering. When I mention shrinking without being specific as to hot or cold, I mean cold shrinking.

In addition to cold shrinking by hand, there are several mechanical shrinkers. Some mechanical shrinkers are manually powered. Some are electrically powered. All shrinkers consist of *two sets* of metal dies—two on the top and two on the bottom. The dies clamp onto the metal, gripping it. The dies then move toward one another, gathering and shrinking the metal between them. This gathering process also increases metal thickness in the same area.

SHRINKERS

Several types of shrinkers are available. Although some are more

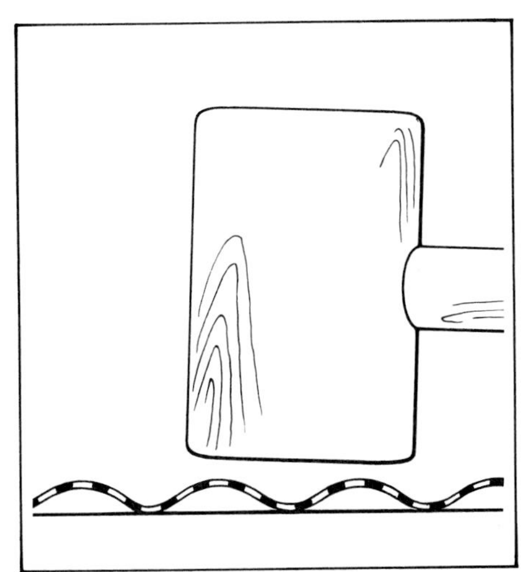

Cold shrinking can only be done using the combination of a hard surface and a soft mallet or hammer. The larger the mallet face, the better. As the metal shrinks, it *gathers,* increasing its thickness.

Shrinking and stretching combine to form a compound curve. English wheel stretches panel in the center. Outside edges were shrunk with a shrinker.

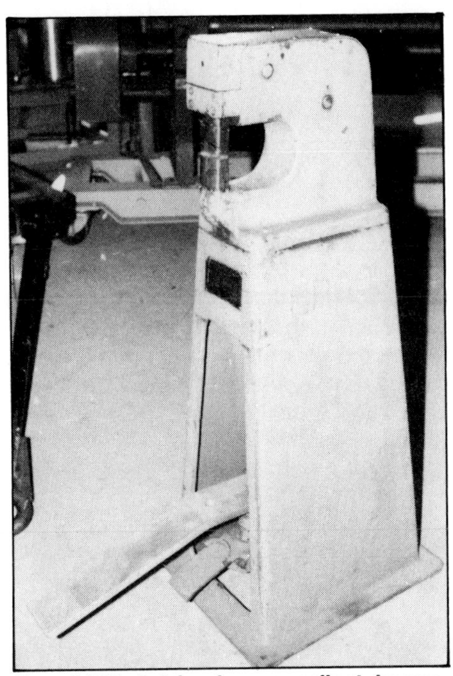

Erco 1447 shrinker is an excellent, heavy-duty, foot-operated shrinker. It's ideal for shrinking steel.

MMA shrinkers are very good tools. Shrinker in the foreground is from Williams Low Buck Tools. I made the other shrinker with the same jaws and mechanism, but with my own frame for working larger pieces.

complicated and more expensive than others, all do the same basic job. A relatively simple, inexpensive hand-operated shrinker will do a fine job of shrinking. The limitation is *throat depth*—how far it can reach in from the edge of metal. It may only reach 1 in. from the edge. Sometimes you need to reach farther to help form some curves.

An MMA shrinker, made in the United States, is a simple hand-powered shrinker with a 1-in. throat depth. Shrinkers available from Williams Low Buck Tools are fine as flange and edge shrinkers.

Another U.S.-manufactured shrinker is the popular Erco 1447. Foot-operated, a weighted pedal is pushed to power the dies. The Erco 1447 has a deep throat, accommodating up to 8 in. of metal. Because it is foot operated, both hands are free to maneuver the workpiece. This gives you greater control of the workpiece. Because shrinking steel is more difficult than shrinking aluminum, this machine is better than the hand-powered models for shrinking steel. However, it can shrink aluminum as well.

A complex and expensive electric-powered *shrinker and stretcher* is the Swiss-made Eckold KF314. Commonly know as a *Piccolo Kraftformer,* its dies can be changed according to the metal being shrunk. Although this super machine is great, its cost is rather high for most fabricators. A KF314 has an on/off switch. A similar Piccolo with a foot-operated control is coded KF320.

The Eckold HF 80 Ch hand-

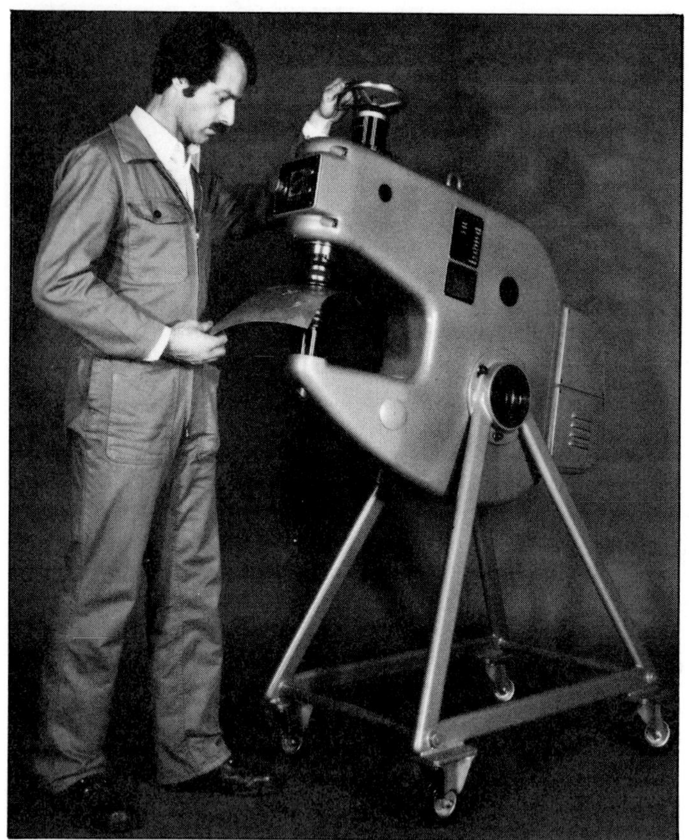

Eckold KF314 Kraftformer—a good shrinking machine. Different types of metal-shaping heads are available. Caster are standard. Photo courtesy of W. Eckold AG.

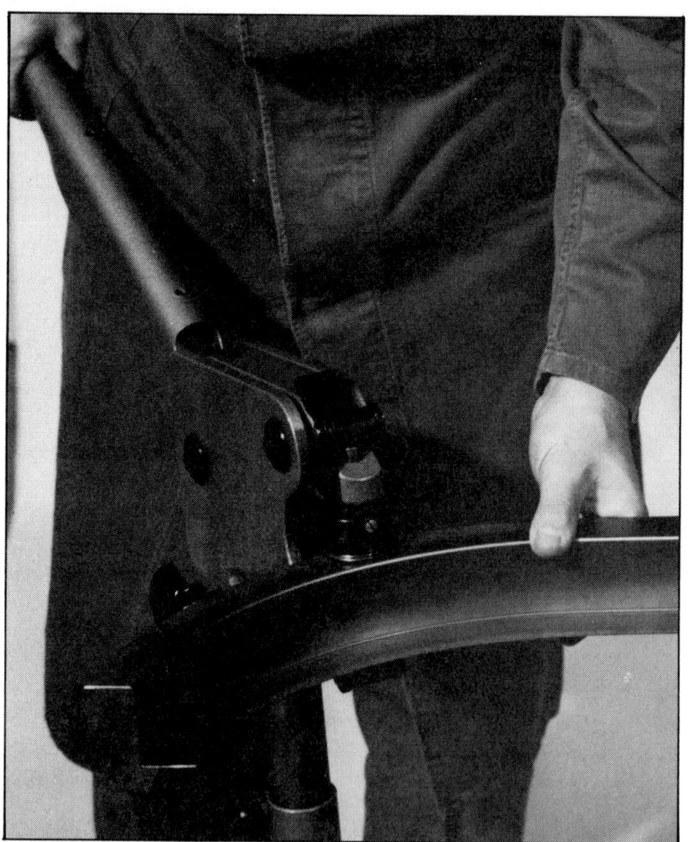

Eckold hand-former type HF 80 Ch is a hand-operated shrinker/-stretcher. Head is mounted on a post and the jaws are easily changed. Photo courtesy of W. Eckold AG.

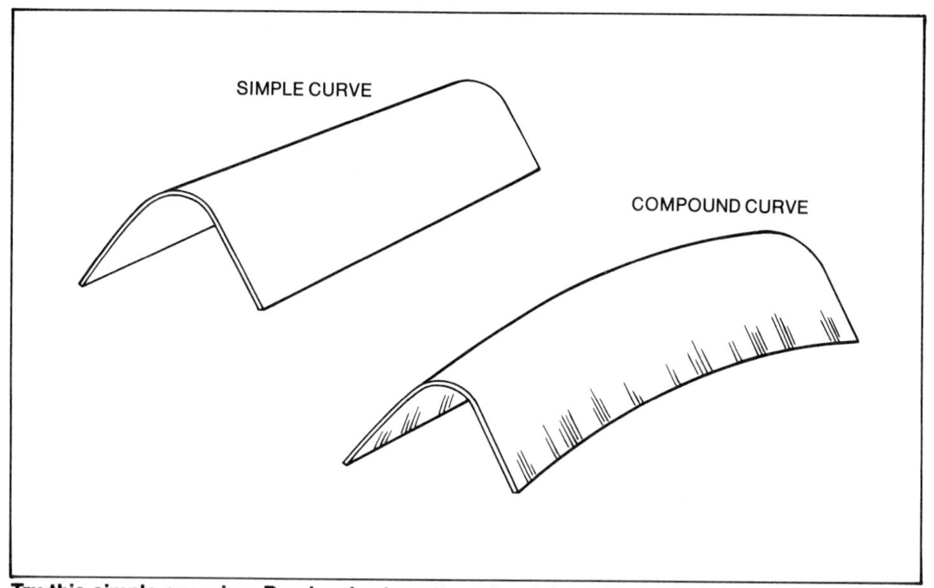

Try this simple exercise. Bend a short section of metal as shown. Shrink the legs to form a compound curve. It's not hard to do and it illustrates how metal forms.

operated shrinker/stretcher is moderately priced. A quality machine, its throat depth is about 3 in. It has a stable foot stand and a long handle for that needed leverage.

COMPOUND CURVES

To understand how shrinking and stretching shapes metal, let's start with a simple curve. Although not all compound curves are made by starting with a simple curve, this example gives you an idea how metal is shaped by shrinking or stretching.

Let's begin with 6 X 12-in., 0.050-in.-thick sheet of 3003-H14 aluminum. See the accompanying drawing. Start by bending a simple 2-in. arc lengthwise in the sheet using a brake and a 2-in. radius die. To form a compound curve from this piece, shrink both *legs*—the two 2-in.-wide, 12-in.-long flat surfaces bordering the arc.

Shrinking the leg edges curves the part up in the middle along the length of the 2-in. bend, or down at each end. This forms a rainbow-like compound curve. If I had stretched the 12-in. length of each leg, the piece would have curved up at the ends rather than down.

Repeat this metal-shaping exercise and you will quickly see what a compound curve is and how shrinking or stretching works. Practice a couple of examples and try for a smooth regular shape. Learn to control the shrinking and stretching *evenly* along the edges. It may end up looking like one of those modern metal sculptures.

A real-life part—such as a race-car nose or a hood scoop—is more complex and requires several compound curves. Therefore, most projects require a *form* or guide—similar to a dressmaker's form—to check the shape for accuracy and fit as the part is being fabricated. Once in a while a simple curved part can be made successfully without the use of a form to check its shape—like a one-of-a-kind hood scoop.

FORMS FOR METAL SHAPING

Made *before* metalwork begins, a *form*—a three-dimensional representation of the inside surface of the part—helps to ensure that a metal part will be shaped correctly. Forms can also be used to ensure accurate duplication of hand-fabricated parts. Auto manufacturers use highly accurate and expensive forms, *bucks* or *die models* made of solid mahogany, or other hardwood.

Two inexpensive types of forms are commonly used for custom metal fabricating: the *station buck* and the *wire form.* They are "skeletons" over which the shaped metal is checked. Station bucks and wire forms are easy to make. You can get very good results using them.

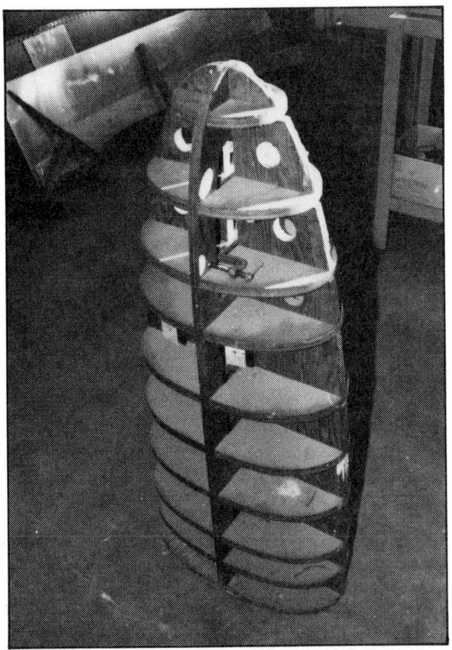

Typical wooden station buck for making an aircraft part includes all the important shapes of the finished part. Holes are for clamping the metal pieces.

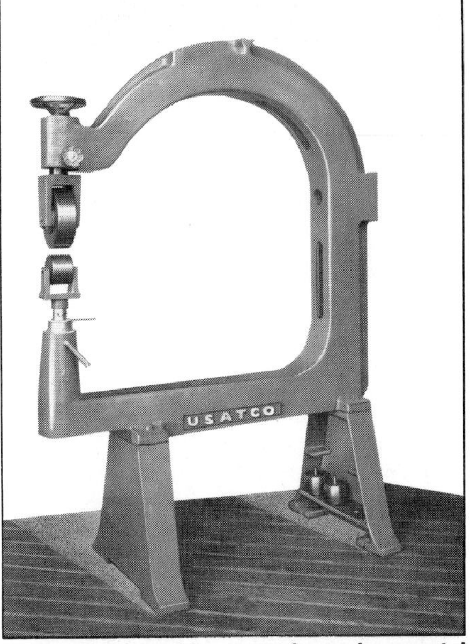

An English wheel is a must for serious metal fabricators. USATCO offers this unit, made of cast steel with a quick release lever that comes with three standard lower wheels. Three other lower wheels are optional. USATCO also offers a table-top model. For their address, see the Supplier's Index on p. 174. Courtesy USATCO.

STATION BUCK

Station bucks are checking devices frequently used in metal shaping. Usually they are made out of plywood sheet in the style of an eggcrate.

The station buck consists of a series of *sections.* It is not a full-surface representation of the part. See the nearby photo. Three types of sections are used: one *vertical center section,* one *horizontal center section* and several 90° sections, or *stations.* Stations are spaced along the two center sections from one end to the other.

Station bucks are most often used when making parts—such as a hood or nose—that fit *over* an automobile frame. The buck must include all of the important shapes of the desired part. By checking the part or parts being formed against the buck, you'll know when it has the correct shape for fitting to the vehicle.

A station buck may also be used as a *tacking jig*—a holding device to allow accurate alignment of metal parts for tack welding. Don't completely weld a part on a buck. You'll burn it up.

Making a Station Buck—To make an accurate station buck you must start with a full-scale drawing of the part to be made. Draw both a top, or *plan* view and a side view.

To make section, or *station* drawings, make measurements 90° to the horizontal and vertical sections every 5 in.—closer if there is a major surface change: where the surface drops, rises or ends. Measure enough points so as they are transferred to paper, they will clearly define each specific section when joined by a smooth line—like a dot-to-dot drawing. Establish the points for each station as you make your measurements. You should end up with a station drawing at least every 5 in. along the length of the part.

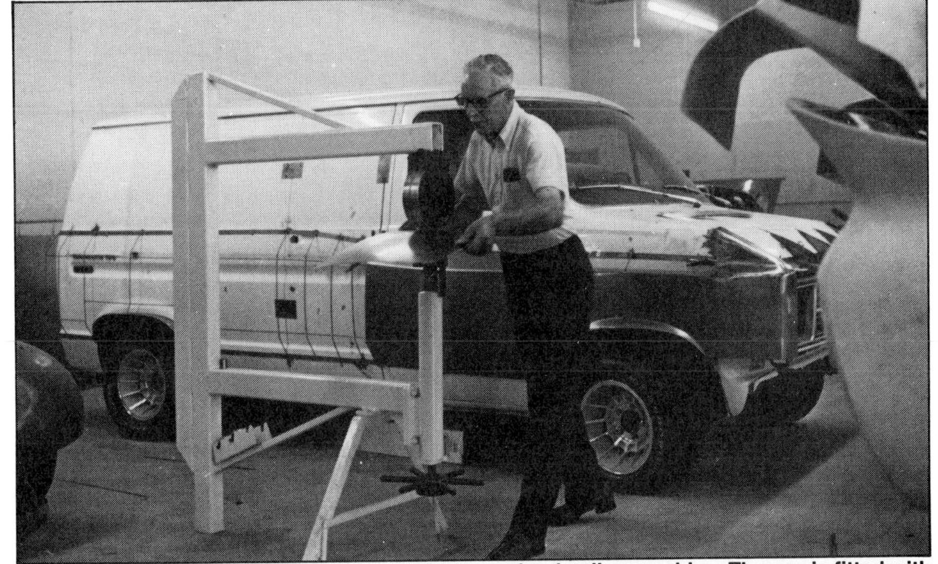

Herb Goode demonstrates his skills on a homemade wheeling machine. The van is fitted with formed metal panels over a wire form. Styrofoam helped develop wire form in front corner area. The form is attached with sheet-metal screws through flat-washers tack welded to form.

Using your drawings, draw each station on 3/8-, 1/2- or 3/4-in.-thick plywood. Using a saber saw, cut out each station carefully. Number or letter code the stations in sequence to keep them in order. Note the codes on the drawings. It's a good idea to provide a *base station*—a bottom surface—that can be clamped to a bench. Drill 2-in.-diameter holes about 1/2 in. from the outside edge of each station, where you think two pieces of metal will join. These holes allow you to use C-clamps to secure each piece of metal to the buck for fitting.

Assemble the buck with wood screws and glue. It must be strong and stable. Take time to make it accurate. The stations should blend together smoothly. If they don't, adjust the buck as necessary after rechecking its

Beginning of the bowl-shaping exercise: pattern and metal blank with a curved-face wood mallet and circular shot bag.

Start by striking the blank hard in the center. Work in a spiral toward the outside edge of the blank. Hit it with many overlapping blows.

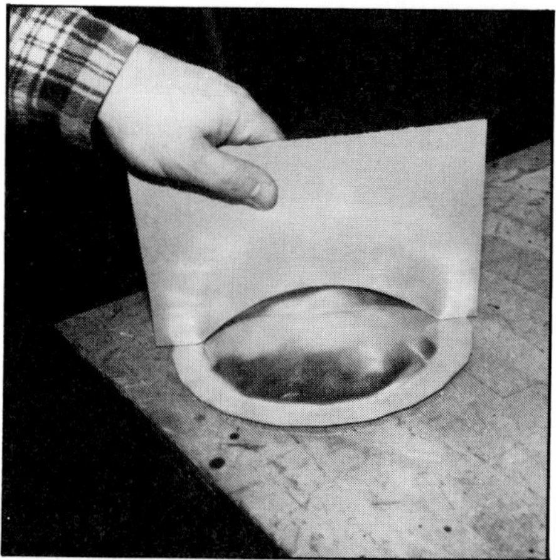

Checking shape and depth with the template to see where additional forming is needed.

measurements against your original drawings *and* the area where the final part will fit. An area on a station that is too high, is reshaped by filing and sanding. Stations that are too low may need replacing. Note: *The metal part will be only as accurate as the station buck.*

WIRE FORM

As with station bucks, *wire forms* are used to check the accuracy of a hand-shaped metal part. Full-size forms are made from heavy wire or light rod: 3/16-in. diameter is ideal. It is easy to form and remains stiff once in place.

A wire form is most often used directly on the vehicle. All of the important shapes and section changes must be reflected by the wire form.

Installing a Wire Form—A wire form is attached to the vehicle by welding small sheet-metal tabs to the vertical wire stations—see photo p. 81. The tabs are then riveted to the body to hold the vertical forms in place. The tabs must be positioned so they'll be under the new metal. You don't want to end up with holes in good body work. After all of the vertical stations are secured, the horizontal wires are welded or brazed to the vertical forms in their proper positions. They must be welded *flush* to the vertical—no overlapping.

As with a wood station buck, shaped metal can be clamped to the wire form. If you're wondering, a wire form is *not* intended to be left on the car inside the new metal part. It is removed before the part is installed. If the form is sturdy, it can be used again on another vehicle.

Once you've gained experience, you may not need a station buck or wire form for simple metal-shaping projects. A complex project will require a form of some kind regardless of your experience. The buck is needed to check the shape of the piece/s as it is being formed to prevent scrappage and help ensure you get an accurate part. So check repeatedly as you shape the metal. You will have a clear idea if the metal being formed is developing in the right direction. With skill and experience, you will be able to achieve the exact shape you want.

METAL-SHAPING EXERCISE

As an exercise in metal shaping, form a circular piece of 3003-H14 aluminum into a shallow bowl 0.050-in.-thick—about 9-in. diameter. Start with an 11-in.-diameter blank. This will leave some extra stock to work with. Doing this will give you a feel for how metal reacts. *Anyone* could grab some aluminum, bang on it—boom, boom, boom—distort the metal and call it a shallow bowl. But you are going to do something different:

You are going to make the bowl an exact depth. You will control the smoothness and curve of the surface. And you'll *stop* beating on the metal when those requirements are met. *You will be in control.*

Make the 9-in.-diameter bowl 1-1/2-in. deep at the center with a smooth and continuous surface. To assist in achieving this, cut a template from thin cardboard such as that shown in the photo. You will use the template for checking shape.

Template—A template is also used to check the accuracy of a part. A minor difference is the template is taken to the work rather than the work being taken to the station buck or wire form.

Make a template like that shown in the photos. Start with a piece of cardboard 11-in. long—9 in. for the bowl diameter, plus 1-in. at both ends—and 3-in. wide or more. At the mid-point, draw a center line 90° to one of the long sides. Make a mark 1-1/2 in. from the edge on the center line. At the same edge, make marks 4-1/2 in. from both sides of the center line. Draw a smooth arc through these three marks—it should be symmetrical about the center line. Now, cut the cardboard out along the arc you drew.

The template will be the final shape and guide for checking the metal as you work. While forming the aluminum, pause often to check your progress with the template. Turn the template in all positions on the bowl to check its shape as it develops.

Hammer—The smoothness you want can be achieved by the *way* you hit the metal. You will need a steel hammer with a minimum 1-1/2-in.-diameter face. The curve of the hammerhead face should always match, or come close to the curve of the finished shape. This will minimize bumps or dents in the finished piece. The hammer-face shape is easily checked by comparing it to the template. A mallet can be used too, if the face of the head is dressed to match the finished shape.

Shot Bag—Finally, you will need a

Shrinking the edge of the bowl. Hold it over a wood surface and strike it firmly with the flat face of a soft mallet, like this wooden one. Rotate the bowl as you shrink the edge.

When the depth is right, planish the bowl with a smooth round steel hammer. Shape of the hammer face should match the shape of the bowl. Be careful not to do any additional stretching.

Check final bowl shape with the template. You can re-cut the template and deepen the bowl for further practice.

12—14-in.-diameter shot bag or sand bag. Using a shot bag is an old and very effective way to shape metal with curved-face hammers, or a mallet. The sand or shot in the bag will conform to the hammer's face and allow the metal to stretch. Place the circular aluminum blank on the shot bag. It's time to start pounding.

Always begin at the center of the metal, using firm and direct blows. Be firm and consistent but avoid excessive force. Strike the metal in circular patterns, working evenly all around. Work from the center toward the outer periphery in concentric rings. Don't try to get to the final shape all at once. Return to the center and repeat the process. Where you need more depth, strike harder. Check and recheck against the template to see where you may need to work more. Conversely: You are not trying to make the world's deepest bowl. You are trying to learn how to make metal move to conform to a specific shape. Excessive hammering won't prove a thing.

It may take more than one try, but when you match the bowl shape to the template, go a step further. Cut a new template, with perhaps a 2-in. depth at the center. To keep the bowl smooth you may have to use a hammer with a more-curved face. Check it against the new template. Rework the same bowl to the new depth. If you can smoothly reshape it to match the 2-in. depth, give yourself a gold star on the forehead. As you work the bowl deeper, you will find that its periphery gets wavy. Shrink it to keep it smooth. This will begin the development of another new skill.

Shrinking by Hand—To shrink by hand, position the bowl edge on a clean, flat, wood or metal surface. The corner of a workbench will do fine. Strike the wavy edge with a *flat-face* wood or rawhide mallet to flatten it. A steel hammer *cannot* be used. A steel hammer used over a steel surface will *stretch* the metal.

Feel the edge carefully to find small irregularities—you'll be able to see the big waves. Work this way until you are satisfied with the smoothness and evenness of the bowl edge.

Making a bowl sounds dull and unpromising. It is a great exercise. You'll find that metal seems to have a mind of its own. To be successful at making this "dumb" bowl, you have to learn the fundamentals of how metal responds to being formed. You are controlling the stretching and shrinking of the metal.

Once you learn the skills needed to make a simple bowl well, you can apply these skills to projects such as putting *blisters* in hoods or flares on fenders—these too require curved and stretched surfaces with shrunken edges. Now, doesn't practicing with that dumb bowl make sense? It's surprising how many shallow bowl-like shapes you will find in a race-car body. Look carefully sometime.

Stretching by Hand—A slapper, used in combination with a shot bag, can be used for stretching metal. How and when it's done is different than stretching with a hammer or mallet. A hammer or mallet is good for stretching small, deep areas—like the stretching done to make the bowl.

Wooden slappers are used to stretch larger areas for forming more gradual curves. Steel slappers are used for metal finishing. Almost without exception, hammers, mallets, and wood and steel slappers are all needed for larger projects—such as a hood. Remember the mallet or hammer works on small deep areas, the wood

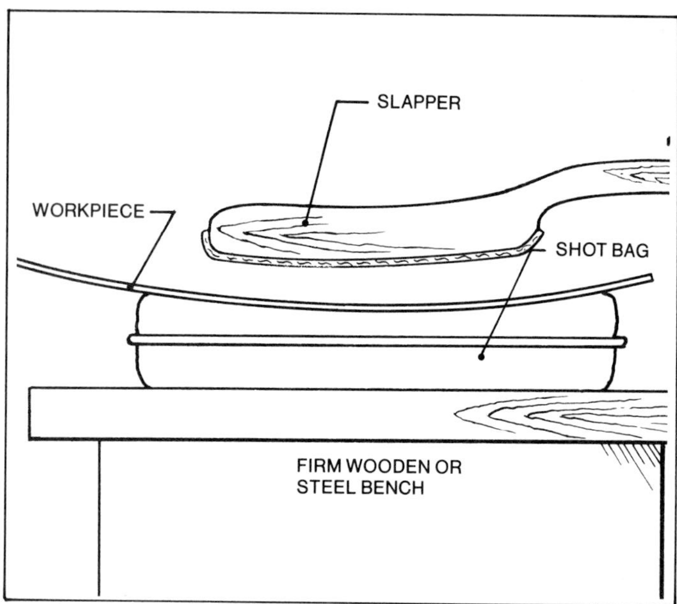

A slapper is good for shaping metal. Each blow must overlap to avoid leaving marks. If you do leave marks, they'll have to be smoothed out later.

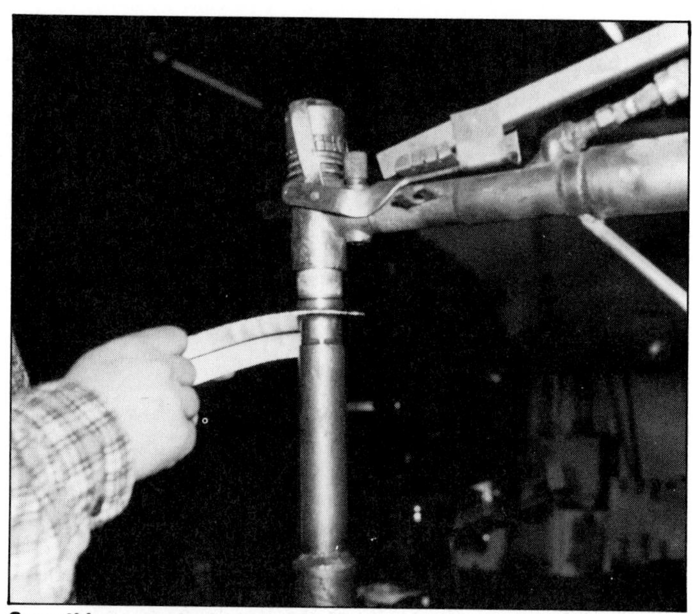

Smoothing, or planishing, with a pneumatic planishing hammer. This is quicker than smoothing by hand.

slapper stretches more gradually over larger areas. The steel slapper finishes the surface.

Slappers—As explained in Chapter 1, page 13, these homemade tools are inexpensive and easy to make. So you will probably end up with several slappers, all with different shapes. Each slapper is good for stretching a slightly different curve.

Making Slappers—Hardwood, such as oak, makes fine slappers. To start, cut out three basic slappers: One should be almost flat. The second should be slightly curved. And the third should have a deeper, more-pronounced curve.

Start with 4-in. square, 14-in.-long hardwood blanks. Draw a pattern for the profile of each slapper on heavy paper or cardboard. Cut out the patterns. Lay each pattern on a wood blank and transfer the pattern to the wood. Cut out the slapper with a jigsaw or bandsaw.

After cutting out each slapper, file and sand its sharp edges and rough surfaces on the handle and face. Remember to preserve the curve of the slapper face. Keep it smooth and even.

To prevent marring aluminum, cover the slapper face with leather. Use a piece of 1/8-in.-thick leather, about 2-in. longer and 1/4-in. wider than the slapper face. Use the extra length to secure the leather tightly to the ends of the slapper with wood screws. With the slapper face covered with leather, it is ready for use.

Using a Slapper—Using overlapping blows when stretching metal with a slapper is even more important than when using a hammer or mallet. Striking the metal too hard *and* without sufficient overlap can put ridges in the metal surface. You not only want to stretch the metal to a certain depth, you want its finish to be as smooth as possible. Remember, both shape *and* finish are produced by the slapper. As your skill in using slappers grows, you'll learn how they form metal smoothly.

Planishing—After metal is shaped it must be *planished*—made smooth by hammering. Planishing removes small irregularities in the metal surface. There are a couple of ways to planish: by hand with a combination of hammers, dollies and steel slappers; or with a pneumatic *planishing hammer.*

Hand planishing requires the use of a dolly that is very close to the shape of the *backside* of the metal surface you want to smooth. While the dolly is held against the backside of the metal, a body hammer or a slapper with very smooth face is used to tap the metal *very softly.* It's important to avoid hitting the metal hard. Hard hammering or slapping stretches or work-hardens the metal rather than smoothing it. So use many light blows.

Steel Slappers—A *steel slapper* is a hand tool specifically used for planishing. As you can see from photos throughout this book, steel slappers are a common tool for metal finishing. They are excellent for smoothing broad-curved metal pieces. The dolly used to back up the metal surface must closely match the length of the slapper face. Planishing with a steel slapper and dolly is about the only way to smooth a body panel already welded to a car.

It's easy to make a steel slapper from an old 12-in. flat file about 1-1/4-in. wide. To make the slapper, heat the file in the center and make a 45° bend. About 1-1/2 in. from the first bend, heat and bend the file 45° in the opposite direction. These two bends offset what will be the slapper face and handle. You can use the cutting torch to narrow the handle end, making it easier to hold. Now you'll need a grinder to dress the slapper face and smooth the handle edges. The slapper face *must* be very smooth. Grind off the file teeth *completely* so the slapper won't transfer marks to the work.

Pneumatic Planishing Hammer—A stationary tool, the pneumatic planishing hammer is quicker and easier to use than a body hammer or slapper because it is powered by compressed air. Hammer power is regulated by a foot control. This arrangement lets you use both hands to guide and control the work piece. The planishing hammer does have a disadvantage. Parts that are too large to handle by hand or those that are attached to a car cannot be planished with a pneumatic hammer. They must be planished by hand.

I made this fender extension with a slapper and shot bag. To complete the piece, I stretched the lower mounting flange with this MMA stretcher. Planishing was done with a pneumatic planishing hammer.

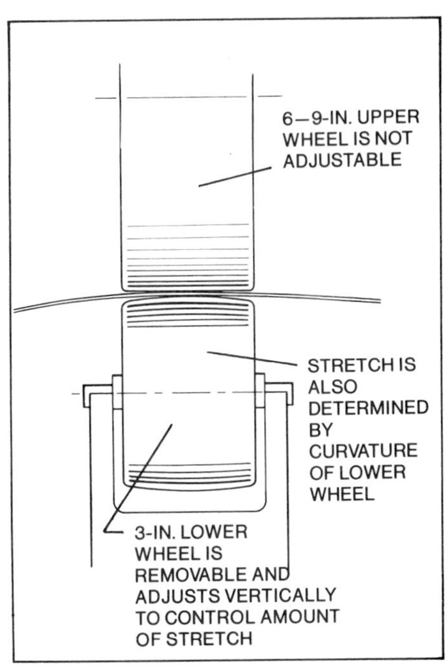

Force applied by upper and lower rollers of an English wheel stretches and shapes the sheet metal. A lower roller with more curve gives more shape. Likewise, so does more force.

Metal is smoothed by guiding it between the hammer's two steel heads, one stationary and one hammering. Using manual planishing as an analogy, think of the stationary head as the dolly and the moving, or hammering, head as the body hammer or steel planisher. The hammering head is above the stationary head.

Pneumatic planishers have interchangeable heads. As with dollies used for planishing, the shape of the lower head should closely match the shape of the material. The lower head should be almost flat for planishing a flat surface and curved similarly to a curved surface. The upper head of a pneumatic planisher may or may not be interchangeable. It should always be flat. The diameter may vary if the upper heads are interchangeable.

A foot-operated air valve controls the force with which the upper head strikes. The higher the air pressure, the harder the head hits. Just as metal shouldn't be hit too hard when hand planishing, it shouldn't be hit too hard with a planishing hammer. The same things happen: work hardening and additional stretching.

Where to Get a Pneumatic Planisher?—My planishing hammer is homemade, but they are commercially available. Sometimes they can be found at surplus air-tool or aircraft-part stores. You can always try used sheet-metal-equipment companies. Some dealers don't even know what an air hammer is, so they'll practically give them away. Otherwise you'll be looking at a $300—500 price tag, depending on the size of the hammer.

METAL-SHAPING EQUIPMENT

So far I've talked about shaping metal with small hand tools. There are larger pieces of equipment which are used specifically for metal shaping. These include the *English wheel,* the Eckold *Kraftformer K665* and *the electric power hammer.* Speed and quality finishing are the main advantages of these machines. The disadvantages are: they are expensive at hundreds to thousands of dollars; and they are hard to find.

ENGLISH WHEEL

Also known as an *English roller, wheeling machine,* or just plain *wheel,* the *English wheel* gets its name from the two steel wheels between which metal is formed. The wheels—one above and one below—are rolled against one another with a piece of sheet metal in between under a preadjusted pressure setting. The wheels roll as the sheet metal is moved back and forth. Shaping results as the lower wheel is forced under pressure into the metal, stretching and forming it.

The wheel is an old and very effective piece of equipment for shaping sheet metal. It is especially useful on large panels—such as a door or roof skin. The wheel works equally well on aluminum or steel. Two men work together to form a large panel: One manipulates the piece through the wheel while the second man supports the panel. They must work together very cooperatively. The wheel can produce very smooth, very complicated curved shapes. Regardless of size or the complexity of the curves, what the wheel can do ultimately depends upon the skill and experience of the operator.

The upper wheel—always with a flat rolling surface—is fixed to the machine. The curve, or crown, of the interchangable lower wheel—the *roller*—determines the shape of the piece being formed. The more curve in the roller, the more curve in the work.

Several different interchangable rollers come with an English wheel. A roller with the *least curve possible* must be used. A roller with more curve than necessary may overstretch and leave roller marks in the metal. So the roller must be chosen carefully.

An adjustment on the English wheel varies the amount of pressure exerted against the metal by the roller. A knob below the roller sets this pressure. In addition to the pressure adjustment, English wheels have a quick-release lever. This allows the operator to drop the roller away from the upper wheel so the work can be removed and inserted quickly without losing the pressure setting. The quick-release feature saves a lot of time and helps to get consistent work.

BUILDING A ROADSTER NOSE

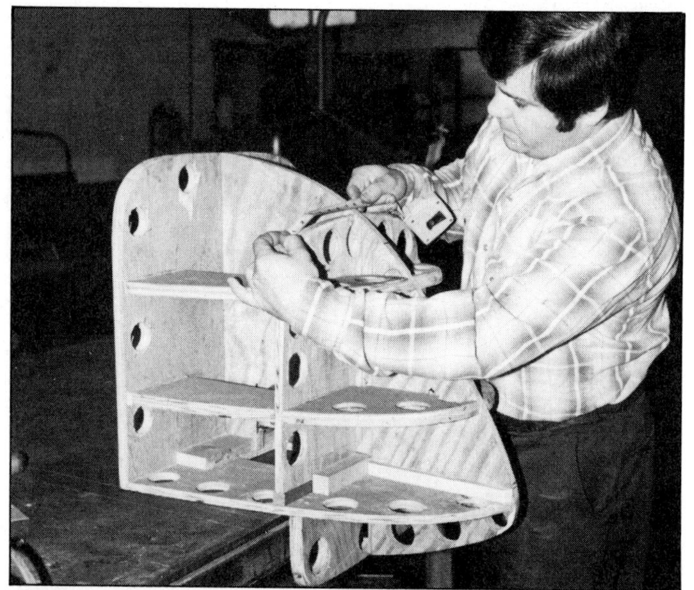

Making a sturdy, accurate station buck was the first step in making an aluminum race-car nose. Buck is nailed and glued together. Holes are for clamping the work.

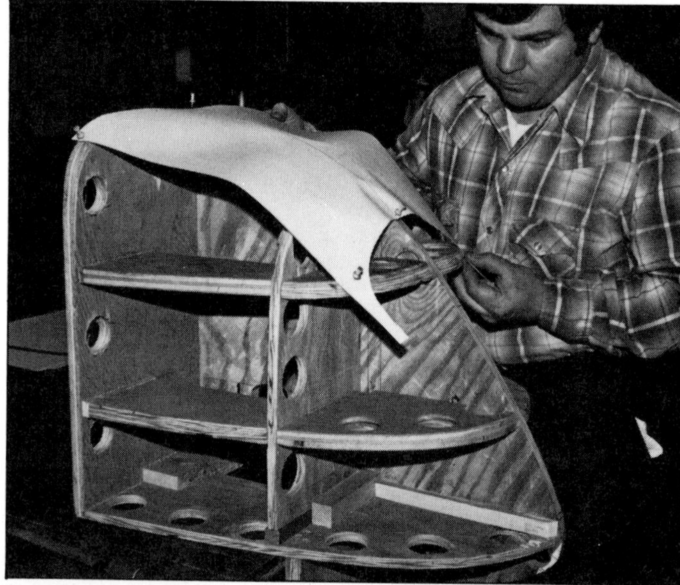

Developing templates to fit a station buck is not hard. Make full-size templates exactly the way you want each metal panel.

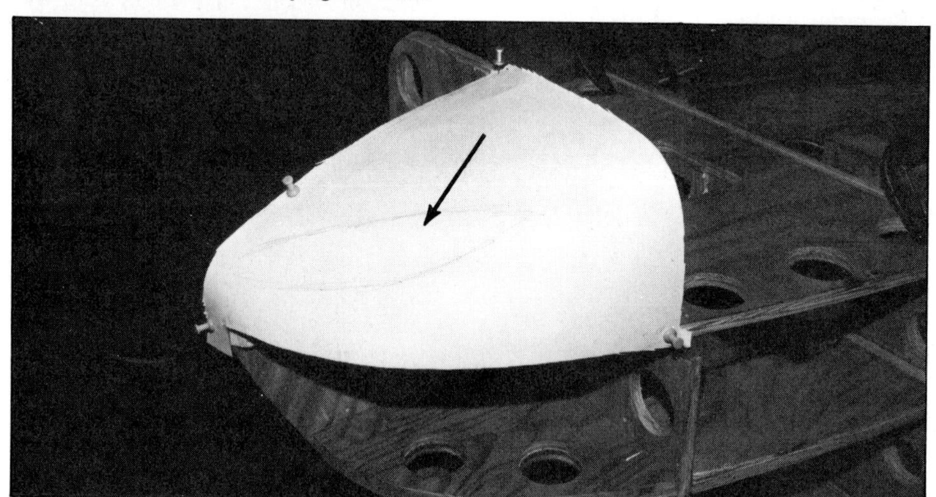

Tape or pin the templates in place and mark areas needing stretching (arrow). A pattern can usually be reversed for the opposite side of the construction. Note the center line on the top pattern for reference.

Start with the top panel. Stretch it a little in the center and shrink side edges. Some shrinking is needed at the front and rear edges. Check the panel on the station buck. This process is continued until it fits the buck. Be careful not to overstretch the panel. When you're happy with the fit, clamp the panel in place.

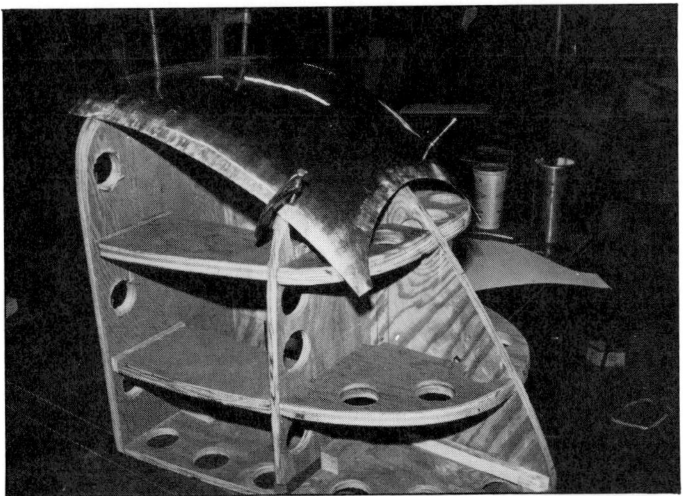

When you're happy with the fit of the center panel, clamp it in place to the station buck.

Side panels are formed by shrinking the top edge and along the grill opening. Panel is wheeled gently with a flat roller. Note how the side panels will join at the center of the nose.

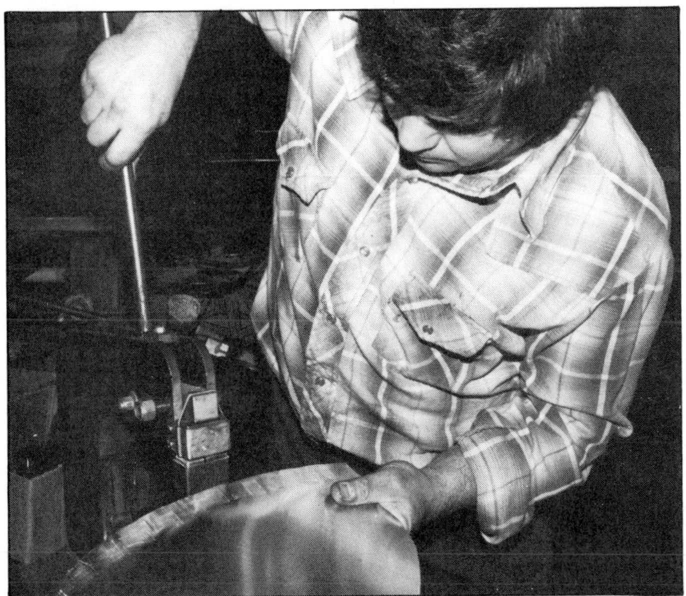

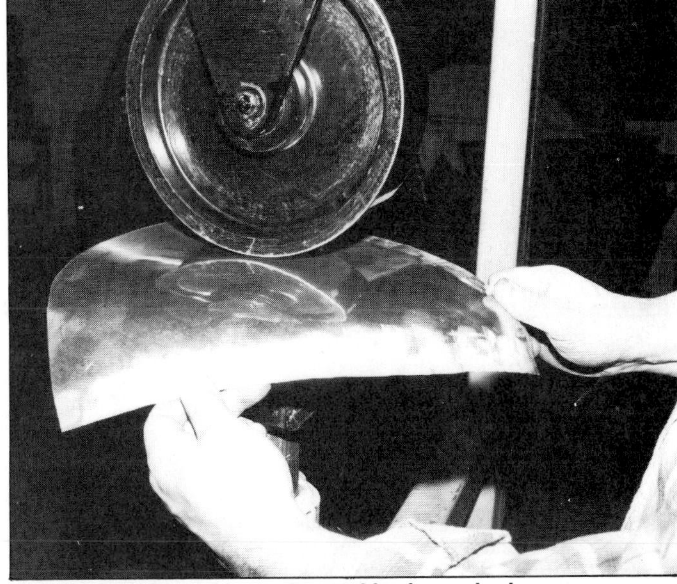

Bottom panel will have the most shape, so it is the most difficult to make. A great deal of shrinking and stretching is required.

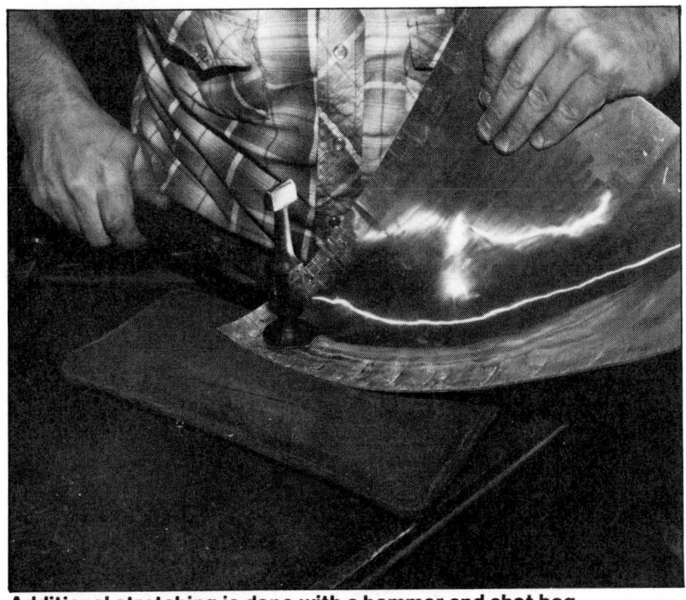

Additional stretching is done with a hammer and shot bag.

Bottom-panel half fitted to buck. Wheeling did final smoothing to make a perfect fit. Follow the same process to form other half.

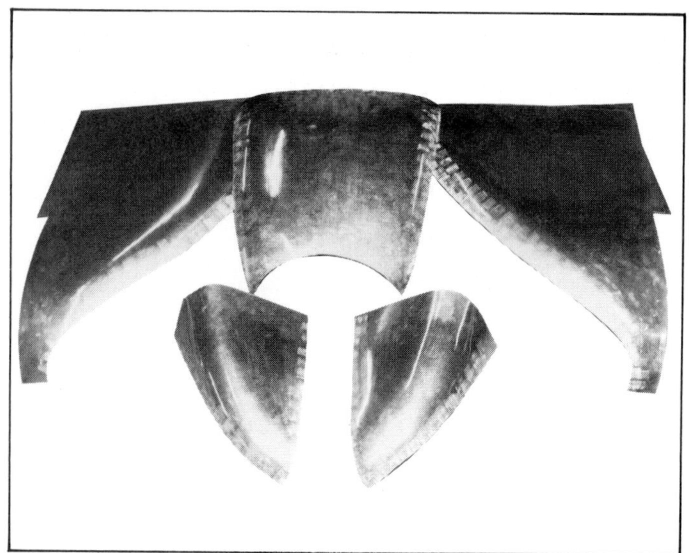

All five panels formed, checked on the buck and ready for joining.

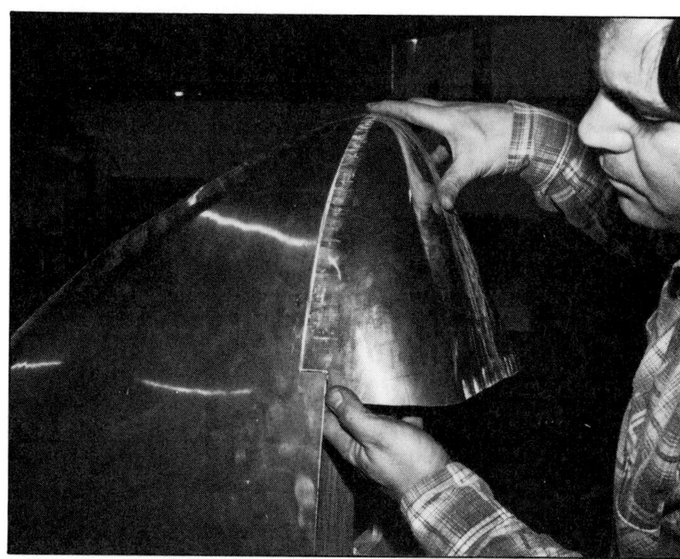

With side panels clamped in place, lower panels can be fitted. Remember: A good fit means good welds.

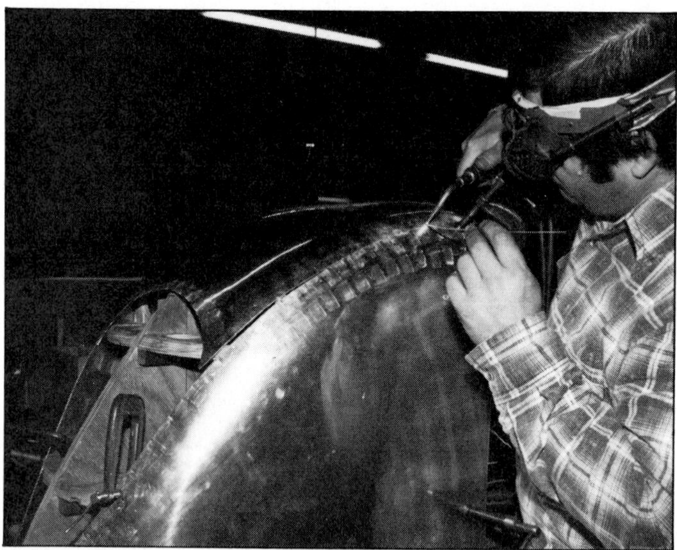

Tack-weld panels while they are on the buck. This helps to retain the shape of the part. Final welding is done with the nose assembly off of the buck.

Tacking the nose causes some warpage. Planish as much warpage out as possible *before* final welding.

Use a hammer and dolly for the tight curves. A slapper is better for the flatter areas.

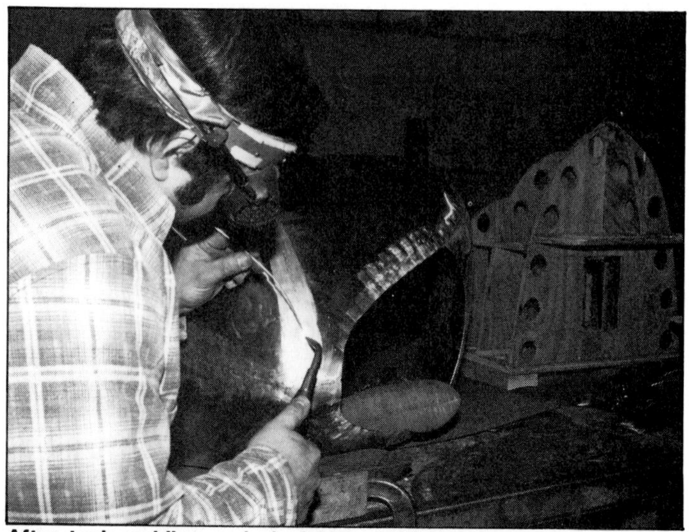

After tack-welding and planishing, hold the nose down with C-clamps or sandbags while you weld. The work has to be held very steady as you weld.

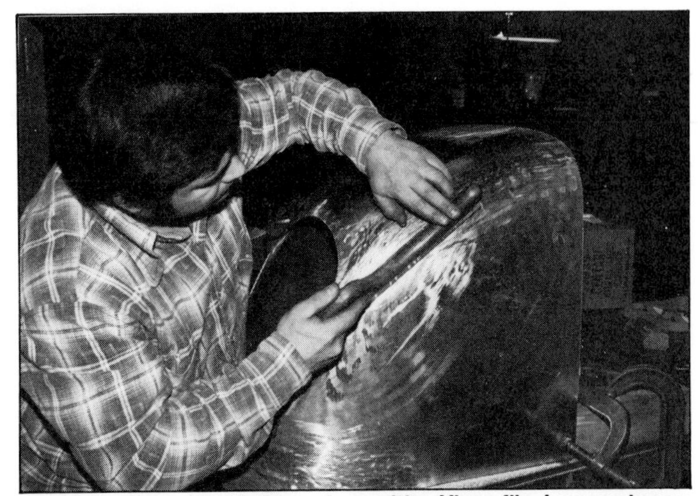

After final welding, file the seams. Remove the high spots with a conventional file, then smooth the surface with a Vixen file. Low spots can be bumped up with a body hammer and filed again.

Roller patterns can be seen on the inside surface of the finished nose. Each panel required different rolling directions. Note the good weld penetration.

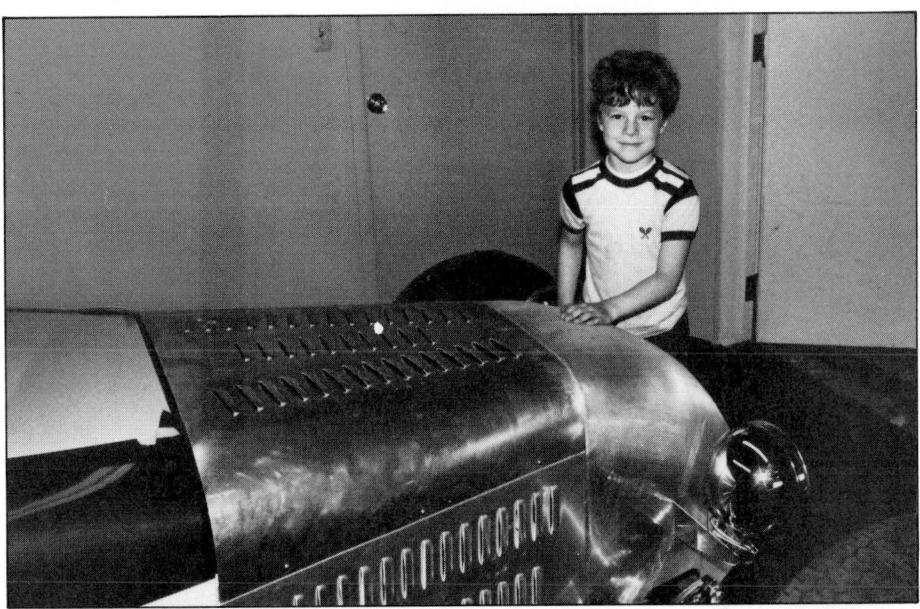

Metal finishing is very important. You never know when a critic will examine your work.

Using the Wheel—An English wheel works best when you follow a certain sequence. First, check to see how much curve you want in the metal panel. Choose a roller corresponding to that curve and install it in the machine. Insert the metal between the wheels. Adjust the roller so it has a slight amount of pressure against the work. Remember: Less pressure is needed to form aluminum than is needed for steel. The panel is pushed and pulled—back and forth—between the wheels. The first few *passes*—movement in either direction—will show if the pressure is right. If it is too light, the work will skid or slip between the wheels. If it's too heavy, roller marks will show up in the work.

How a metal panel is moved between the wheels is *very important.* Each pass of the metal through the

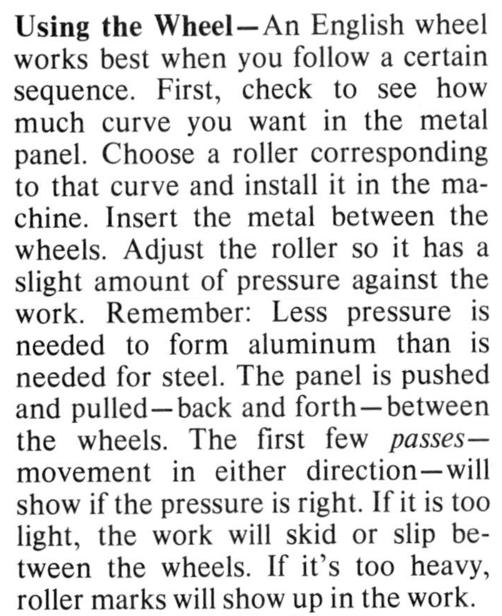

Finished nose installed on a '27-T roadster. It gives the car a lean, mean sprint-car look.

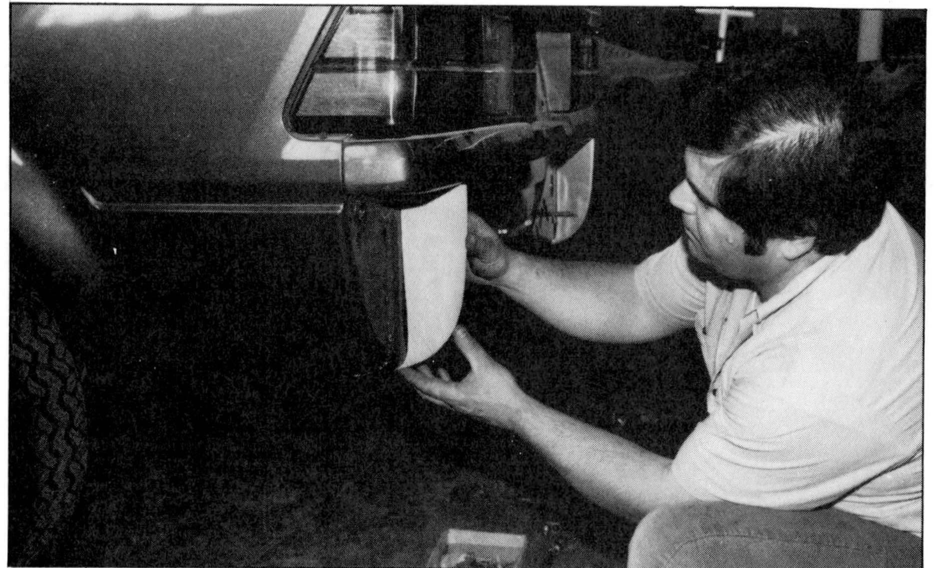

Mild-steel valance panel began with pattern development. Panel width was dictated by car width, but side profile was developed. Cardboard pattern represents this profile.

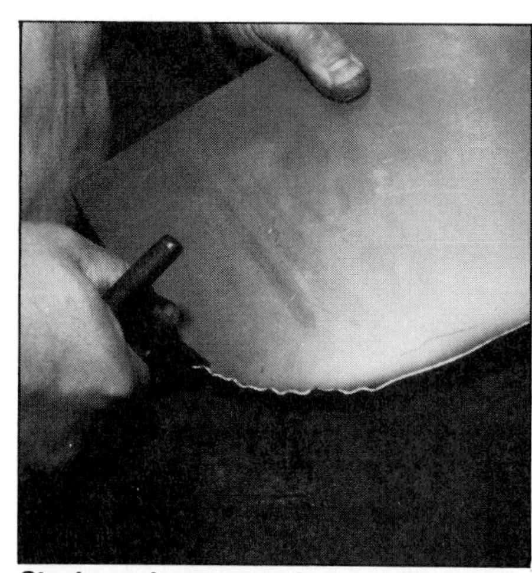

Steel panel corresponds to profile pattern —right and left panels were required. Flange was made first. I tipped the bend as far as possible with the beader, page 38. I gathered the flange further with a small *fork*.

With the flange backed with a dolly, material was gathered by striking it on the opposite side with a mallet. Dolly is moved along the flange until the flange is smoothed.

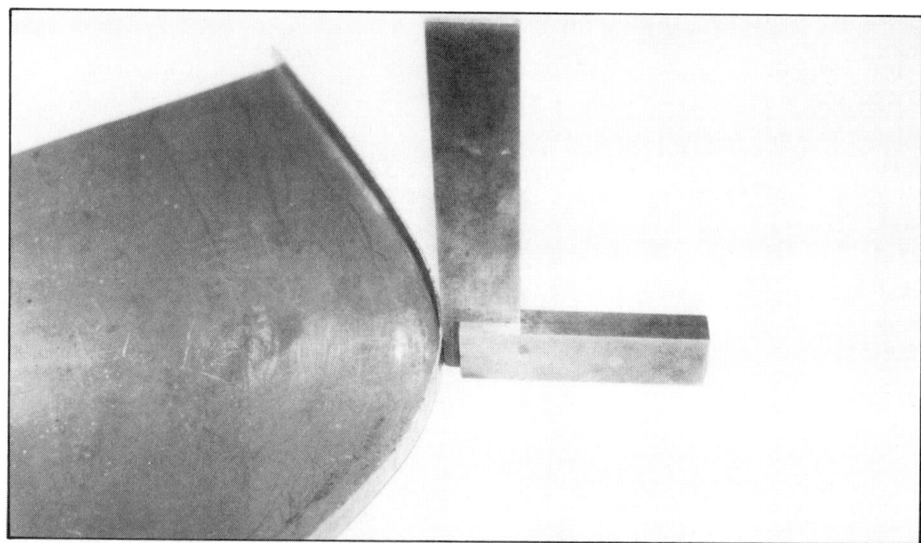

Finished flange: Square and wave free.

Mounting flange is bent on leading edge of each panel so it can be riveted to the car. Clecos hold them in place.

wheel must be so close to the previous one that they nearly overlap. Otherwise the panel will stretch unevenly, making its surface wavy and irregular. So keep each pass very close to the previous one. If you don't, you'll miss areas, making the shape of the metal uneven.

Because the English wheel is primarily a stretching machine, you must avoid running the edge of the work through the wheel. A stretched edge requires the extra work of shrinking—and shrinking is not easy.

If you haven't guessed, English wheels are imported from England. An English wheel can be fabricated or they can be purchased from a tool supplier. Two names to look for are F.W. Edwards, Ltd. and Kendricks—they are expensive.

Mike Archer and I wheel a long, sweeping curve in the panel. Long overlapping passes through the wheel give the panel the desired shape.

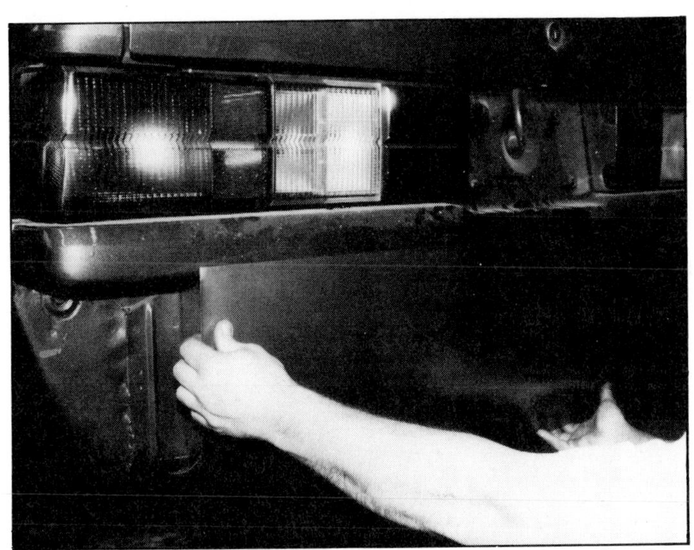

Checking the panel against the car. This should be done often to see how a curve is progressing. This check showed us we needed more shape on the lower end.

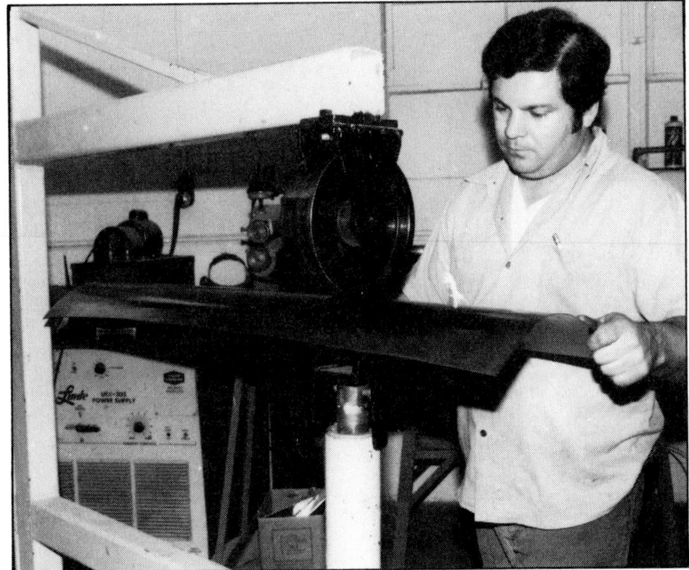

Back at tbe wheel working the panel crossways to make it curve under for the lower back of the car.

Fitted panel clamped in place. Next job is to develop the end pieces.

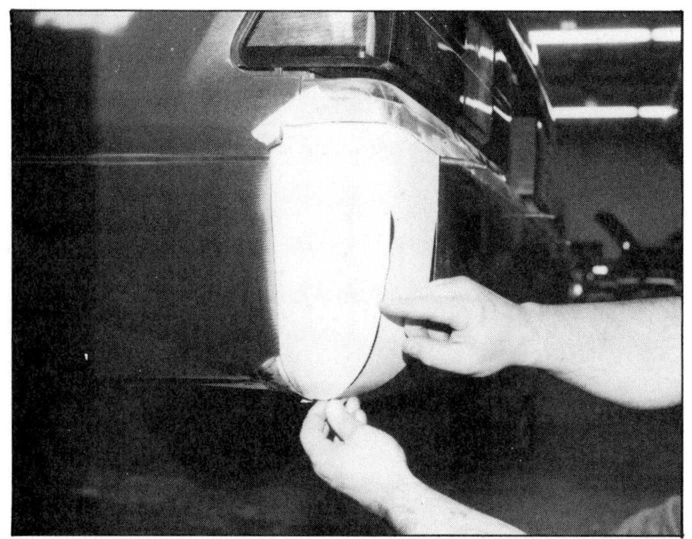

Developing the end in cardboard.

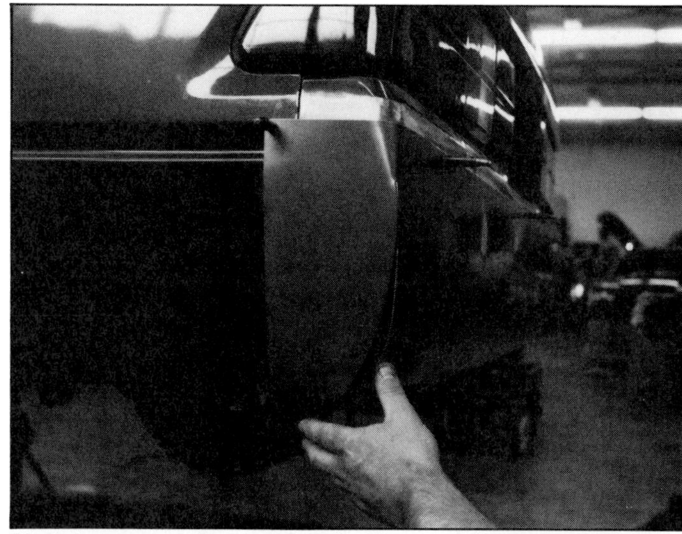

Blank is cut according to the end pattern and Clecoed to the car. A rough fit: I just wanted to see how the blank should be formed and to see where to remove metal.

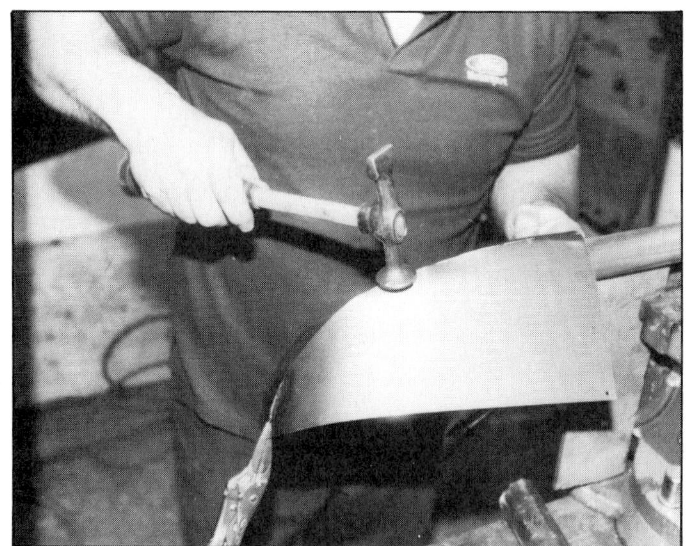

Hammering the edges together over a large, curved-end T-dolly.

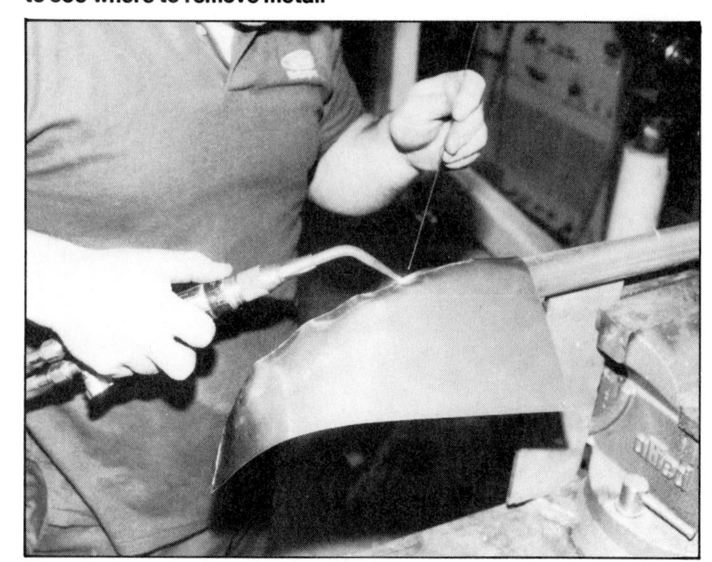

The edges are tack-welded together.

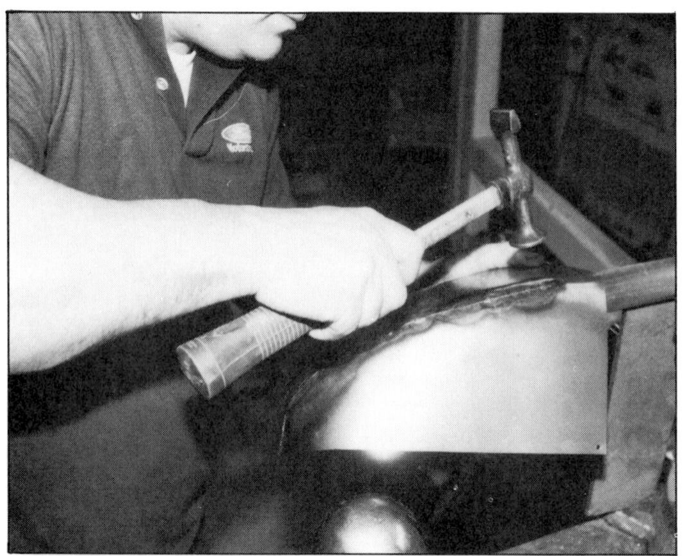

The edges are hammered again for more tack-welding. This process was repeated until the whole seam was tack-welded.

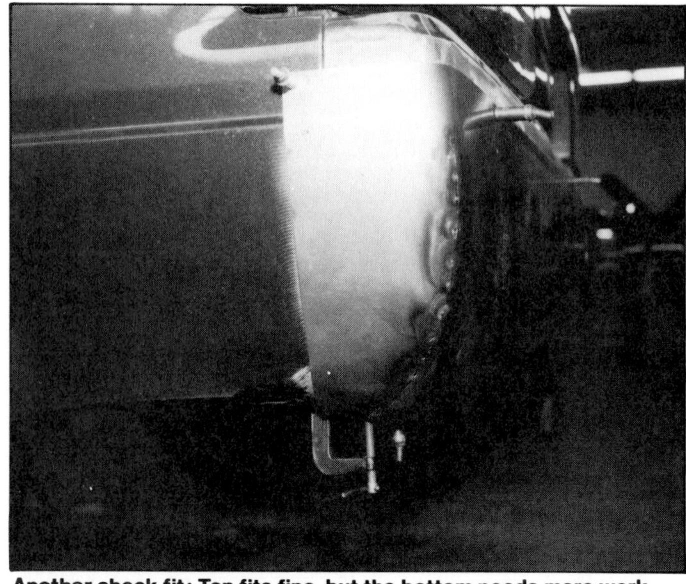

Another check fit: Top fits fine, but the bottom needs more work.

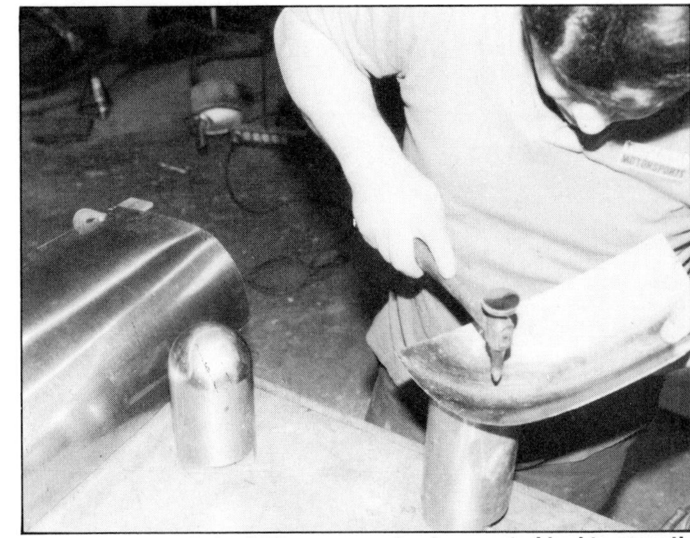

Smoothing with a body hammer and large domed free-standing dolly—another homemade tool. There were also low spots I had to smooth from inside by tapping gently over a flat surface.

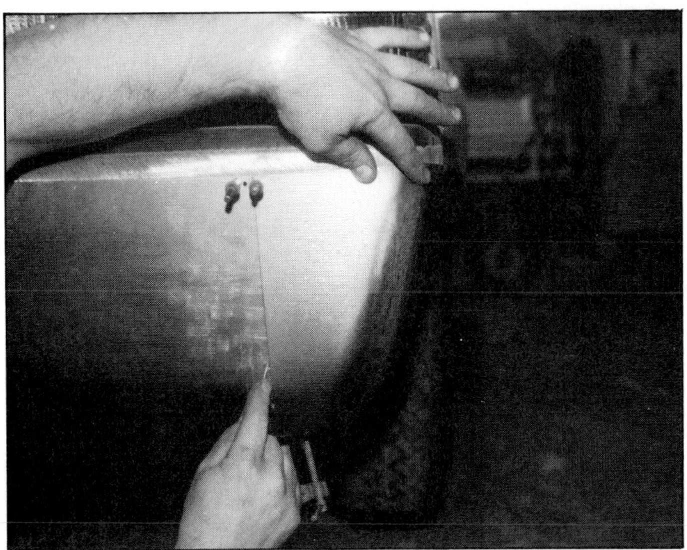

One more check for fit after smoothing showed a tight fit. I clamped it in place. A cut line is scribed where the end piece meets the center panel.

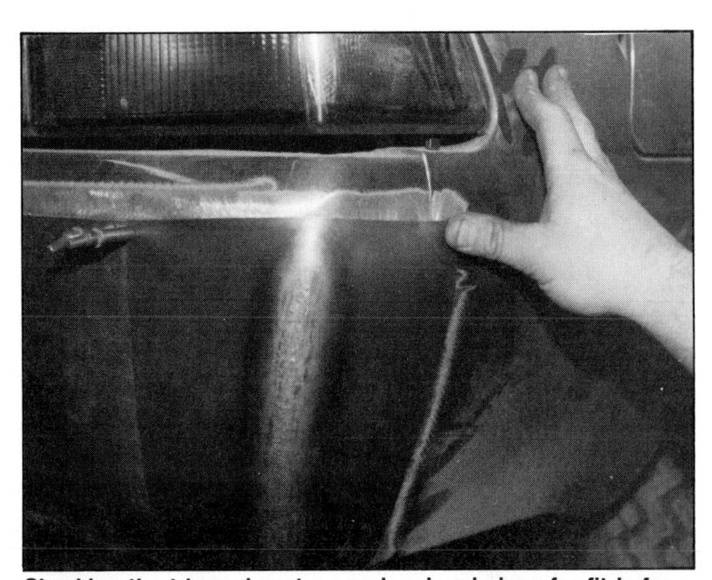

Checking the trimmed center panel and end piece for fit before welding.

After cutting the center panel, the end piece is butt-welded into place. Oxyacetylene welding allows the welds to be metal-finished more easily.

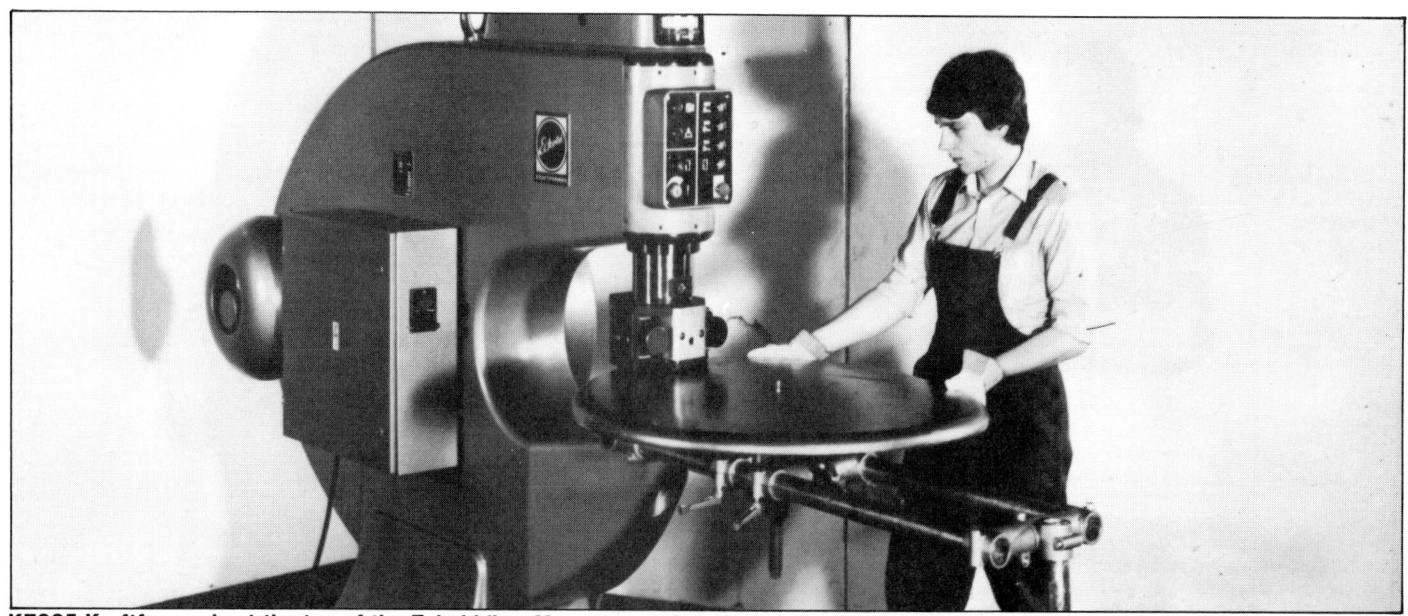

KF665 Kraftformer is at the top of the Eckold line. Many types of shaping can be done with this machine by changing its tooling. Photo courtesy of W. Eckold, AG.

KRAFTFORMERS

Included in Eckold's fine line of metalworking equipment are three of their electric-powered Kraftformers. The KF665 heavy-duty Kraftformer is the largest and most complicated of the three. It can perform the most metal-shaping operations. This is mainly because it has more interchangeable tools than the other two Kraftformers.

The KF665 Kraftformer accepts metal up to 26.5-in. (675mm) wide into its throat. It stretches and shrinks steel and aluminum easily. Additionally, this Kraftformer can shrink or stretch sheet metal *without* marring the surface when the type LFA or LFR die set is used. This means that little or no metal finishing is required on a piece stretched or shrunken by this machine using these dies. It saves time and effort, and results in a beautiful finish.

In addition, the KF665 Kraftformer can planish metal. It can also do other jobs with the PFW *doming tool*. This tool curves or shapes metal panels. It can also be used to form indentations or rises in metal.

The KF665 machine has an on/off switch and three working speeds. This speed control makes the machine perform very well when using the TPS planishing tool. A foot control regulates the speed of die movement, leaving both hands free to control the work.

The KF460 Kraftformer is smaller than the KF665, however it shrinks and stretches metal well. The

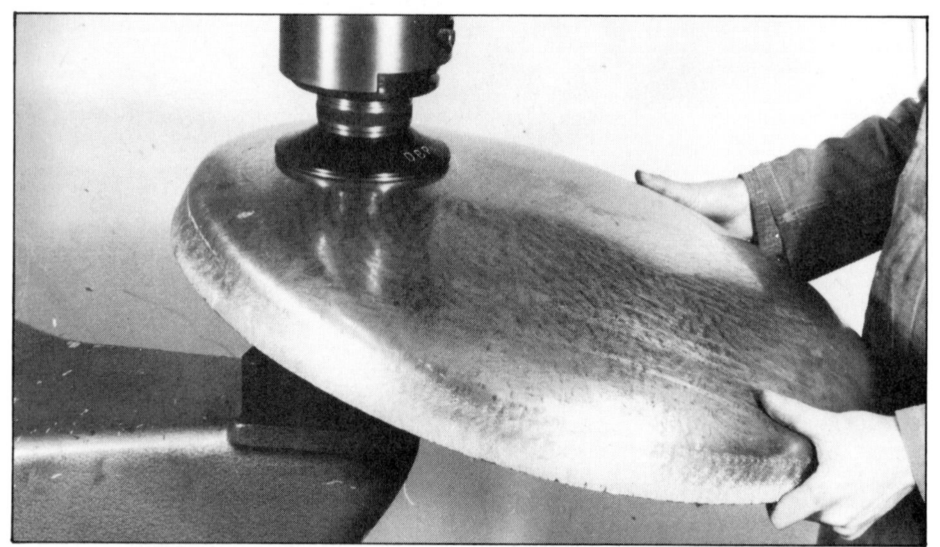

Doming tool PFW being used in the KF665 Kraftformer to form an aluminum tank bottom. Photo courtesy of W. Eckold, AG.

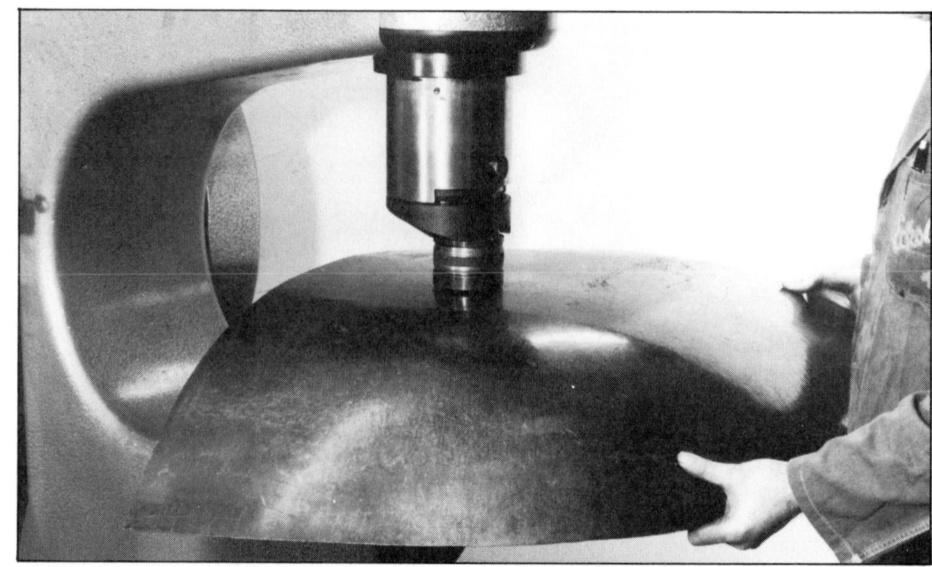

KF665 Kraftformer with tool TPS planishes this body panel easily. Photo courtesy of W. Eckold AG.

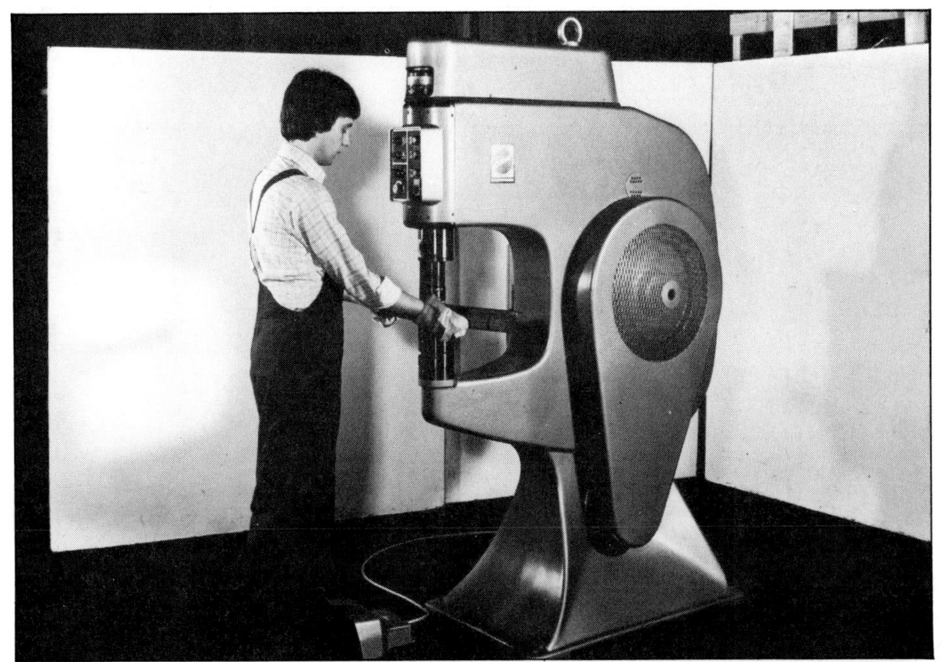

KF460 Kraftformer is smaller than the KF665—460mm (18.1 in.) throat depth as opposed to 665mm (26.2 in.). The KF460 has many interchangable tools. Photo courtesy of W. Eckold, AG.

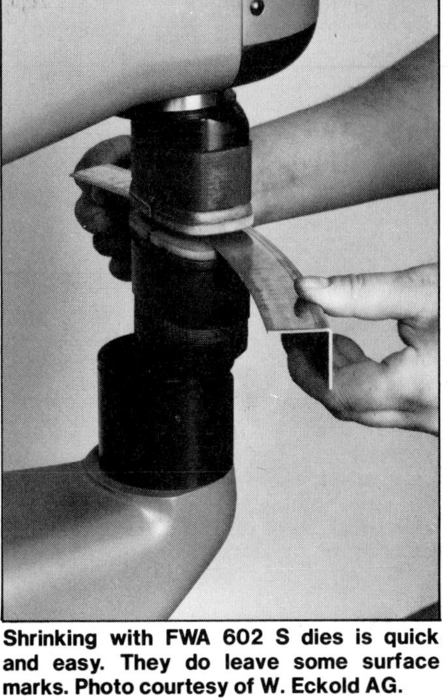

Shrinking with FWA 602 S dies is quick and easy. They do leave some surface marks. Photo courtesy of W. Eckold AG.

KF460's throat depth is 18.1 in. (460mm). The KF460 Kraftformer has a foot control.

The Piccolo models are mounted on wheels and can be moved to the job rather than having to bring the work to the machine. This will save those extra steps and lost time.

I've gone on at some length to inform you about Eckold Kraftformers for good reason. Although expensive, Kraftformers are important tools because they are durable and can handle such a wide variety of metalwork. They can be considered many tools in one. And because they are powered, time and energy is saved.

I had the chance to use the KF665 for a period of two years. I became familiar with—rather, awfully spoiled by—the capabilities of the machine. Its versatility is remarkable. If you shape metal frequently you should be familiar with these machines and how they work. If you can afford a Kraftformer, it will work well for you.

GO EASY **sign taped to this power hammer is there for a purpose: Hammer exerts tremendous force with each blow.**

Power-hammer dies of many sizes and shapes. Each die forms a different shape.

Electric Power Hammer—Another power tool made specifically for forming sheet metal is the electric power hammer. A very large piece of equipment, the electric power hammer is ideal for shaping large aluminum or steel panels. Also, it greatly eases forming oddly shaped parts.

The foot-controlled electric power hammer uses a 3-HP electric motor to deliver 120-lb blows in rapid succession. Like the air-powered planishing hammer, the lower die is stationary and the upper die moves. Also like the air hammer, the lower die is shaped and the upper die is flat.

While more and more Kraftformers and English wheels are popping up around the country, the power hammer is becoming scarcer. And very few people can operate a power hammer skillfully simply because it requires more skill than other types of metalworking equipment. Because they are no longer produced in the United States, only used electric power hammers are available. Power hammers are mostly found in shops specializing in automotive custom metal shaping.

To Summarize—Shaping metal involves two processes: stretching and shrinking. Before you start any metal-shaping job consider this carefully: The ability to shrink a part is much more important than stretching. Shrinking is difficult to do and it

Large panel takes two men to support it while shaping at the power hammer. Each must be skilled. Fellow wearing the cap is operating the machine with a foot control while the second man helps guide the panel.

Partially built Indy roadster. Most race-car bodies of this type were 3003-H14 aluminum formed with a power hammer. Photo by Denny Hassell.

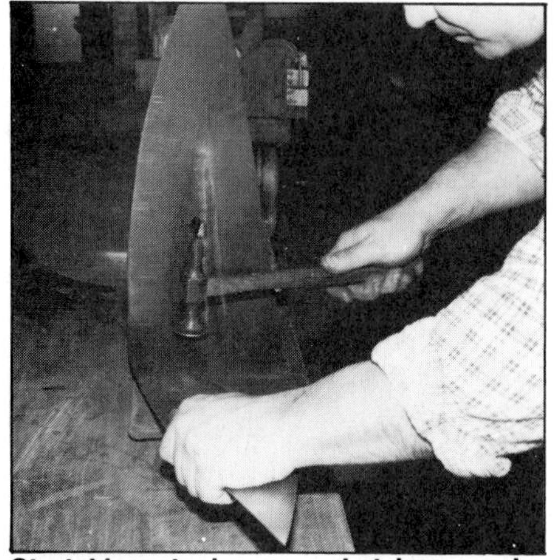

Stretching steel over a shot bag can be done with steel hammers, as long as they're smooth and you use overlapping blows.

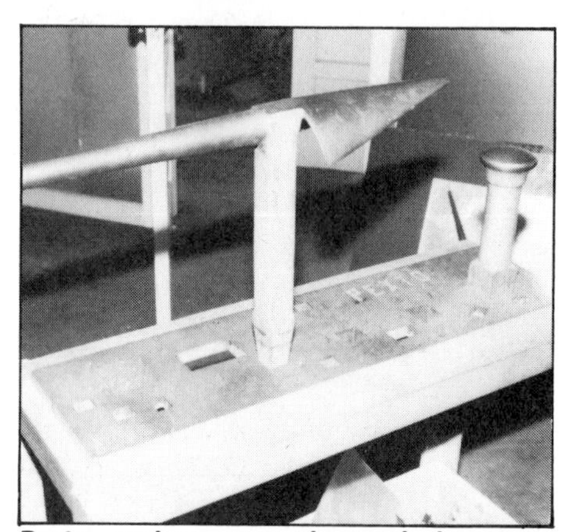

Pexto makes several good hammer-forming tools. These two *stakes* are mounted in a special table.

Fender-flare top panel required considerable stretching. Lower panel was shrunk along the top of the wheel opening so it would curve in and meet the upper panel with a smooth curve.

strongly determines *where* the metal can be joined. On the other hand, stretching metal is much easier. Metal can be stretched merely by pounding on it. Take great care not to overstretch metal. Overstretching your work will make the final product more difficult to metal-finish and will probably require addititional shrinking.

Metal shaping is a skill that must be developed through practice. Eventually, you should be able to create custom designs in unique ways on complex projects. So be brave, dive in and give it a try. Start with a simple project. Then move into more difficult jobs. Your projects may become very professional and very individual. At the very least, you'll learn what *not* to do. You'll definitely learn to recognize and appreciate quality metalwork. You may even get hooked and become a real "metal man!"

In Chapter 9, I discuss sheet-metal add-on's. I will demonstrate how metal shaping is used to produce a given project and how excellent the results can be. Believe it or not, all of the practice making precise bowl shapes and compound curves will lead into making air dams, hoods with performance blisters, spoilers and other body parts. In Chapter 10, I cover roll bars and roll cages. Then there are chapters on metal tanks and sheet metal interiors. First, though, you should be familiar with another special metal technique—hammerforming.

8

HAMMERFORMING

Hammerforming can be especially useful to make duplicate parts—whenever several identical pieces or parts are needed. Hammerforming is actually easier than it looks. Photo by Michael Lutfy.

Of the many ways to shape metal, one of the most-economical methods is the way the old masters did it—*hammerforming*. In this chapter I tell how to hammerform metal and when to use it. Although not complicated, hammerforming takes time and skill to do right.

Hammerforming is done by hand. It requires a few simple tools: a selection of hammers, mallets and some homemade forms. You'll also need some C-clamps and possibly a vise. Chances are, you already have some of these tools. The tools you don't have can be obtained at relatively low cost. You can make the forms.

Hammerforming is done by sandwiching a sheet-metal blank between a wood, phenolic-resin or metal form and a clamping device. The sheet metal is forced down over the form by hammering on it. The clamping device simply prevents the sheet metal from shifting on the form as you work. Such shifting would distort the final product.

DUPLICATING HAND-MADE PARTS

Duplicating parts when making them by hand isn't as difficult as you might first think. You should be able to master the skill with practice. It is surprising how many times you may need to make duplicate parts. The method discussed is one I use whenever several identical pieces or parts are needed. It is a simple and effective way to make a small number of identical metal parts without expensive dies.

If you are making the top and bottom of a tank, hammerforming may be the best answer. If the tank has the same required shape for both top and bottom—often the case with water surge tanks—you can make each *end* from the same hammerform. When both parts are made from the same hammerform you can be sure they will be identical. Duplicating the

Metal is hammered with body hammer and homemade forming tools. C-clamps can be repositioned out of the way as the metal is worked down over the form. Check clamps frequently to make sure they are tight.

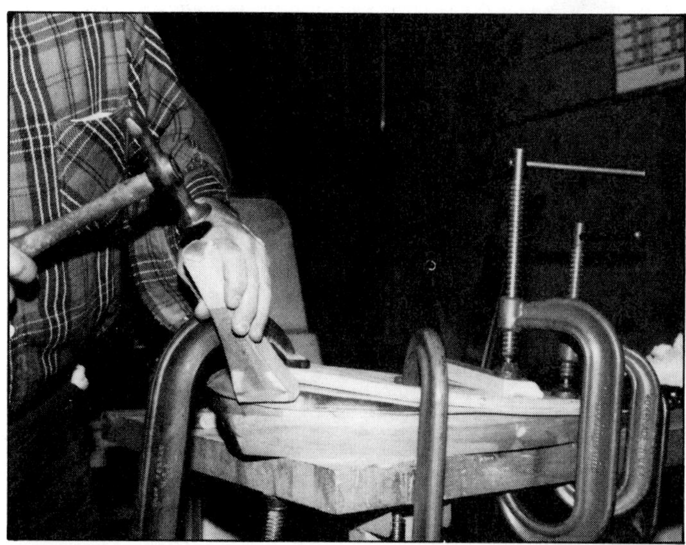
Wood forms and clamps are the main ingredients in some hammerforms. C-clamps and upper form must keep the metal from shifting on the bottom form as it's being hammered.

Solid wood form serves two purposes: It represents the shape of the desired part, and metal can be hammered directly on it.

ends without a hammerform is not easy. Another example: When building a wing, eight or ten identical wing *ribs* may be needed—the inner pieces to which the wing *skin*—or outer sheet metal—is attached. These ribs can be duplicated with a minimum of variation by using hammerforming.

SYMMETRICAL HALVES

This technique can also be used to make *symmetrically opposite*—mirror image—halves of a part. You can make the halves of a given part, such as a hollow brake pedal. The advantage of this is you can get a strong, light part. Hammerformed brake-pedal halves joined in the middle by welding make a sturdy part. A brake pedal with the same strength, but made from a single thick piece of the same material will be much heavier. This is why hammerformed parts are frequently used—to save weight without losing strength.

Brackets—I frequently hammerform brackets to save weight. You can begin with a thinner material and achieve the same strength as a much thicker solid bracket. You *hammer in* strength by putting the metal where it counts. Material toward the center of a bracket adds little to its strength, but it does add weight.

To illustrate this, compare the weight and strength of a solid bar to that of a commonly used tube: A 1-3/4-in.-diameter, 1/8-in.-wall steel tube is 66% stronger than a 1-in.-diameter solid-steel bar in bending, but it's 19% lighter!

Because no welding is done to form the bracket halves—compared to a similar part made from flat pieces welded together—the part is made quickly. Not only is a hammerformed bracket stronger and lighter, it will look good and give a professional touch to a project. When you think of the number of brackets used in a car, it takes little imagination to realize the weight hammerformed brackets can eliminate.

MAKING A HAMMERFORM

Hammerform Base—The first step in hammerforming is constructing the *base shape* or *hammerform*. A hammerform must be of the same shape you desire to reproduce in sheet metal. It must conform exactly to the *inside* sheet metal surface of the finished part. If the hammerform matched the outside surface of the intended part, the resulting part would be larger by the thickness of the metal. So you must allow for sheet-metal thickness when building a hammerform. The hammerform shape plus the gage of metal used will determine the outside surface of the sheet metal.

For example, let's say you want to hammerform ends for the tank illustrated on page 149. The 6-in. diameter ends are to be fabricated from 0.063-in. stock metal. In this case, the hammerform base diameter must measure 6 in. minus *twice* the 0.063-in. (1/16-in.) metal thickness, or 5-7/8 in. If you follow this formula, the outside diameter of the finished piece will measure 6 in.

Remember to allow for metal thickness on every hammerform, regardless of its shape. If you don't, the finished piece will be too large from side-to-side by twice the sheet-metal thickness. It won't fit!

Hammerform Top—The top, or upper part of a hammerform is only for holding the work. It must be the same shape as the base, but *slightly smaller*—1/8—1/4 in.—around its periphery to allow room to hammer the metal over the edges of the form. This small difference in the overall size of the top is essential. *You must clamp all of the sheet metal securely to the form except at the edge where it will be hammered.* This is to prevent distorting the final part, thus helping to produce an accurate part.

Base Material—A hammerform base may be made of hardwood, phenolic-

Aluminum hammerform for forming a single steel part or many aluminum parts. Top and bottom forms of 7075-T6 aluminum were used to hammerform 3003-H14 aluminum. I'm making a hood reinforcement.

Hammerform and workpiece is clamped in a vise. C-clamps keep the outer ends tight. Bolts at the ends align forms and workpiece.

resin, steel or aluminum. The material you use depends on how many parts are to be fabricated with the form. The form must be durable enough to withstand repeated hammering without being damaged. Any damage to the form will be transmitted to subsequent parts. They could easily end up being the wrong shape or size.

Two factors will decide which material to use for the base: blank material and number of parts. The harder and thicker a blank material is, or the more parts that are to be fabricated means the base must more durable. For example: Sheet steel automatically means you can forget about using a wood base. Even the hardest wood won't hold up when steel is pounded over it. Steel sheet requires a steel base for hammerforming. If you only need a few pieces, you *might* get away with using a T6 aluminum form if the steel is less than 0.035-in. thick.

The number of parts to be formed is virtually unlimited with a steel form. As for aluminum, oak is generally fine for forming up to six 0.063-in.-thick parts. If you want more 0.063-in. aluminum parts, a thick aluminum hammerform base is needed.

With these two points in mind, start by choosing the sheet metal and decide how many pieces you want to make. That pretty much decides what material you should use for the hammerform base. You're ready to make the base itself.

Making the Base—Remember to allow for metal thickness when measuring and sketching the base. If

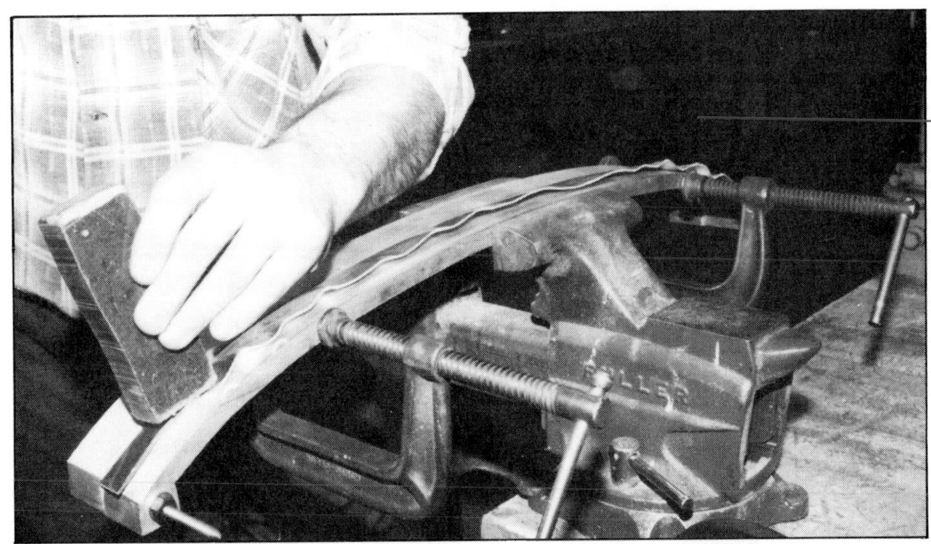

Metal is hammered down equally along its length. To keep the bend tight against the hammerform, strike along the base of the bend.

Finished hood reinforcement and hammerforms. It is strong and light. Equal-quality part would have needed an expensive die to make it any other way.

Two-piece hardwood hammerform was used to form this NACA-type duct.

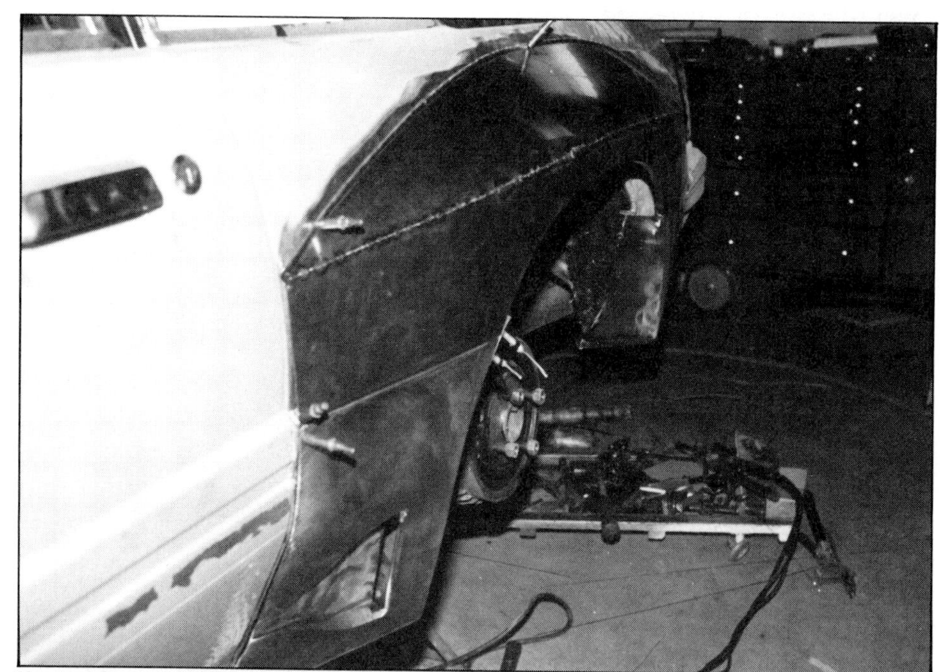
Hammerformed NACA duct is welded into quarter panel. Duct supplies cooling air to the rear brakes.

Steel disc acts as a hammerform base and steel bench provides the "top" for forming a curved flange on this panel. Rawhide mallet is the tip-off that I'm shrinking the flange.

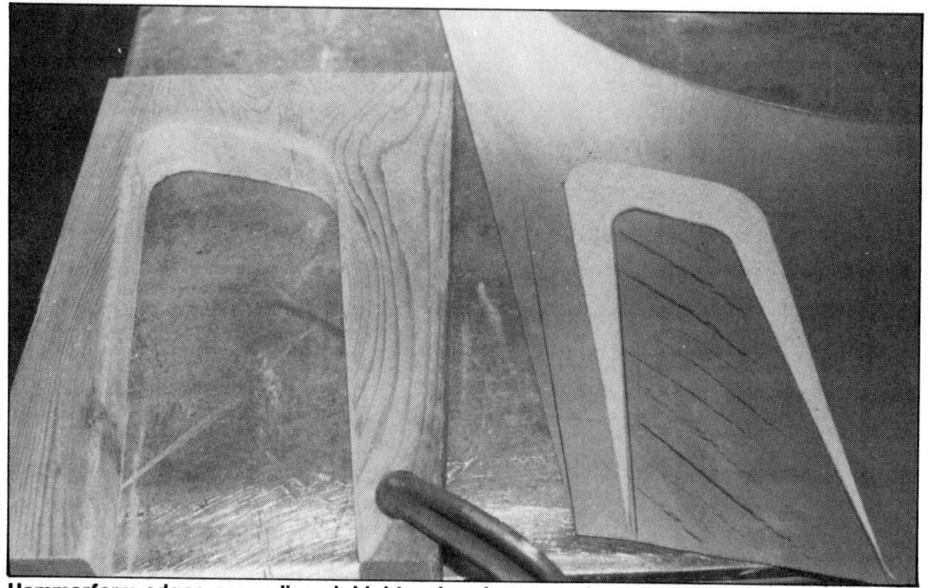
Hammerform edges are radiused. Light-colored area on the quarter-panel blank indicates where metal will be stretched. Crosshatched section was removed.

Cutting out a hammerform for an air scoop. Scoop fits custom quarter panel for a Fiat X1/9. Hammerform was made from a full-scale drawing.

the shape is complicated, cut out a pattern, then transfer this pattern to the base material. Cut out the basic shape of the base according to your measurements. Then carefully smooth the base so there are no rough areas or sharp corners. Remember: Imperfections in the hammerform will be transferred to the sheet metal. This could ruin the appearance or performance of the final part. Aluminum sheet metal is especially vulnerable to this kind of damage.

Radius Edges—All edges of a hammerform must be radiused: 1/8-in. minimum. Metal formed over a sharp edge—no radius—can break or tear at the edge as you pound it. The edges of the hammerform must be slightly rounded and very smooth. This will let the metal bend down and around the edge so it remains undamaged. The final part will have a nice appearance inside and out. And it will be stronger.

Making the Top—The upper half of the hammerform must be made of the same material as the base. Although you don't pound on it, the top must be as durable as the base.

The top should be smaller than the base by 1/8—1/4 in. as measured from the *bending point.* The bending point starts where the edge radius ends, or becomes *tangent* with the top surface of the base. Therefore, if you measure 1/8-in. back from the bending point and a 1/8-in. radius is used, the edge of the top should be 1/8 + 1/8 = 1/4 in. back from the edge of the base. There is a limit to this rule. *Don't make the top more than 1/2-in. smaller than the base.* If the top is too small it won't keep the sheet-metal surface flat.

Like the base, the upper half of the hammerform must be very smooth. If it's not, imperfections may transfer to the metal when the top is clamped down. Remember: A smooth form is needed to make a smooth part.

Metal Blank—A sheet-metal blank for hammerforming should be the same general shape as the base, but larger. How much larger the blank should be is determined by the length of material needed at the edge of the part. For example, blanks for the 6-in.-round tank ends on page 148 measured 6 in. *plus twice the depth, or length of the side wall.* If the tank-end diameter is 6 in. and it is 1/2-in. deep, the blank must be 7-in. diameter: 6 in. + 2 X 1/2 in. = 7-in. diameter. Even though the part is irregular in shape, *the blank must always allow for the depth of the part.*

Limits—It is a bad idea to try to hammerform anything more than 1/2-in. deep. If you try anything deeper you'll have trouble shrinking the edges. It's tricky enough shrinking 1/2 in. or less over an edge and getting it right. It is extremely difficult shrinking more than a 1/2-in. length of sheet-metal over an edge. I hesitate doing it unless it is absolutely necessary.

Secure the Blank—Carefully position the metal blank on the base. The blank must overhang the base *equally* all around. Place the upper half of the hammerform on the base, against the blank. Clamp these three pieces together.

This "sandwich" consisting of a hammerform base, sheet-metal blank and upper form must be clamped together very tightly. I use C-clamps when they fit easily. Sometimes I clamp the assembly with a vise and C-clamps. You may even be able to put the whole thing in a vise. It is nearly impossible to clamp the hammerform and blank too well. *Make sure it is tight.* If it comes loose you could spoil the part. Regardless of how the hammerform and blank are held together, the work surface must be strong and firm. I rely on a heavy hardwood workbench most of the time.

Check to make sure the blank doesn't shift after you clamp it. Sheet metal must overhang the form equally all around before you do any hammering. If it doesn't, the turned-down edge will be uneven: long on one side and short on the other. So if the blank is out of position, loosen the clamps and reposition the blank and the top.

MAKING A PART

Mallets—It is time to pick up a mallet. Choose a soft, smooth-face wooden or rawhide mallet. Just as a rough hammerform can mar the metal blank, so can a rough mallet. You want a nice, even metal piece

Blank is clamped tightly between hammerform top and base. It's ready for forming.

Additional stretching is done with body hammer on shot bag. Shape of curved-face hammer matches scoop shape.

Wheel-well flange was extended before fitting new quarter panel.

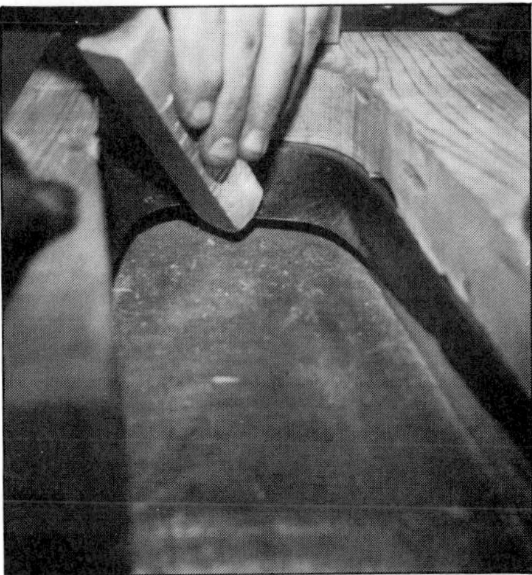

Aluminum forming tool is rounded on the end. Metal is stretched downward, starting at the top and going around the edge repeatedly.

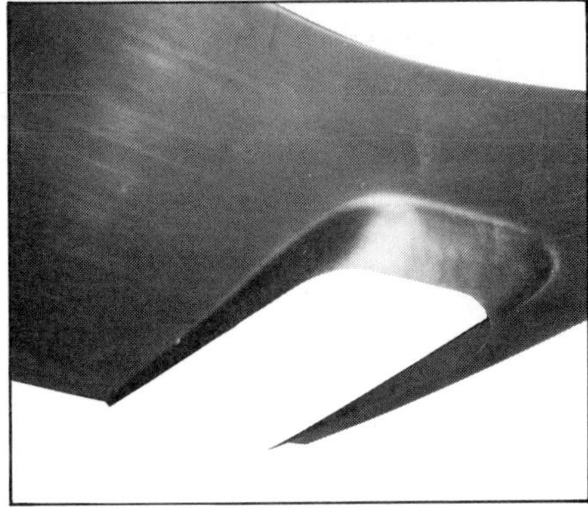

Panel is ready to be fitted to the car.

Panel is Clecoed into place. One last fit before the top edge is bent and the bottom edge rolled.

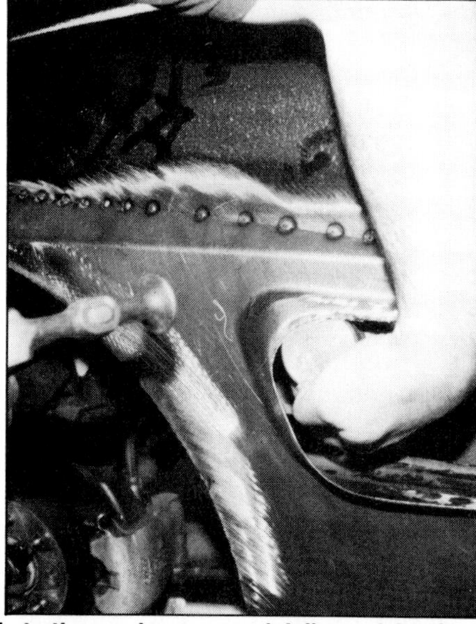

After finished quarter panel was tacked securely to the car, hammer and dolly work begins. Narrow dolly is used to form leading edges. A large round dolly is used at the back. Hammer is Proto's 1427.

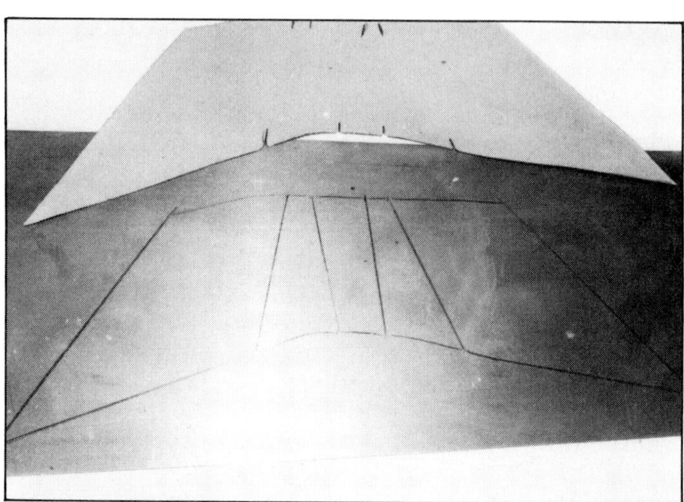

Pattern for scoop inner wall taped in place. Marks on pattern and lines on the metal indicate bend *tangent points*—where a bend begins and ends.

when you're done—no scratches or dents. Look over the mallet face and dress it smooth if necessary. A little sanding on the mallet face now will reduce the possibility of damage to the metal.

How to Hammer—Your first instinct may be to pound like mad in one spot. Don't do it. It's not wild hammering that counts. You have to move the metal *gradually.* What you are actually doing is gathering and shrinking the metal around the base. Gently urge the metal over the edge and down the sides a little at a time by tapping evenly all the way around. Then go around again, tapping evenly and moderately. It's this series of consistent hammer blows that does the magic.

If you concentrate in one area, it won't be long before you run into trouble. The pounded area will lay down OK, but the adjacent metal will flare up, making it harder to shrink down over the form. It will want to fold over on itself. So use the "easy-as-you-go" method. You'll get much better results.

As you form the metal down, another area of the edge may raise up. This is more likely to happen on parts with edges longer than 1/4 in. Otherwise there should be no problem. However, if you find areas that raise, concentrate on hammering them down. As for low spots, they will almost always take care of themselves as you work. The low spots are pushed between the top and bottom halves of the form and blend into the highs that you're pounding down.

Shrinking—Shrinking is a term used to describe how metal is formed over a convex surface. Forming metal over a circular hammerform base—like the tank ends described—requires only shrinking. The deeper the circular

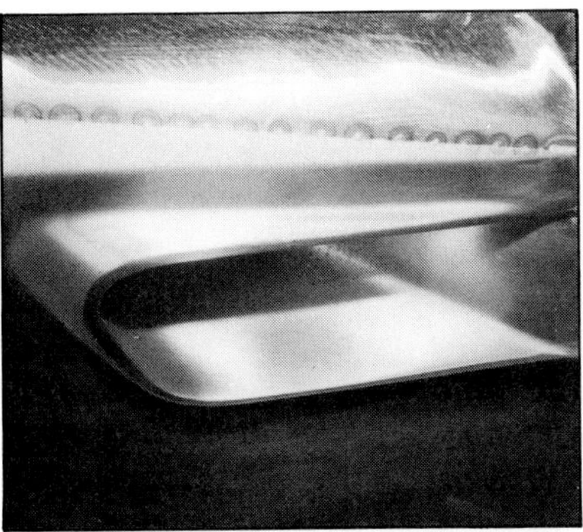

Metal-finished scoop and quarter panel is ready for final welding.

Brad Spitz primes the quarter panel before painting.

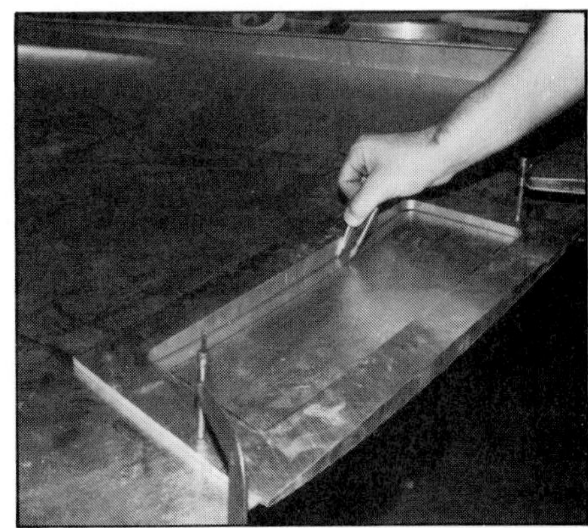

Pencil dividers mark width of metal that will be formed down over the hammerform base. Metal bordered by mark will be removed.

Painted scoop not only looks good, engine cooling is improved. Wider quarter panel allowed for wider wheels and tires.

Make hardwood or aluminum forming tools to suit the job. A forming tool should match the curve in the hammerform like this. Work the metal slowly, starting at the bend and working out to the edge.

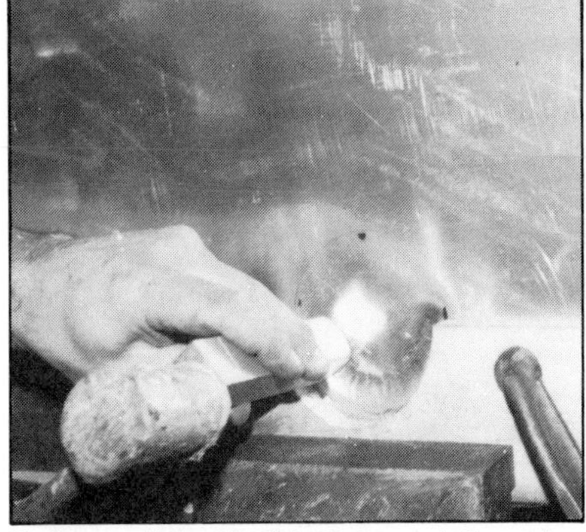

To form a bulge in this flat panel, I'm using a half-moon-shaped hammerform. After half of the buldge was started, the hammerform was repositioned 180° on the panel and the other half was partially formed. This back-and-forth process continued until the desired bulge was achieved.

edge, the harder the shrinking job. It is usually very difficult.

Stretching—Other hammer-formed parts, such as brackets, may involve both convex and concave edges. On a concave edge, metal has to be stretched. The metal must be stretched and pulled over the form. Stretching is relatively easy, but there is one big danger. When you stretch metal it gets thinner. As it gets thinner, it may tear or split. So be careful not to overstretch or damage the metal. The key is to stretch metal slowly and use *many very light* hammer blows. Go easy and you'll avoid damaging the metal. A torn or split part is junk.

HINTS

As you do more and more hammerforming you'll pick up tricks of the trade. You will learn that some curved edges on a mallet face help form some kinds of parts better than other edges. Sometimes I get better results by using the mallet to hit a specially made *wooden forming tool.* This is a piece of hardwood sanded smooth in a particular curved shape. The shape varies according to the job. The mallet hits the wood tool and the tool telegraphs the blow to the metal. These kinds of forming tools are good for either shrinking or stretching. They are especially useful for stretching because they cushion the force of the hammer. This lets you stretch intricate shapes more gradually with less chance of damaging the metal.

Learn as you go. Begin with something simple and you'll build your hammerforming skills for harder projects. A shallow tank top or bottom is a good start. As your skill grows, so will your use of hammerforming. Once you've mastered the basics, don't be afraid to experiment either. You may be surprised at the quality of work you produce.

SHEET-METAL ADD-ONS

9

Custom-built fender flares with wide wheels and tires give the impression of a much-widened Capri. Flares added only 3 in. to overall body width! I detail the construction of these flares beginning on page 118.

Here's where you begin to produce some good-looking parts by applying those welding and metal-shaping skills you've developed. You will find that practice pays off.

In this chapter I discuss how to add special custom parts to an existing vehicle. These *add-ons* include fender flares, hood scoops, air dams and spoilers. Sheet-metal add-on's are custom fabricated from sheet aluminum or steel. They improve an existing car's appearance or performance—or both. Added parts can be practical: fender flares to enclose wider wheels and tires or a *performance blister* to clear dual four-barrels sitting on a tunnel-ram intake manifold.

Add-on's are not only practical, they add a certain element of style. They can give a car its own personality. But not all add-on's are for appearances only. Some are functional. Spoilers, for example, not only look nice but they enhance a car's aerodynamics and handling. Let's look at add-on's.

HOOD SCOOPS & BLISTERS

Hood scoops are probably the most common add-on. You can buy any number of fiberglass or plastic hood scoops at your local hot-rod shop. Installing one of these scoops will give your car some degree of individuality, but it'll be the same scoop everyone else has installed on his car. But you now have the skill to create a totally individual look.

With tape, thin cardboard, or chipboard, and some imagination, you can create a unique hood-scoop design. No one else will have a scoop like it. Once you have a design you like in cardboard and tape, make it in metal. It's a relatively simple process if you follow some basics.

An air scoop directs air into the engine compartment. This air is usually ducted to the engine's induction system, resulting in greater engine efficiency. A scoop may also provide clearance for some underhood component.

A hood-scoop can have its opening at the front or rear. A rear-opening

Racy-looking spoiler was designed using thin cardboard and masking tape. Cardboard pattern was then transferred to metal. Start-to-finsh photo series on this project begins on page 113.

Addition of supercharger meant modifying Camaro's hood. Scoop is very tall—too tall for safe visibility on the street. Air is drawn in through rear. Only two pieces were used to fabricate this scoop!

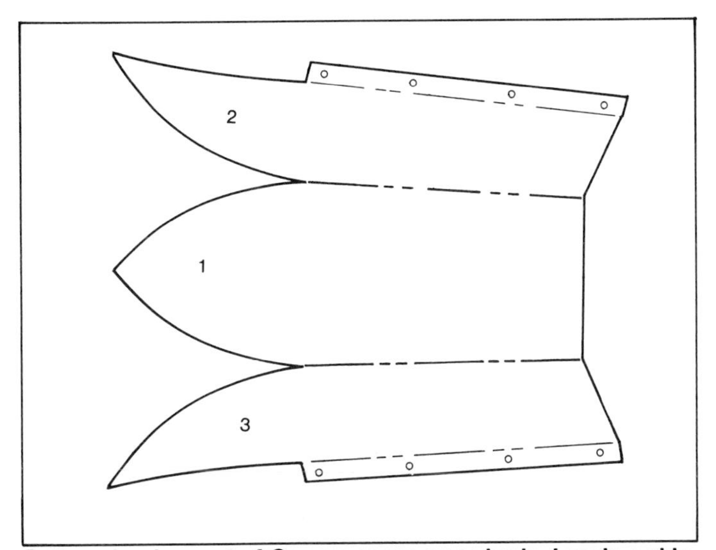

Pattern development of Camaro scoop was simple. I made a side pattern and a top pattern, then combined them into one piece.

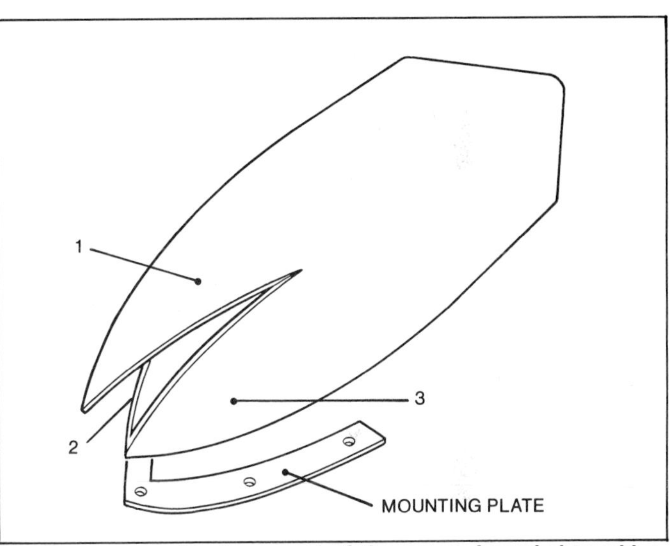

Bending was used to form most of the scoop. I needed to add a small mounting flange at the pointed end. This piece was welded on after scoop was fabricated.

scoop must be close to the windshield for maximum benefit—no closer than 1 in. Otherwise air may exit from the engine compartment through the scoop rather than enter. Just make sure the hood will open. Except for the very front of the hood, the rear of the hood is the point of highest air pressure. A front-opening scoop provides more air, but it also creates more aerodynamic drag.

A performance blister is a rise on a hood—like a blister on your hand. It provides clearance for some high-performance part such as a turbocharger or supercharger. Both air scoops and performance blisters are constructed in a similar way.

There are two kinds of scoops or blisters: those which function and

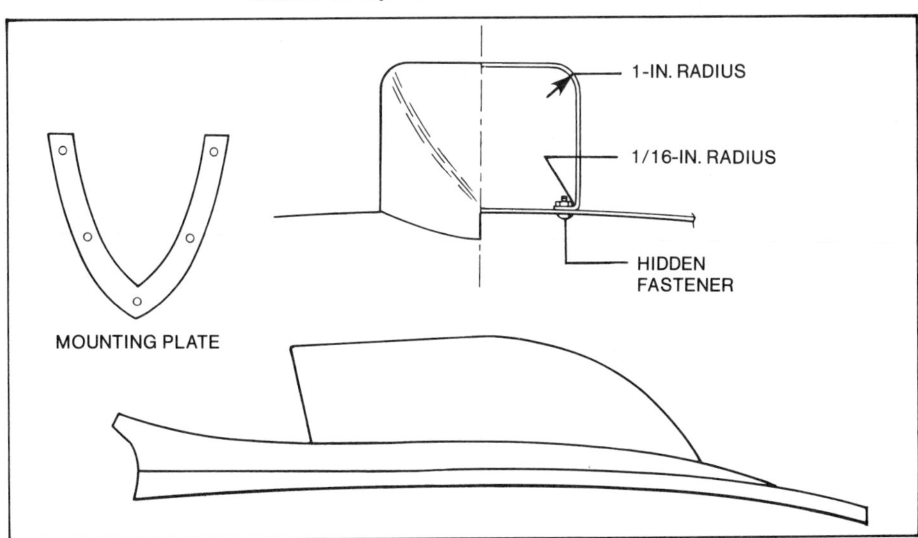

Scoop nose required hand work. I used a hammer and dolly to reduce the seam radius gradually where the scoop sides joined the top. This made welding easier.

Fabrication of a new hood begins by removing the old hood skin. Original skin is removed from the hood inner panel by grinding through the outer edges of the hood and prying the two pieces apart.

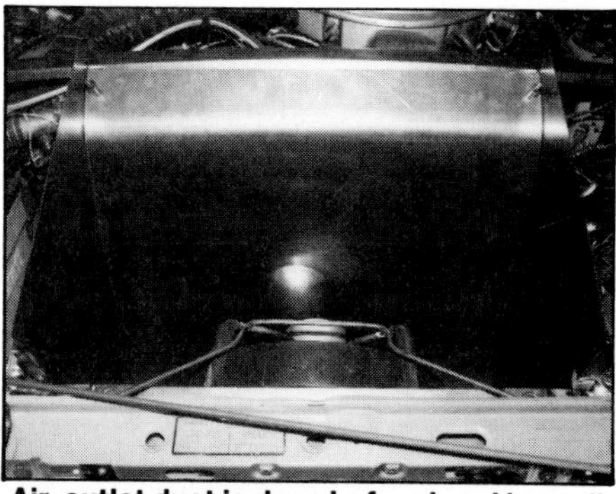

Air-outlet duct is done before hood is modified. Making hood patterns is easier to do with ducting in place. The sides of this duct are blank lengths. They were trimmed later for radiator-cap clearance. Bulge in the middle of the duct—shown being formed on page 103—is for water-pump-pulley clearance.

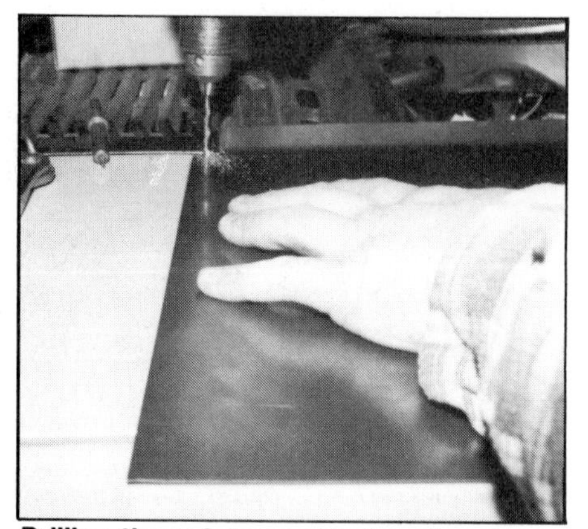

Drilling through hood side for Clecoing so it won't shift out of position while other pieces are fitted. Hood sides are simple rectangular pieces, matching the curve of the fender.

fakes, or those for appearance only. A fake scoop just needs to be attached to the hood and must be low enough to see over. You don't need to cut a hole in the hood or worry about function, so fake scoops or blisters require less work. I'll describe the construction and addition of a functional scoop or blister.

Clearance—Suppose you're building a rear-opening hood scoop to enclose an induction system similar to that for the supercharger illustrated. There are some very important things to consider. The general shape of the component—in this case a supercharger—will determine the general size of the air scoop. There must be enough clearance around the unit for air to flow in and down around the engine. There must also be enough clearance to ensure that engine components won't contact the scoop.

There must be at least 1-in. clearance between the scoop and the top of the air cleaner or intake. Side, front and back clearances are also important—particularly the sides if rubber engine mounts are used. Rubber mounts allow the engine to rock from side-to-side. It will rock to the right under power and to the left under deceleration.

The throttle linkage and fuel line/s make side clearance very important. Insufficient clearance can cause a jammed accelerator or a broken fuel line—a dangerous situation. So pay special attention to side clearance. If it's possible, run the engine and blip the throttle to get an idea of how much side-to-side movement there is in the plane of the hood. Take a measurement while someone else operates the throttle. Use this figure, *plus 1-in.* for side clearance.

Note: The higher a component is above the engine, the more its side-to-side movement will be. So take this into consideration when checking lateral engine movements to allow for scoop clearance.

Measuring—To design the hood scoop, start by cutting a hole in the hood. Don't make the hole bigger than necessary. To determine the size of hole and where it should be, make some measurements. Follow this rule: "Measure twice and cut once." It is much easier to trim excess metal away than it is to put some of it back once you've cut it off.

Measure from the back of the hood opening to the front and back of the supercharger, carburetor or whatever. Transfer these measurements to the hood. These points will establish the front and rear cut lines for the hole in the hood. Find side cut lines by measuring in from the side of the hood opening.

Using these measurements, draw four lines to describe a rectangle. Cut a hole in the hood using these lines as a guide. You can now do a trial installation, trimming material as necessary for clearance. After trimming to fit, bolt or pin the hood in place and measure for the scoop.

Measure the tallest point above the hood line: a velocity stack or air cleaner, for example. Measure the length and the width of the unit at its longest and widest points. Add 1 in. to the height, and provide at least 1 in. of side clearance for the fuel line/s and throttle linkage. You have now determined the minimum size of the hood scoop and where it will go on the hood. The basic shape is a cube. Remember to allow for side-to-side engine movement.

Styling—Here's your big chance to be a stylist. The box shape you need for minimum clearance is pretty ugly. You must round corners, soften lines and develop an aerodynamic shape. *It is OK to make the scoop larger for styling and aerodynamic improvements, but never make it smaller.*

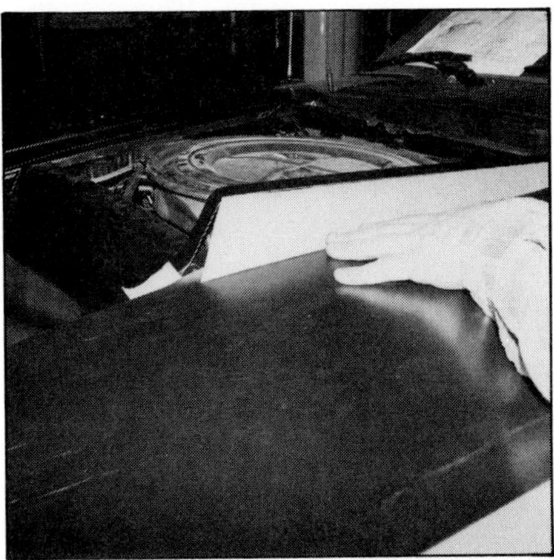

Test strip and side pattern help establish carburetor clearance and hood-blister and radiator duct fit.

Performance-blister pattern was developed using test strip, cardboard side and top shapes. Bend lines and their order are noted on pattern.

Draw front and side views in full scale on a large sheet of paper. Then soften harsh angles with gradual curves. Follow the basic styling theme of the car you're adding the scoop to. Remeasure your drawings to ensure that you have allowed for correct clearances.

Cardboard Mock-up—When you really like what you've drawn, transfer the sides to cardboard using your drawing as a guide. Cut the cardboard. Assemble the sides, top and front with masking tape. You have a cardboard hood scoop.

Set the cardboard scoop on the hood and check it for fit and looks. Make any necessary clearance and styling changes. When you really like the shape of your cardboard scoop, you're ready to develop the final *working pattern.* A final pattern is a cardboard template you lay over the metal to use as a guide for cutting the blank.

Try to put as many surfaces of the final part as possible into the template. Also, fabricate the scoop from the *fewest* number of separate parts. Make a bend rather than a weld seam. The extra time invested in working out a pattern requiring fewer weld seams will result in a more attractive scoop. It's also a lot less work.

The procedure I've just described is 99% the same whether you're making a big scoop, small scoop or whatever. The size and shape will vary according to your needs, but the basics remain the same. The general process of measuring, making drawings and developing a pattern is the same. Clearance is the crucial factor.

Whether the scoop opens in the front or in back is a relatively unimportant fabrication consideration. And styling is entirely up to you. That's what makes it a custom scoop.

The same thing applies to designing a *performance blister.* Clearance is just as crucial. Use the same rules for establishing clearances as you would a scoop. Measure, draw, make cardboard templates and remeasure—the procedure is identical.

Test Strip—A good way to avoid scrapping a metal blank for a scoop or blister is to make a test strip using 1-in.-wide aluminum. Use the same thickness as the material you are using for the scoop. Using these strips, bend up a *sample section* of the planned scoop. By sample section I mean a typical cross section of the scoop. If your scoop tapers in the top view, make two test strips—one for the widest section and one for the narrowest.

These strips will give you a last chance to find and correct any unsuspected clearance problems. They will also help with locating radius-bend *tangent points*—where a radius stops and a flat surface or another bend begins—and mounting-flange angles accurately.

Each test strip should have a *marked* center line that coincides with the metal blank. By bending these strips you can see how well your planned bends will work *before* bending the blank. This will let you spot a mistake and easily correct it prior to shaping the final part. Trial-fit the test strips over the protruding object and check clearances at the top and sides.

Trial fit of bent hood blank. This points out any irregularities in bend angles so adjustments can be made early. Note how inboard edge of hood half is bent up to meet the blister.

Pattern Adjustment—It takes time to make a good pattern. Don't be discouraged if you have to change the pattern because the test strips revealed some problems. That's *why* we call them "test strips." It is less trouble and less expensive to make changes in cardboard, or even toss the whole pattern out and start over, than to ruin metalwork. Only when you are positive the pattern works can you begin to work in metal. Don't hurry!

A good scoop pattern can also include a lot of time and trouble-saving features. The pattern itself can include mounting flanges.

Because of its importance I'll

1-in.-diameter T-dolly will be used to round the corner at the weld joint. Gap will be closed as the corner is hammered over the dolly.

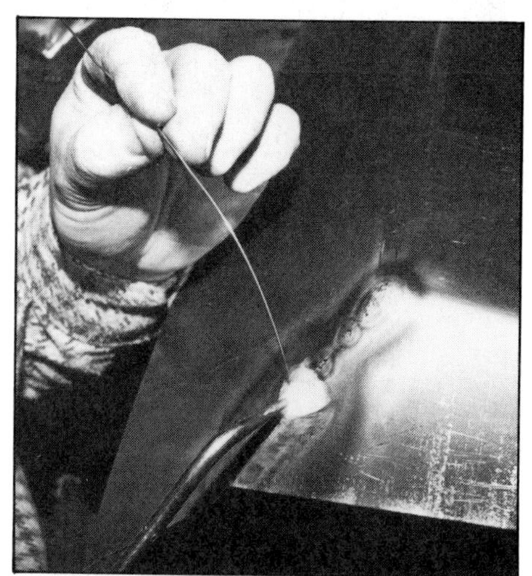

Tack-welding the corner seam. Hand-held dolly backs the corner while many gentle taps with a flat-faced hammer does the smoothing.

repeat: It's best to bend one large metal blank rather than to weld several smaller blanks together to make the same basic piece. Either way the scoop has the same shape, but bending avoids welded seams that must be metal-finished. And it is quicker and easier to bend rather than to weld and metal-finish.

Again: Spend time developing a pattern which requires *as few pieces as possible.* Fewer pieces means you'll spend less time fitting, welding and finishing—*simpler is better.* Fabrication will be easier and the finished product will look better.

Flanges—Always include mounting flanges as part of the blank rather than adding them later. Include them in the blank and bend them. You'll eliminate the time needed to weld them to the blank. Although it doesn't matter whether the mounting *steps,* or flanges, turn in or out, I always try to bend mounting flanges to the *inside.* See the drawing. Flanges bent inward allow hiding the mounting bolts under the installed scoop or blister.

Marking the Blank—It's time to use the patterns to make a metal blank. Lay the patterns on sheet metal and trace around them. Use an HB lead pencil when marking aluminum. *Do not use a scribe on aluminum. Scratches will weaken the metal, possibly causing it to crack at a bend line.* Using the test sample strips as guides, locate and mark the center line on the blank. Transfer the bend lines to the blank too. These marks will be a "road map" for cutting and bending the blank.

Number the bend lines on the blank *in the order the bends will be made.* The test strips will be your guide for establishing the proper bending sequence. Follow it. Otherwise you will "paint yourself into a corner," preventing the possibility of making some other bends.

All it takes to understand the need for doing this is to violate the bending order once, particularly if the blank has to be scrapped. So plan ahead and follow a logical bending sequence. You'll save time, money and headaches.

Cutting the blanks—With the cut lines and bend lines on the sheet aluminum you can cut out the blank/s. Use shears, a bandsaw, saber saw or tin snips: whichever you find works best. File any sharp or rough edges smooth.

Bending—As you'll soon discover, bending is the most difficult part of fabricating a scoop. You already know the importance of making bends in the proper sequence: Don't complicate matters by violating the sequence. It affects how easily and how well the metal takes shape.

Depending on the size of the scoop, you may or may not need a brake to do some of the bending. A large scoop—over 20-in. long—requires a brake. On the other hand it's often easier to bend a smaller scoop—about 14-in. long—with a workbench, large C-clamps and homemade radius dies. Scoops smaller than this can be bent by hand.

Avoid making long bends by hand because they tend to deflect at the center. The uniformity of the bend radius is then lost. However, short bends, like those across the width of the scoop, can be made by hand. They can also be made with C-clamps and homemade radius dies on a workbench.

Use a T-dolly for forming bends. It often takes some forming to get the desired radius where two surfaces must be welded.

Tack-Welding—After you have the sheet-metal blank/s bent up into the shape of a scoop, carefully tack weld the piece/s together. Then do the final welding.

Whether the scoop is steel or aluminum, *gas-welding is the best welding technique.* A gas-welded joint will be soft and easy to metal-finish. This is extremely important on exterior parts where surface appearance is clearly visible.

Metal-Finishing—Metal-finishing is the final touch applied to your work before priming and painting. Welded or shaped areas must be finished as smooth as possible. An irregular surface must be filled with lead or plastic filler to even it out. Fillers are OK, but should be used sparingly. Metalwork quality can be judged by the amount of filler used. A smooth part needs little filler. More filler is needed when the basic part has many surface flaws and irregularities.

There are times when you can't avoid using filler. For instance, when a low spot in a fender cannot be brought up because of an obstruction behind it, you need to use filler. That situation is probably the most

Forming a slight curve in the blister by shrinking each side.

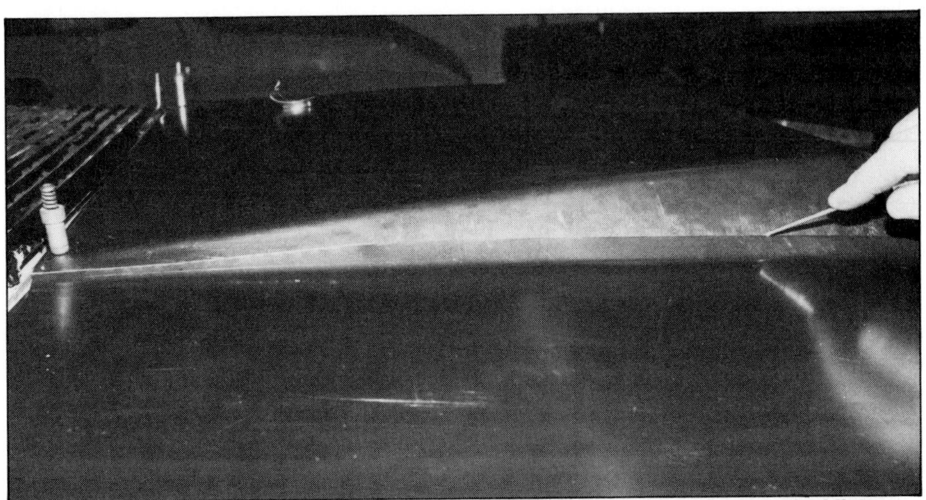
Scribe line shows where the side-panel edges should be trimmed for joining to the blister.

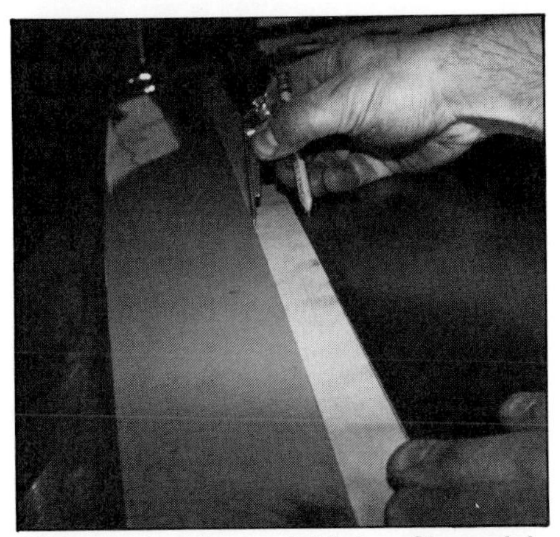
With hood sides and blister Clecoed in place, the outer edges of the hood skin are marked for bending. The 2-in.-wide tape lined up with the inside edge of the fender guides the pencil compass. Compass is set at 2-1/8-in. to give a 1/8-in. *margin,* or hood-to-fender gap. Edges are bent first 90° to stiffen the hood. Some shrinking was required.

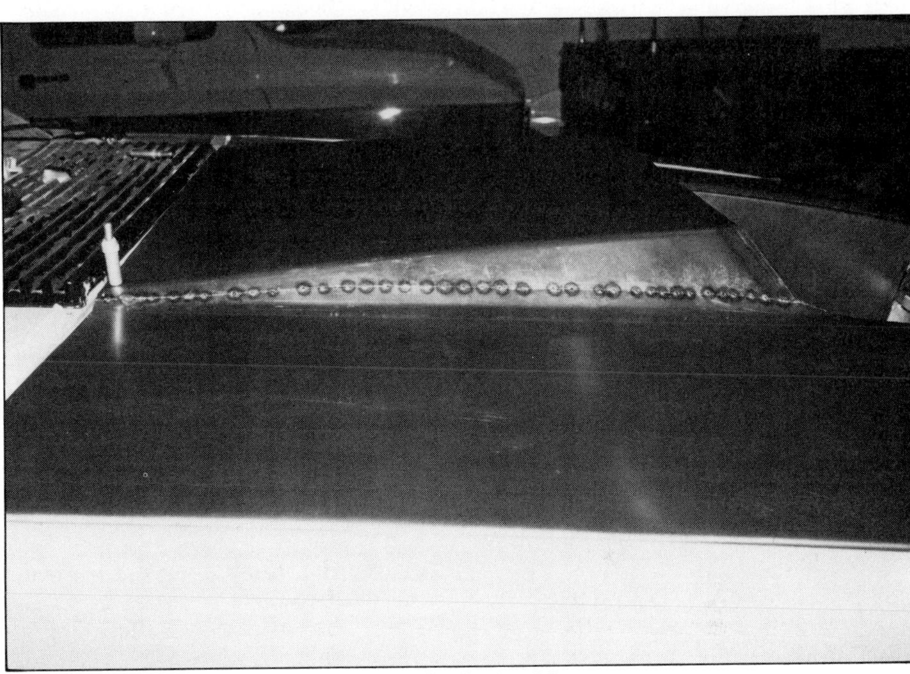
Tack-welded performance blister is ready for final welding.

common *valid* reason to use plastic or lead filler.

Hammers, slappers and dollies come into play to smooth metal-surface irregularities. The first metal-finishing step is to tap the metal gently with a hammer or slapper while backing it with a dolly. The idea is to move high areas down and low surfaces up.

One of the most important metal-finishing "tools" is your hand. *Feel* the surface often during the tapping process by running your hand across it. You'll be able to detect high spots, low spots and other irregularities. I've been accused of "massaging" aluminum by people who have watched me metal-finish. When you're really happy with the evenness of the area, move on to the final metal-finishing step: filing.

Filing—You'll need a special file for metal-finshing your scoop—a *Vixen file.* It is a long, wide file with teeth designed to remove small amounts of material, without catching or snagging, as it moves across aluminum or steel on each pass. Use long, smooth strokes with the Vixen file.

Bar hand soap rubbed over the surface of the metal makes the file move easily over the work. Soap also stops the file teeth from *loading,* or clogging, with metal shavings. If the file loads up, use a file card to clean the teeth.

Continue to feel the surface for smoothness as you file. Work in a well-lit area so you can see problem spots.

Front, center hood section is ready to be constructed now that the blister is fully welded.

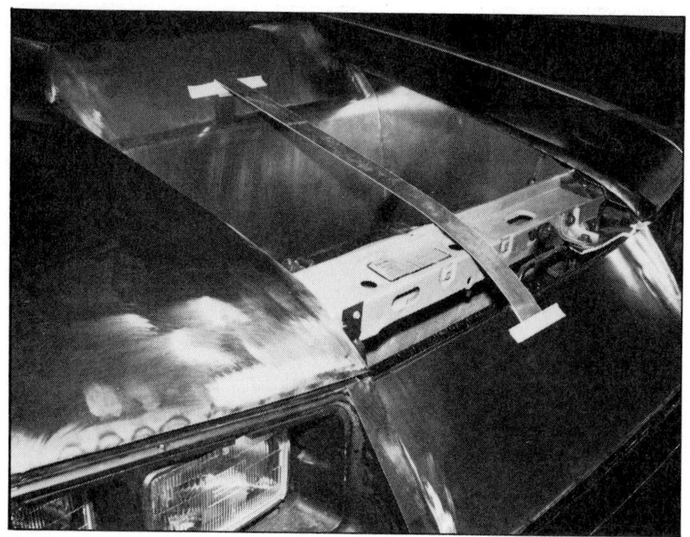

Test strip establishes clearances and shape of hood center section.

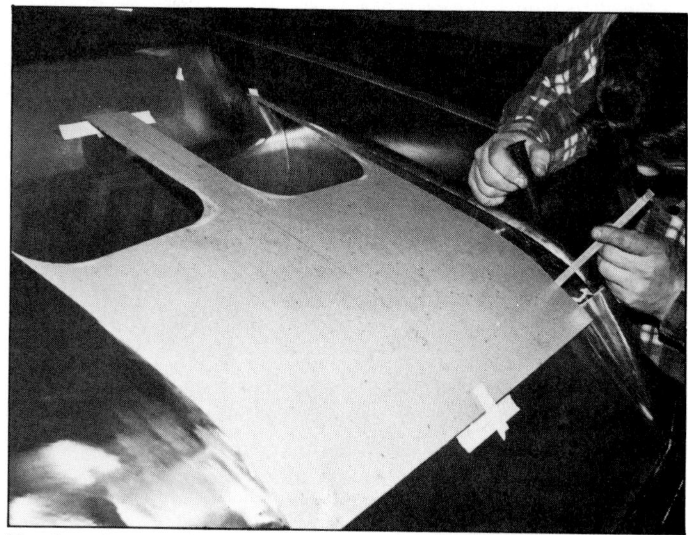

Hood center-section pattern developed and fitted to car. Measuring center-section-to-side-panel gap. Material will be added to center-section blank to close this gap.

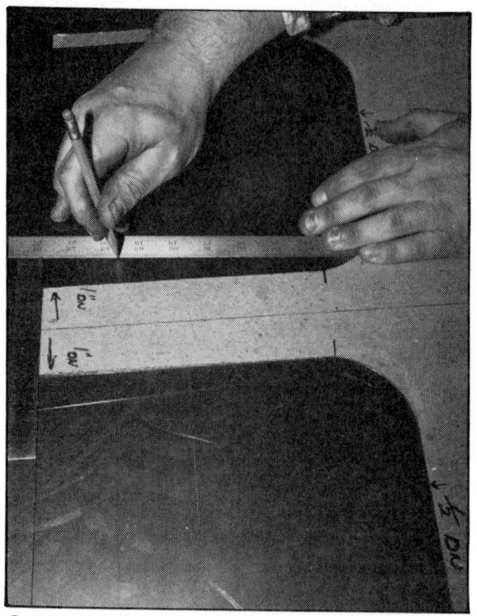

Special features noted on pattern are transferred to metal blank. Use a soft pencil to mark the metal.

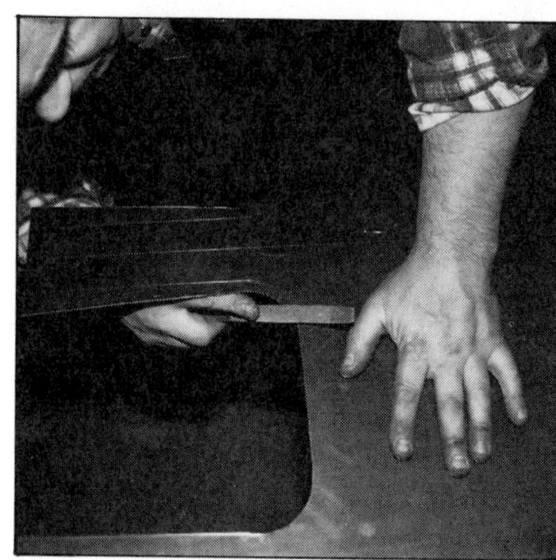

Filing metal edges in the corners. This reduces the chance of tearing the metal when the flanges are formed.

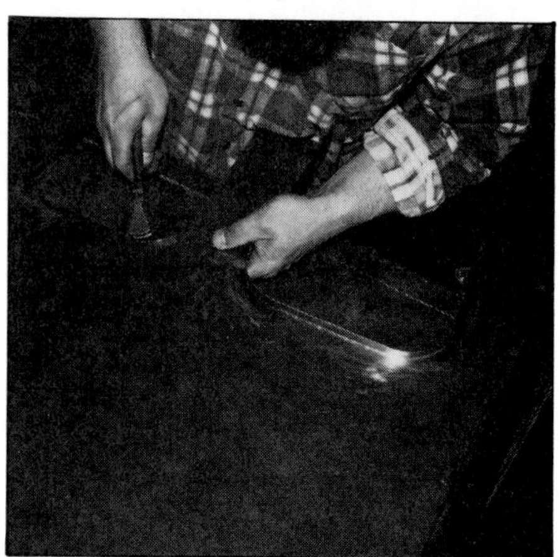

Begin flanging in the corners. I'm using curved glass pliers.

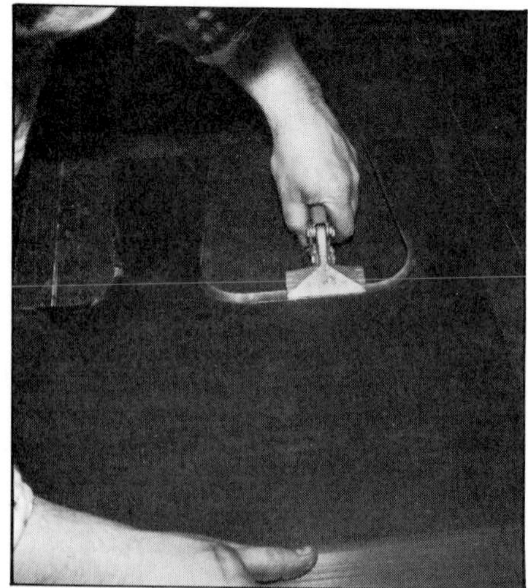

Hand seamers turn the flange along straight sections. Don't bend flanges more than 45° with glass pliers or hand seamers.

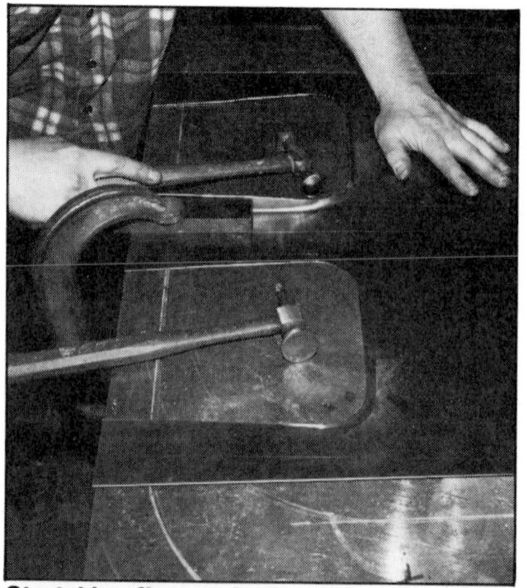

Stretching flanges further or bending them on a sharper radius is done with body hammers.

Installation—When you are satisfied with the shape and finish of *your* scoop, it is ready for installation. If you turned the mounting flanges in, the scoop can be attached with the bolts or rivets hidden from view.

In the case of a metal hood, the need to weld the scoop to the hood is eliminated. Such welding can be a big problem. Welding on a large, relatively flat piece of sheet metal, such as a hood, will distort it. Much extra work will be required to restore the hood to its correct shape in the weld area. However, if you must weld the scoop on, be prepared to do a *lot* of metal-finishing to the hood. Another advantage of bolting the scoop on is you'll only need to paint the scoop, not the

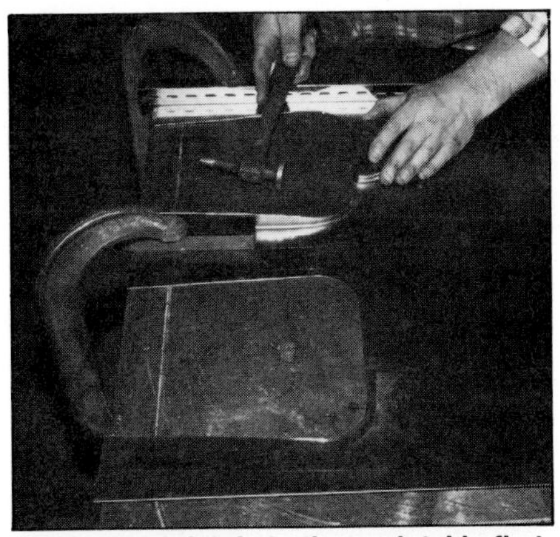

Clamp panel tightly to the work table first. Note how holding block functions as a dolly. Short straight sections of the flanges can be smoothed with a hammer and hand-held dolly.

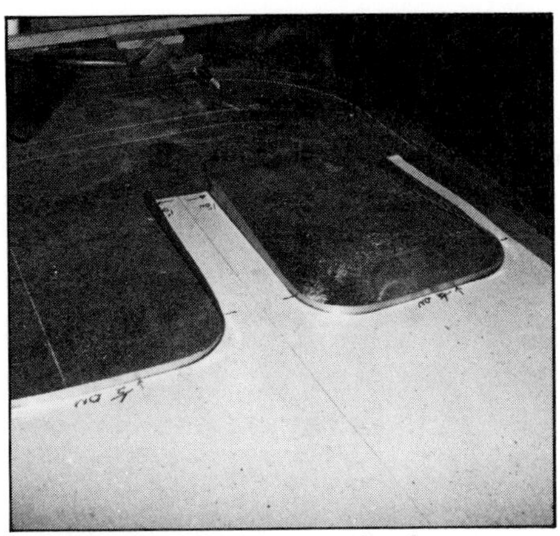

Checking hood-center section for accuracy by laying the template over it. This is a very important step.

Top side of finished hood center section with turned-down flanges. Outside flanges have not been bent.

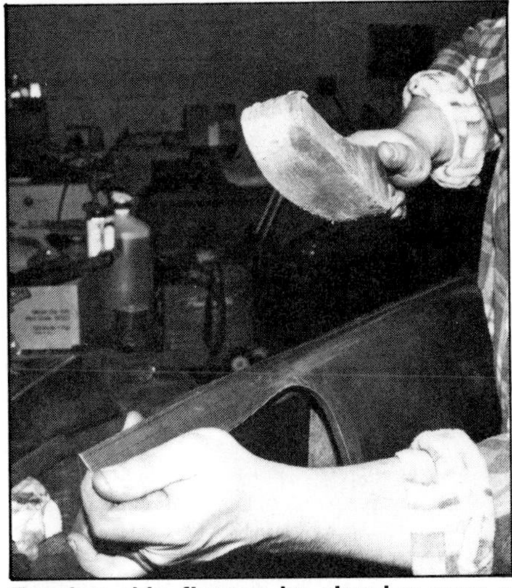

Forming side flanges by slapping over a large T-dolly. This brings the outside flange into correct position for joining to side panels.

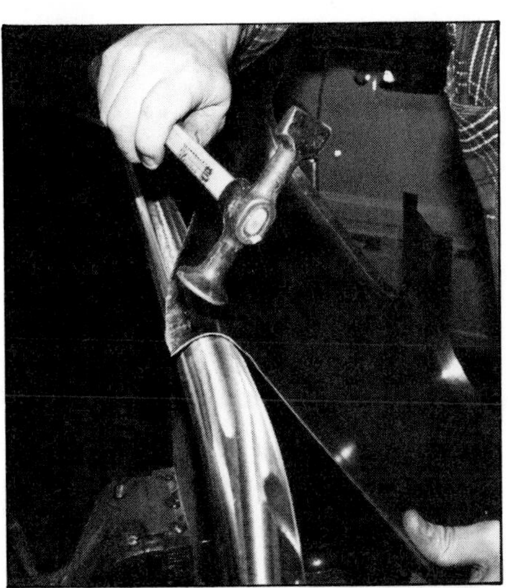

Stretching side flanges curved the center section to match test strip. In addition to the hammer and T-dolly, a stretcher was used to curve panel.

Additional curve is formed with the English wheel.

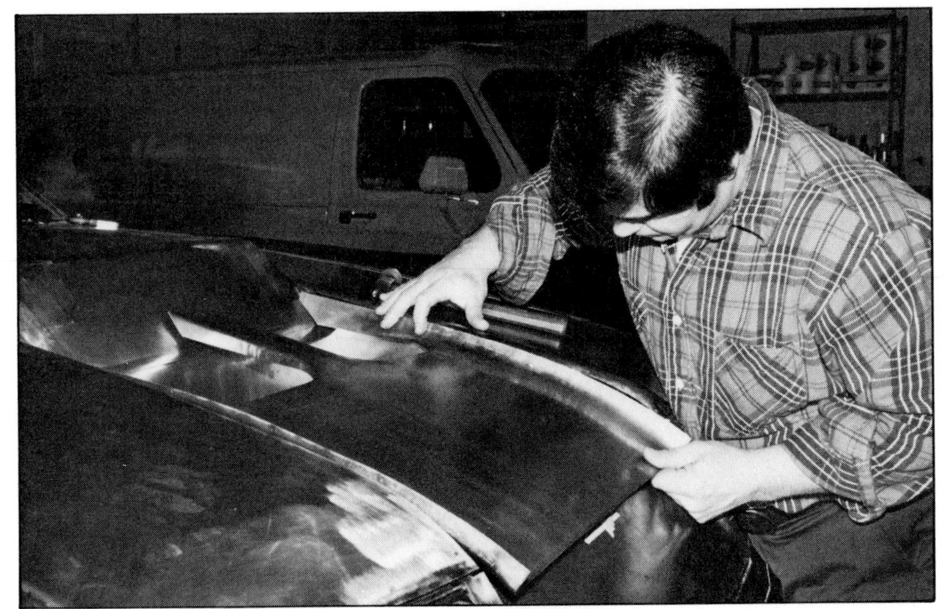

Trial fit of hood center section showed it needs more curve.

Front mounting flange is gradually formed in a beader. A slight curve in the panel meant I couldn't use a brake for this flange.

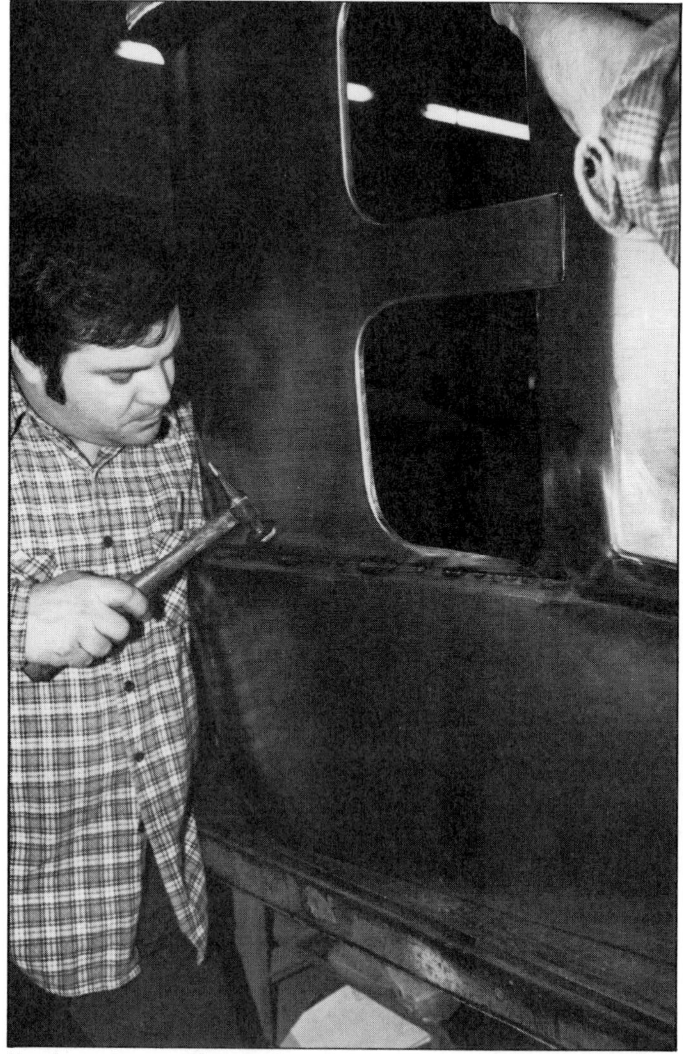

Hood center section tack-welded to hood sides and blister. Hammer and dolly are used for smoothing.

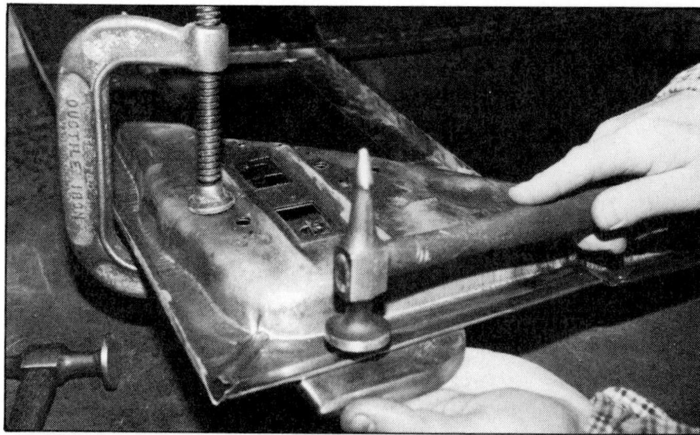

Edges of hood are carefully folded over hood reinforcement with a hammer and dolly. Dolly is held firmly against the skin to keep from denting it.

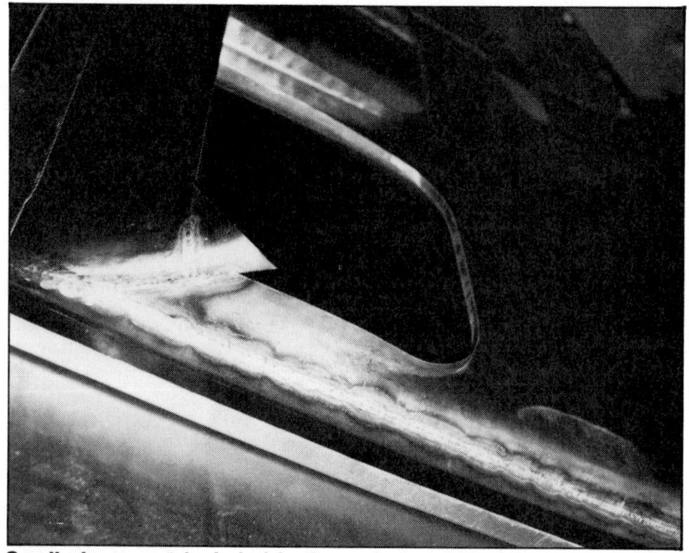

Small pieces not included in the main hood blanks are fitted and welded.

Assembled hood with performance blister is ready for final welding and metal finishing. Keep margins even and in correct alignment with fenders, cowl and front end.

whole hood. Welding means you must paint everything.

SPOILERS

Spoilers, or *air dams,* are another common exterior add-on. Like scoops, fiberglass or plastic spoilers are available. However, if you want something different, make one in metal. The basic rules for fabricating and installing front and rear spoilers are similar. The main difference is in function: how a vehicle's aerodynamics is changed.

Front Spoilers—Mounted under the front bumper or to the lower edge of a car, the main function of a front spoiler is to prevent air from flowing under a car. If done correctly, a spoiler reduces aerodynamic drag by *spilling* air off to the sides. This can increase the top speed of a car and make it more stable at high speeds. It also improves engine cooling.

Finished spoiler blends in with the design of the Mustang. The following pictures show the construction of the spoiler.

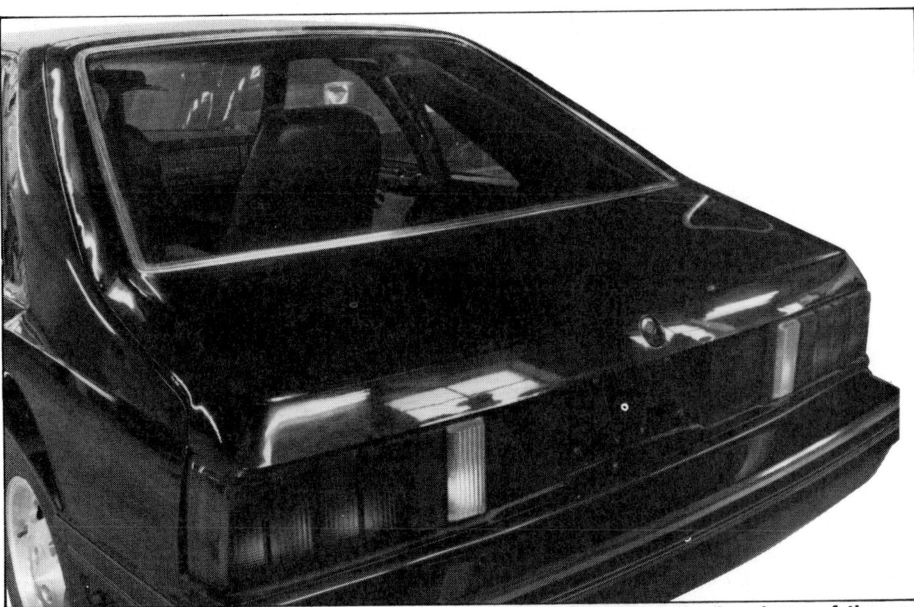

Original Mustang before the spoiler was added. Take time to study the shape of the car before designing the spoiler. It should blend with the rest of the car.

A well-designed front spoiler increases air flow through the radiator, improving engine cooling. It also reduces front-end lift that was previously caused by air flowing under and *packing* under the car. In many cases, front-end lift is reversed and becomes *downforce.* This increases the cornering ability of the front tires. Finally, a front spoiler offers a good place to install intakes for front-brake cooling ducts.

A front spoiler must have adequate ground clearance. That's important whether it is for a race car or street use. Otherwise, there's a good chance you'll bend or break the spoiler off every time you go up a ramp, such as a driveway, or park facing a curb. A spoiler isn't much good lying on the street.

Let's look at rear spoilers and their design and fabrication next. When fabricating a front spoiler, use the same basics as for a rear spoiler.

Rear Spoilers—Mounted to the trailing edge of a *deck,* or trunk lid, the rear spoiler is also an aerodynamic aid. The results are similar to a front spoiler, but at the rear of a car: Appearance and aerodynamic drag and downforce are changed. Similar to a front spoiler, downforce on the rear improves high-speed handling by increasing tire traction at the rear.

Two common mistakes are made with rear spoilers: making the rear spoiler too high or making the leading surface of the spoiler at too "severe" of an angle—more than about 45° from horizontal. Either of these mistakes may defeat your purposes.

If a spoiler is too high, drag will be proportionally higher than the downforce gained. Similarly, if spoiler angle is too severe, drag *gain* will be excessive. *The most efficient spoiler will gain the most downforce with the least increase in drag.* With this in mind, design a spoiler not over 4—5-in. high and with an angle of 30—40° from horizontal.

Spoilers can be attractive as well as functional. When aerodynamics is *more* important than appearance—like the spoiler on a short-track stock car—the spoiler may be a large piece of sheet aluminum or plastic riveted to the trunk lid. This kind of spoiler is not very attractive—particularly a *big* one; up to 2-ft high—but it can provide a lot of downforce! These big spoilers create excess drag—not an important consideration on a short-track car. Improved cornering is more important, considering the time spent in the corners and the relatively low speeds.

A great deal of care must go into the design of a spoiler when appearance is important. So let's look at the process of constructing a rear spoiler by starting with design.

Design—Carefully study the lines of the car. Draw a profile of the car body. Draw a line that extends at an angle up from the rear edge of the deck lid. This line represents the cross section of the front surface of the spoiler. For deck lids with high openings like the Mustang shown in the photos, the lower edge of the spoiler's rear surface looks best if it matches the rear lip of the deck. Draw the rear surface of the spoiler with a line running from this point up to the "peak" of the spoiler.

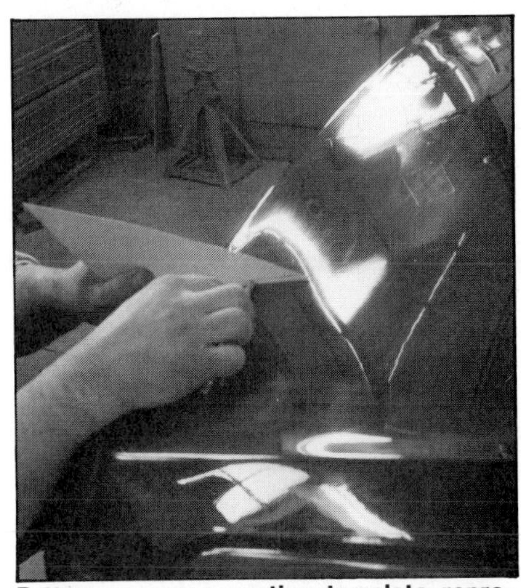

Pattern or cross-section template represents a cross section of the spoiler shape. It will become a rib to support the spoiler panels.

Keep drawing, erasing and drawing until you have it right.

Cross-Section Template—Cut out a cardboard template using your drawing as reference. When it's finished, the template will represent the cross section of the spoiler.

Begin making the template by transferring the body line to the bottom edge of the cardboard. Hold the cardboard against the rear edge of the deck and draw a line on the cardboard as shown in the nearby photo. This line should follow the surface of the deck. Trim the cardboard so it fits exactly against the deck surface.

Don't be afraid to experiment. Using the template, your drawing, a pencil and some scissors, decide

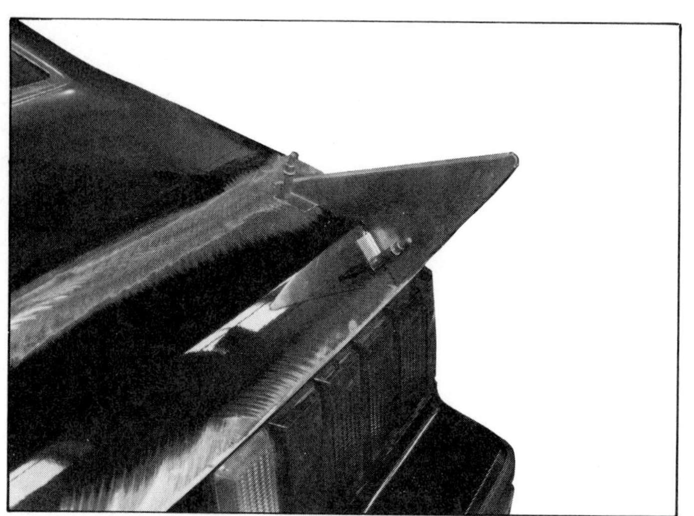

Make one rib and Cleco it in place for a trial fit. Note that the rib does not line up with the rear edge of the trunk lid. This allows room to mount the rear spoiler panel.

When you are satisfied with the fit and shape of the rib, make several more and rivet them into place. Space them equally across the width of the trunk. Rivets are used rather than Clecos. Clecos would protrude and interfere with the development of the exterior spoiler patterns.

Spoiler top and rear patterns can be developed quickly. Forward trim line is dictated by the intersection of top pattern with the trunk lid. Rear trim line is parallel to the forward trim line. Rear spoiler pattern must match the trunk lid at its bottom edge. Top trim line matches the trim of the top panel where they form the peak of the spoiler.

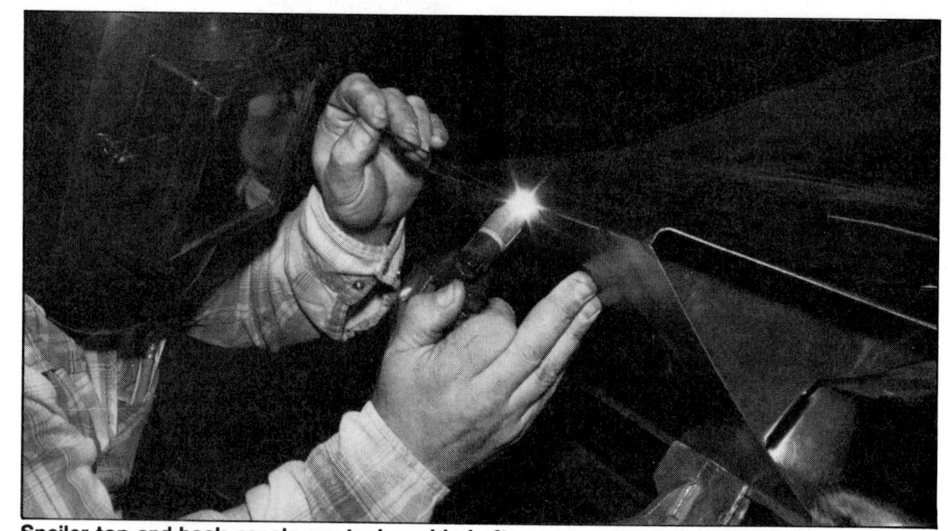

Spoiler top and back panels are tack-welded after seam edges are formed with a hammer and T-dolly. Lower flange is formed with a beader or hand seamer. Because it is curved, flange cannot be bent in a brake.

where the spoiler should begin, end and what angle appears to be best. It is easy to draw, erase, and redraw a pencil line. Cut the cardboard following the general shape you drew.

Tape the template to the deck so you can stand back and view it. The spoiler's front surface should not look awkward or take off from the deck-lid surface at an abrupt angle. Play with the general shape until you're pleased with the height and the angle of the spoiler.

Refit the template to the car and "eyeball" it. Trim and recheck if you're not satisfied. Start over if necessary. Cardboard is cheap. When you are totally satisfied, stop. You now have a cross-section template.

Deck-Lid Template—Here's where we make the spoiler look natural. Looking at the deck lid from overhead, you'll notice the shape put there by the stylists. Make your spoiler follow that same shape so it will blend in with the styling of the car. Otherwise it may look like an ugly duckling.

To pick up the body end curve, make a template the full width of the deck lid. You'll need a big piece of cardboard for this. You'll also need someone to help hold and fit this template. Because of its size and flexibility it will be difficult to control and hold steady.

Trace the body curve across the back of the deck lid onto the cardboard. This line represents the peak of the spoiler. Cut out the curve. Determine the *width* of the front, or upper, surface of the spoiler—the distance from the peak of the spoiler to its intersection with the deck lid. This distance should be the same as the front edge of the cross-section template. If, for example, this dimension is 10 in., duplicate the body curve by drawing a line 10-in. forward of the *peak line.* If the shape of your car's deck lid is curved like the one shown in the photos, this template will look like a giant banana.

Spoiler into Metal—You should now have two templates: a cross section and an upper spoiler surface. It is now time to begin metalwork. I used 0.035-in.-thick mild steel to fabricate the spoiler shown in the photos.

Transfer the cross-section template to metal. The average spoiler requires four cross sections spaced evenly across the back of the deck lid. I added flanges to my metal cross-section pieces: one at the front, one at the back, and two smaller ones for mounting. You should do the same.

Cut out the cross-section pieces and bend their flanges. I used a brake to bend the long flanges on each section 90°. The flanges were bent to the right on two pieces and to the left on the other two. I used a hand seamer to bend the short mounting flanges *in the opposite direction* of the long flanges. I did this so the long flanges wouldn't be in the way when I was ready to Cleco and rivet the pieces in place. Rivet the cross sections to the deck at equal intervals across the rear.

Spoiler end-cap patterns. Make everything fit exactly. Scratch on body (arrow) correlates to a body styling line and forward point of cap.

Rear Template—Once the cross sections are attached, lay the large top template in place across the width of the deck. Now is the time to develop the template for the rear spoiler surface. You can now see the actual spoiler shape. The rear template is formed by tracing the top edge of the upper template onto cardboard. Complete the template by locating the bottom ends of the sections. Connect these four points with a smooth line that follows the rear body line. When you have the shape drawn on cardboard, cut it out.

Make a trial installation of the rear-surface template. It must meet the upper edge of the spoiler's top template and intersect the body. Trim the bottom template *and the top template* for exact fit before continuing—*no gaps allowed.*

Blanks—Cut out metal blanks using the templates. The peak where the top and bottom spoiler surfaces meet must be formed to *soften* it. Use a large T-dolly and a wooden flat-faced mallet. Roll the edges on both pieces where they meet so the front and rear surfaces are joined with a smooth radius. You can see this being done to the end caps in the photo, page 116.

I bent a 1/4-in. flange along the lower edge of the bottom panel so it would fit flat against the body. This flange made it easier to mount the spoiler by creating a natural surface for clamping and spot-welding along its lower edge.

Tack-Welding—Tack-weld the two panels in place on the deck and together across the cross sections. Once these are in place, you can develop

Closely nesting spoiler end-cap-patterns saves sheet metal. Blanks are marked and cut out.

Tipping 3/8-in. flange to ease tack-welding panel to body. Beader easily follows gradual curve. Also formed with the beader, one metal-thickness (0.035-in.) step facilitates joining panels.

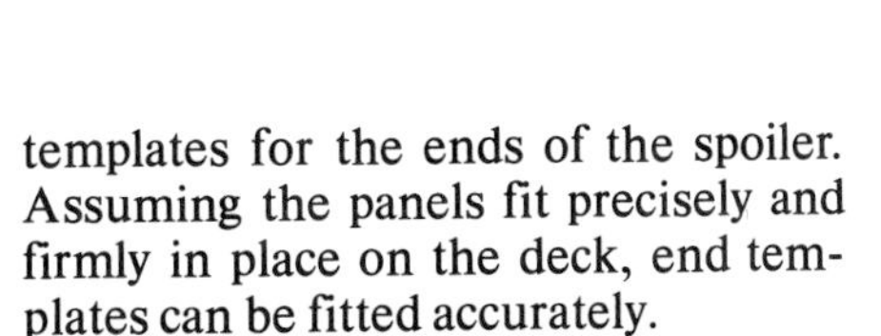

templates for the ends of the spoiler. Assuming the panels fit precisely and firmly in place on the deck, end templates can be fitted accurately.

The beautiful part of following this sequence is that it lets you see exactly what the completed spoiler will look like *before* making it in metal. It is a preview of the spoiler and helps determine an attractive design for the sides.

Fitting End Caps—Templates for the end caps are made so they wrap around to the side of the body. One important point to keep in mind when fabricating pieces that must be the "same," but on opposite sides of a body, like the end caps for a spoiler: The two end caps and the pieces that form the caps will be *symmetrically opposite.* The separate pieces will be the same in the *flat,* but forming must be done in *opposite directions* on pieces that go on opposite sides.

You will need one end panel each for the top and for the bottom to form each cap. Draw, cut, fit and tape the cardboard templates to the car so the end cap will blend into the spoiler and the body in a pleasing way. Repeat this process until you find a shape you like. It should be apparent to you by now that it's considerably easier to modify a design in cardboard than it is in metal. When your design is "perfect," *mark* on the car with a horizontal scribe line where the forward end of each cap ends.

End Blanks—Cut out blanks for the caps using the templates as guides. I added 3/8-in. flanges to the edges of the blanks that join the body. These flanges are similar to the one I added

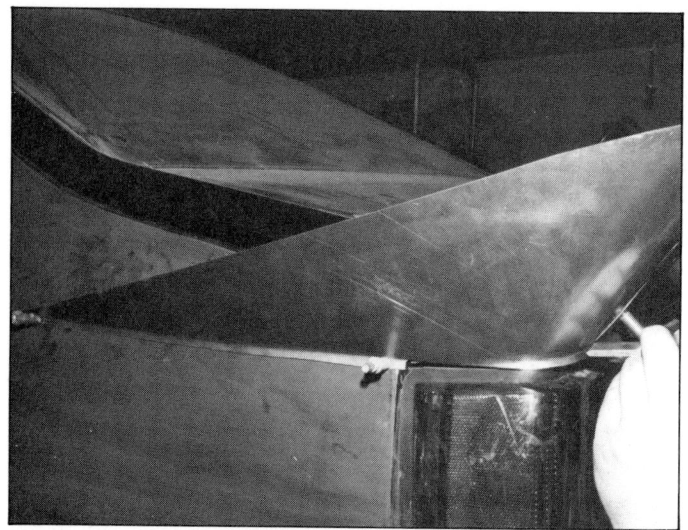
Lower end-cap panel Clecoed in place.

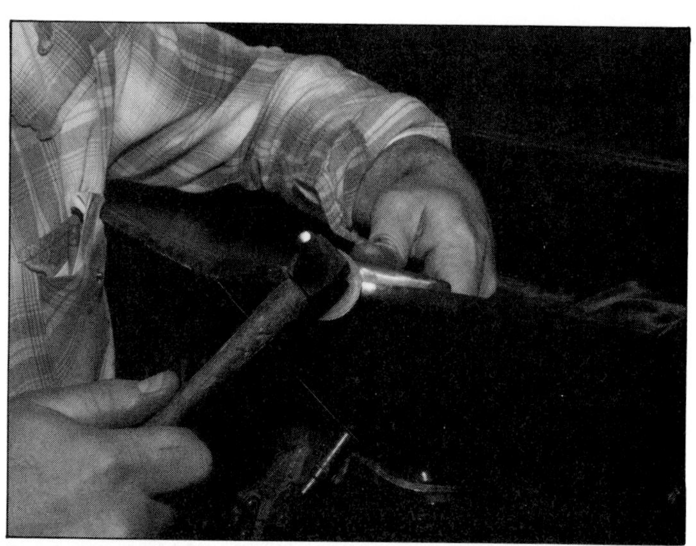
Hand-held dolly and hammer are used to blend lower cap panel into center spoiler panels.

Smooth seam edges before tack-welding.

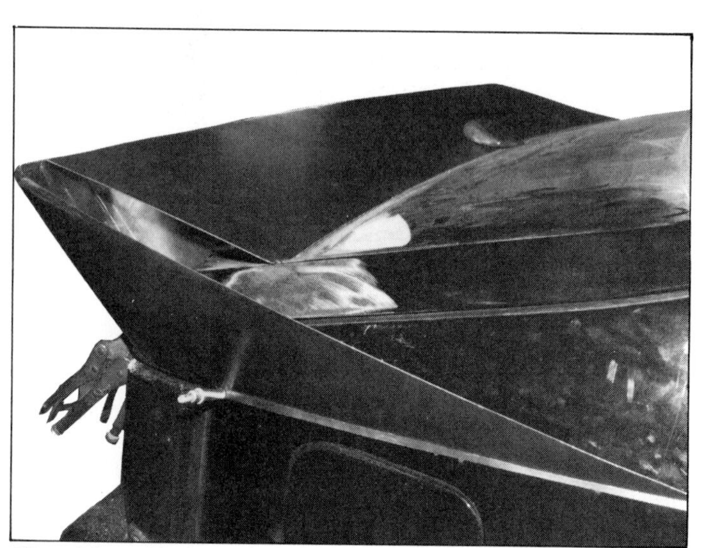
Fit and form lower end-cap panel completely before fabricating top panel.

End-cap-top blank is cut out and trimmed to fit.

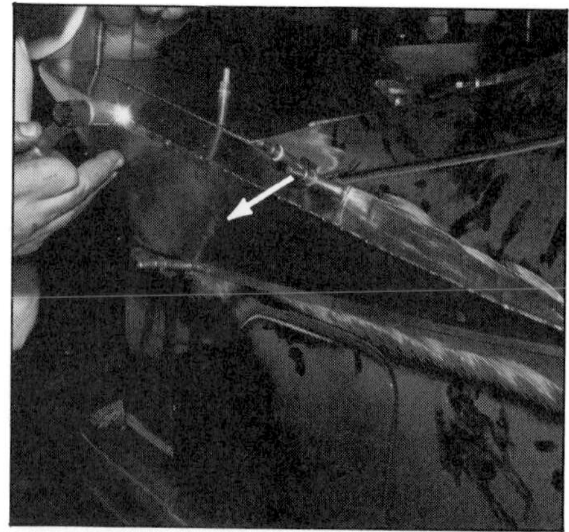
Tack-welding end-cap pieces. Note center-punch marks (arrow) to indicate where end cap must be cut to allow hatch to open. Spoiler ends can be removed and cut after tack-welding without losing the correct fit.

to the lower edge of the rear spoiler panel—and for the same reasons: holding and joining. I also added a step to the ends of each cap piece that joins to the upper and lower panels. See the nearby photo.

After cutting out the pieces, form them to a curve or bend matching the cardboard development. The bend can generally be done by hand. Check against the car several times to see if the bend is correct. The pieces *must* fit extremely closely.

Side construction—Match the pieces to the scribe line on each side of the body. Drill some holes in the end-cap pieces, through the body, or through the upper and lower rear panels while holding everything in place. Note: As

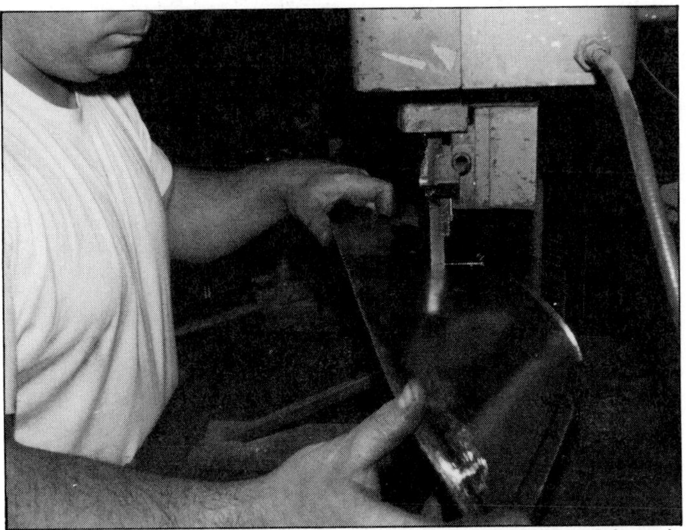

Cutting spoiler in a bandsaw to avoid warpage. It is quick and easy.

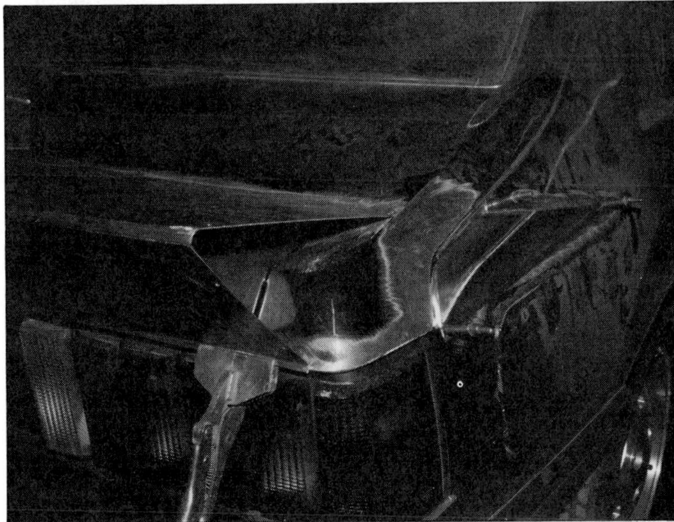

Front half of end cap is Clecoed to the car and it's end capped as shown.

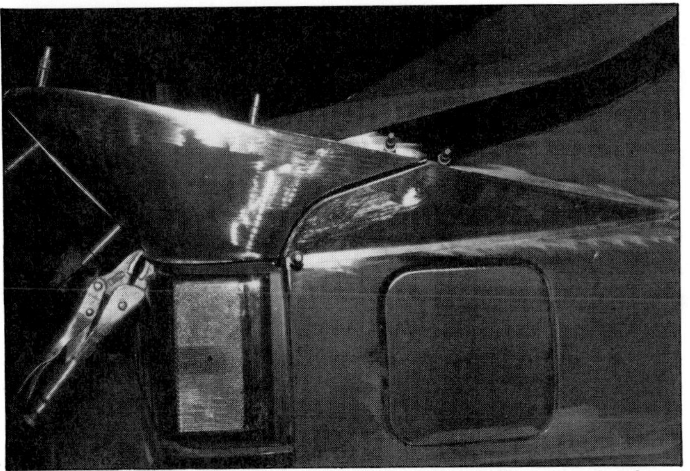

Rear half of end cap must also be capped. Margin, or opening between front and rear cap must be even and equal the hatch-to-body gap.

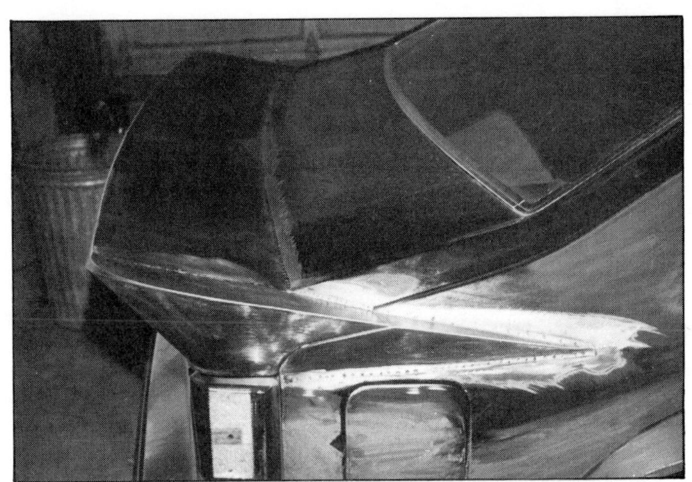

After the spoiler looks good and fits well, do the finishing touches. I heliarc-welded this spoiler to minimize distortion. Then I metal-finished the seams.

you drill each hole, install a Cleco *before drilling the next hole.* This will ensure that the holes in mating panels will match after they are all drilled. With everything Clecoed in place, tack-weld the cap pieces together.

Do not tack the sides to the large panels at this time. The spoiler must have a split, or gap, at each side so the deck can be opened. Notice that the pictured end caps overlap the hatch opening. Finish welding the side pieces together.

Determine where the caps must must be cut to allow the deck to open. Follow the natural curve of the deck opening. Mark this line with a soft pencil or a series of center-punch dots. Remove the end-cap assemblies. Make the gap in the caps the same as the deck-to-body opening.

Cut the caps with a bandsaw along the lines. Cleco the forward half of the caps to the car. Develop caps for the

Minimum amount of body filler and some paint put the finishing touches on this spoiler, the only one like it in the world.

Finished Mustang shows the results of the hood project beginning on page 107. Fender flares were also added.

Rear fender-flare during construction shows how the project was developed: from an artist's rendering. Photo of rendering is in quarter window.

Fender flares I added to this Capri give it an IMSA-GT look. Deadline pressure meant project was done without a rendering or full-scale drawing. Although I didn't have either, I'd prefer a full-scale drawing from which to work.

Fender-flare development starts here: Flare top piece narrows at the ends of the fender. It is widest at the center of the wheel well: 1-1/2-in. wider than stock. Wheel-well strip under the fender is much wider at the bottom.

end caps using the template technique and weld the open end of the caps closed. *Don't weld the end caps to the body yet.* Repeat this process on the rear half of the caps.

Finishing—Cleco the welded caps to the car and check for side-to-side alignment. The deck-opening gaps should be the same. Do not weld these pieces to the car until they fit *exactly.* Now is the time to make adjustments. The caps can be realigned, if necessary, while Clecoed to the car.

Weld the caps to the body and the upper and lower panels when the fit is right. Grind all welds smooth and clean. Use your hand to check for smoothness. This is where the metal man turns it over to the painter. Take a bow. You're done.

FENDER FLARES

One of the most popular and dramatic exterior add-ons is the custom fender flare. They were first used in racing to cover big racing tires. Fender flares can also be very attractive and they often add an exclusive look to a car. Flares are often featured on show cars. The look is racy, muscular and bold. Flares often catch the viewer's eye and imagination. But before you get excited and chop out a wheel well or fender to try fabricating flares yourself, read this section first.

I've built many custom fender flares, both for show and racing. So I can attest to the fact that the necessary metalwork *is not* simple or easy. It can be greatly simplified if you follow certain rules and a set sequence of development. Regardless: Fabricating metal fender flares is *not* a good project for a beginning metalworker. Consequently, if you haven't already done so, start with something easier: a hood scoop or spoiler. Flares should wait until you have developed your skill, confidence and experience. Patience and caution are a must. If you have any reservations, leave this project to a pro.

Design—Because fender flares involve most of the car area, it is very important to begin with a set design—not in your mind, but on paper. This design must include more than the looks of the fenders. It must take into account other aspects: wheel and tire sizes, vehicle ride height, placement of additional lights . . . The aesthetics of the design may involve not only the fenders, but an additional side scoop. You may end up *restyling* the side of the car body. Therefore, this is a big job both to design and build. *The design must come first.*

Simple cardboard pattern cut and taped to fit the top panel and wheel-well strip.

Large metal blank cut out, marked and shaped to produce the desired curves. Panel is tacked and welded to the fender top and wheel-well strip, *not* to the car at this time.

A design begins with sketches. Sketches are then used to make a precise rendering and scale drawing. So work with a skilled automotive designer or stylist to develop a clear picture of what you *think* you want. Your ideas may change afterwards. The rendering should show an accurately proportioned view of the car with the proposed changes. You'll be able to see what the finished project will look like *as* you work.

The Mustang in the accompanying photos was built using an 8 X 10-in. color photo of an artist's rendering for reference. It would've been far easier with a scale drawing to work from. Unfortunately, the deadline required that I fabricate the flares with only the photo as a guide. Fortunately, I found the photo to be very accurate after measuring it with dividers. Only then did I begin to work from it.

Take time to study your design before committing to the build. Measure, look carefully at the body lines, and consider all influencing factors. All these elements must be acceptable before you pick up the first tool.

Caution—Unfortunately, a design may not work. It may overlook body proportions or tire and suspension considerations. Stylists frequently take liberties that can't be worked on an actual car. I find this happens when a designer disregards the style of a car. The basic style of a car must not be ignored. It is the foundation on which you must build.

I have had designs that didn't include a car's major styling feature. One designer ignored an existing shape which ran the length of the car. Even though his design looked super, it was impossible to reproduce on the vehicle.

Another time I got a drawing which looked fine on paper. I checked the proportions of the car to the drawing with dividers. Much to my surprise, the designer had conveniently left out about 4 in. of body height. Again, this made a fine drawing, but it was totally useless as a construction guide.

These examples illustrate that it is extremely important that you always check the design for accuracy. Only when it is carefully done can you make it work in metal. Even then, it is a challenging project for an experienced person. I don't mean to dampen your enthusiasm. I mean to help you avoid the bear traps. Stick in there. I'll be the guide.

Although not of my choosing, the Capri in the photo was modified without a scale drawing *or* a rendering! It happened because the owner had a rendering that wouldn't work and he was short on time. So the owner trusted my judgment and I did it differently. I started out by having him describe exactly what he wanted. I then sketched repeatedly until we found something he liked and *I knew would work*. We then agreed that I would use my own discretion to build something close to the final sketch. It wasn't easy, but it worked.

I suggest that you don't attempt the

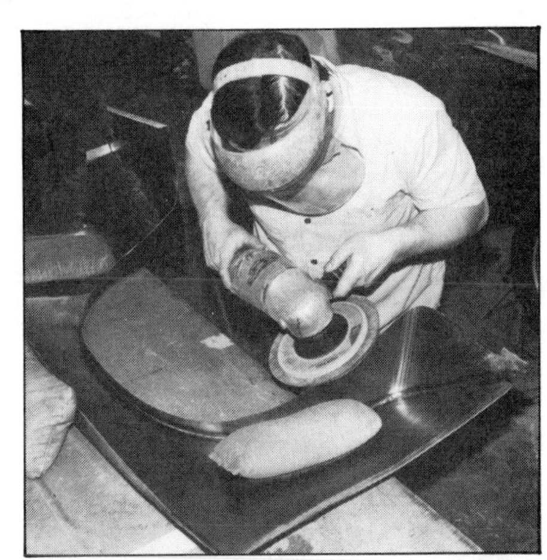

Grinding sharp edges smooth. Metal-finish all welded seams. Sand bags hold fender firmly without clamping and marring the metal surface.

After metal-finishing is done, fender flare is welded to the car. Although it could be welded with heliarc or gas, I'm using a body-shop spot-welder to attach the rear edge of this fender. It made mounting the fender much easier.

Rear wheel-well strip is the beginning of this rear fender-flare project. Note partition parallels top of existing rocker panel. Note also how flare extends forward over door.

Larger tires meant providing tire clearance. Mark wheel-well opening according to how high a tire can go when it is past its highest possible jounce, or bump, position, then trim it.

Weld the wheelhouse to the outer fender. This is particularly important with a unitized-body car such as this. It uses the wheelhouse and quarter-panel tie-in for strength. I used oxyacetylene to do the welding because the paint and undercoating made heliarc welding impossible.

Cut the outer fender. Stretch the wheel house up to meet the outer fender. I'm using an oxyacetylene torch and body hammer to do this.

method I just described unless you have a strong background in metalwork. It is difficult for the professional and virtually impossible for the amateur.

Patterns—Begin developing cardboard patterns for the flares once you have an established design. Because the panels are big, you'll need a lot of cardboard and masking tape. You also need a lot of time and patience.

When developing these patterns, remember they are flat. Compound curves won't be formed until you get to the metal pieces. This means material must be added to some panels to allow for forming the compound curves. So you must use your imagination and compensate for this when fitting the templates.

One set of patterns can be used for both sides of the car. I generally start with one full set of metal parts for one side, set them up and modify the parts as necessary. I then make adjustments in the patterns to make the metal parts for the opposite side. Doing this saves time and mistakes by learning from the first side. The first set of metal parts tests the patterns.

Tires and Ride—There are two important things to consider before you make patterns: You must know the sizes of the tires and wheels to be used on the finished vehicle. All the wheels and tires move up and down, and the fronts steer. So keep in mind that the new body work must allow clearance for these tires and their movement.

You must also take into account any change in the car's ride height. A car that is lowered usually looks better if the fender openings are raised a corresponding amount. Also consider tire clearance. Lowering the car may cause the tires to hit the original body work as the wheels travel up. So consider tire clearance as a necessary element in your build. See the photo of checking tire-to-body clearance.

It's best to make a tire-clearance check after first removing the suspension spring, if possible. With the

Flip-flopping the pattern for one wheel well produces a blank for the opposite side. Right and left pieces are mirror images.

Before I installed the fender flare, I constructed this rear-brake air duct. Doing the duct afterwards would've been considerably tougher. No, you're not seeing things: This is the driver's side of the car, not the passenger's. Photo sequence jumps from side to side because that's the way flares are constructed—not one, then the other.

wheel and tire installed, it's easy to run the suspension all the way up against its *bump stop.* Turn the wheels full right and left and check clearances. If the tire doesn't hit the existing fender, you're "home free." Otherwise, you'll have to clearance the fender similar to that shown on the facing page.

Wheel Wells—Determine how much wider the fender will be at the vertical center line of the wheel. Start developing a wheel-well-opening pattern with this measurement. It will be a lateral projection of the wheel opening. The pattern will usually be fairly narrow at the top—perhaps 3 in. It will be much wider where it finishes at the bottom—both in front of and behind the tire.

Using your pattern-making cardboard, cut an 8-in.-wide strip as long as the periphery of the existing fender opening. Add material for the front and rear where the flare will blend into the bodywork. Tape the cardboard to the wheel opening. Put a mark at the top center of the fender opening and a matching one on the pattern.

Draw a line on the pattern where the cardboard meets the existing body. Let's call this line A. This and the center mark will be your references for reattaching the pattern in the exact place. Using your rendering or drawing as reference, draw another line where you want the flare or new outer fender to end. This is line B. It shows where the existing wheel-well opening meets the new fender.

Remove the cardboard. Cut *on* line B. Next: Cut on a line *1-in. outside* line A to make the pattern 1-in. wider than if the cut was made on the line. The 1-in. additional width will become the mounting flange.

Transfer the pattern to metal. Cut out the metal and roll it to the shape of the existing wheel opening. Cleco it in place using the 1-in. mounting flange.

Top and Sides—You're now ready to develop additional patterns. You'll need a pattern for the top and the side. Develop the fender-top pattern next. The pattern should be widest directly above the center of the wheel. It should gradually taper and intersect the body at points indicated by the designer's rendering. Use a scriber to mark these points on the body for easy reference. Letter-code them for quick identification.

The top pattern will form the upper part of the flare as the fender is viewed from the top. The pattern will be wide at the center—maybe 2—4 in.—and blend in to the existing sheet metal at its foremost and rearmost points. Whatever the shape, follow the basic design. Now is not the time to experiment.

Cut a piece of cardboard as long as the planned flare. The pattern shown in the example is as long as the existing fender. Cut the pattern to match the widest point on the proposed flare. This usually coincides with the point-A-to-point-B distance.

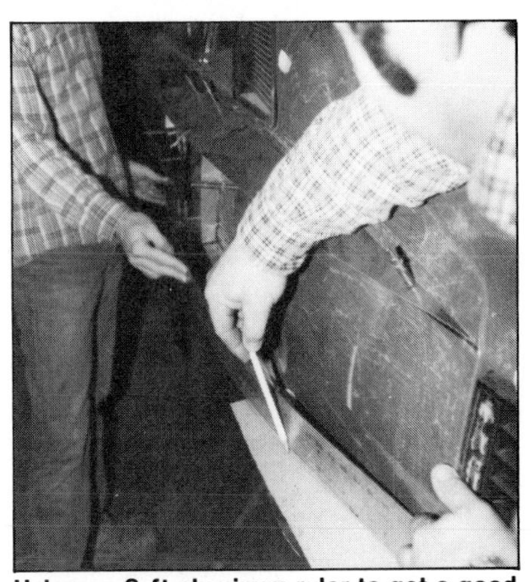

Using an 8-ft aluminum ruler to get a good measurement and a smooth flender-flare contour. I enlisted the help of painter Bob Fehan to help hold the ruler while I took measurements.

Hold the pattern in position against the fender surface. It should lie exactly where the top of the flare will be. Transfer a pencil line from the body onto the pattern, picking up the shape of the fender. This line represents the intersection of the top of the flare with the fender. Cut the cardboard on the line. Carefully tape the cardboard in position to the fender.

Use the rendering or scale drawing to determine the outer edge. You must freehand this line. Take your time and draw until the line is smooth and it matches the line on the rendering or drawing. When it is as close and as smooth as you can make it, cut it out. Trim if necessary. This will complete the upper pattern.

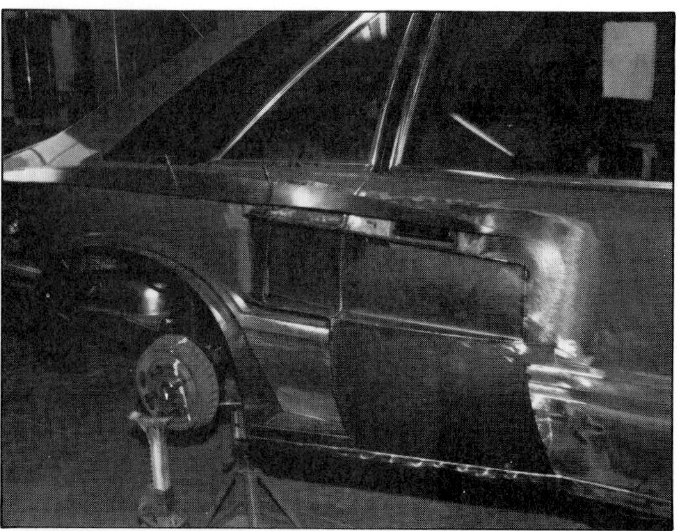

Installation of air scoop was followed by development of lower door section. It attaches directly to scoop and rear door-pillar extension. Shape of door-pillar extension was determined while measuring with the 8-ft ruler.

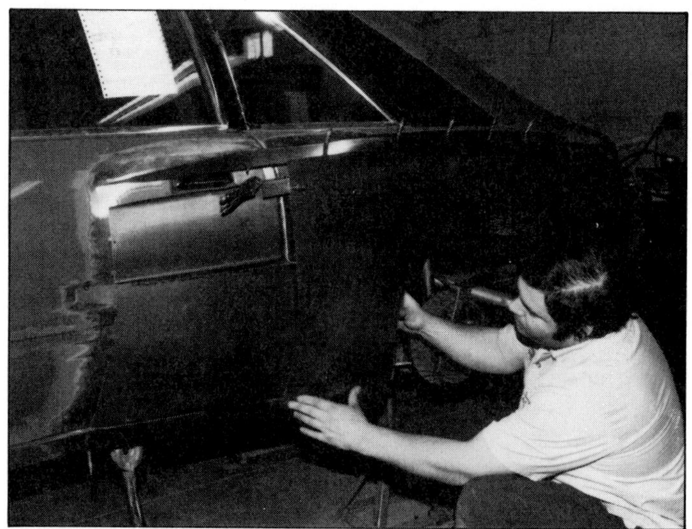

Checking fender-flare-skin shape. A similar outer fender-flare pattern is used for the front to maintain a consistent design. Development sequence is the same.

Constructing air duct was relatively simple. Two large-radius 90° bends in a single blank were made using this "hand-operated" brake: a 1-3/4-in. tube with welded-on angle brackets secured to table with C-clamps.

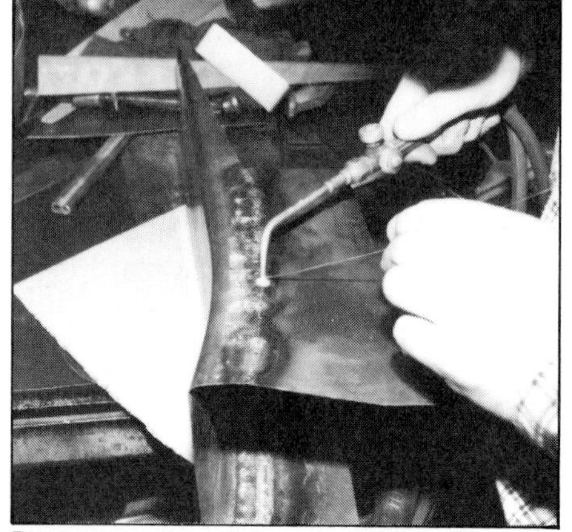

Duct was curved by stretching the two legs. Welding on leg extensions to meet car body. Longer legs could have been put in original blank, but stretching them to make the curve would have been impossible.

Add a 1/2-in. flange to the inside (car body) edge of the pattern if the mating surface of the fender runs close to vertical. A turned-up flange will make it easier to attach the top panel to the fender. If this surface "lays down," you can Cleco the panel directly to the fender without a flange.

Go to Metal—You must have firm, steady surfaces to work to if you expect to make accurate patterns. By making patterns, then metal parts and temporarily attaching them as you go, the patterns and resulting parts will fit better. *Fit is the name of the game.*

Now that you have the pattern for the upper panel, transfer it to metal *immediately*. If shaping is required, do it before Clecoing it in place. Continue developing patterns as you make and attach each metal piece.

Side Pattern—The pattern for the fender-flare outer surface is made from a large piece of cardboard. It is developed by laying the cardboard against the edges of the metal upper panel and extended wheel-well opening.

This pattern may be so large that you'll need to tape two pieces of cardboard together. Hold the big piece of cardboard against the fender and the edges of the two panels. Trace a line following the top panel onto the cardboard. Cut the cardboard along this curved line. Tape the pattern in place against the upper panel so the cardboard hangs down.

Hold the cardboard against the fender and mark the vertical cut lines: the front edge and rear edge of the fender. Tape the pattern in place on the fender. Mark the wheel opening by pressing the cardboard firmly against the edge of the wheel-well extension. The edge of the sheet metal will make an impression in the cardboard. Cut the cardboard by following this impression.

It is important to tape the pattern in place as you cut it. *Draw, cut and tape as you go.* This will guarantee a good fit in the final metal part. Don't let the pattern droop or stretch. Tape it frequently and firmly. The side pattern must be very accurate to ensure that this large panel fits closely.

Cardboard is very flexible, however it can't do the things metal can do. For example, the shape in metal can be creased along a line that follows the fender. Cardboard won't hold such a crease. So mark where you want a crease line on the cardboard so it can be transferred to the metal panel. Similarly, cardboard won't stretch or shrink. Metal will.

Indicate on the pattern *where* you want to stretch or shrink the metal. Write a note on the pattern to indicate where special work must be done to the metal blank. For instance, a flange can be noted. A side scoop opening can be drawn on the pattern and cut out. I also note if metal needs to be left in a certain area, such as that needed to form the inner edge of a scoop opening.

Unlike some patterns, this one is more than an outline of the part. It is a map and a set of directions. It is worthwhile to take time to make a super pattern. Because of its importance, I'll say it again: It pays to devote considerable time and thought to making a pattern. The result will be

Fender is welded to top strip after flanges are bent and final fitting has been double-checked. Note how door handle will be enclosed by fender flare. There's an access hole cut in the scoop.

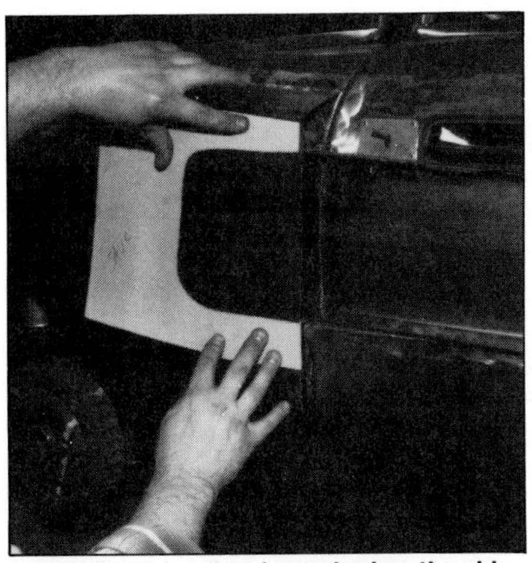

Cutout for the scoop is marked on the skin. Pattern is marked SAVE **because it can be reused for the other side of the car.**

Flanging fender 90° to meet outer wall of scoop is done with a curved-face hammer. Result is a nice rounded edge at the air-scoop opening.

a better metal part. This is particularly important when a major part of the car is involved. Sometimes a pattern is simple and quick. More often it is gradually developed and involves great care. *Don't rush the pattern.*

Carefully remove the side pattern when you are satisfied that the resulting metal part will work and will accurately represent your design.

Side Panel Into Metal—Mark the metal blank for the fender-side panel using the pattern as a guide. Use a soft-lead pencil—*never a scriber*—where stretching, shrinking or special work is required. Cut out the blank and shape it using the rendering as your guide. Check the panel frequently by holding it against the car. Your eyes and hands will guide the shaping of the metal. Learn to trust your senses.

Form compound curves using whatever shaping tools you have: dollies and shot bags, a power hammer, English wheel or Eckold Kraftformer will all work. Each has its advantages. Work to get a smooth, wrinkle-free surface. It pays to be patient.

Welding—The ideal way to weld a large metal panel is with heliarc. As opposed to oxyacetylene welding, a heliarc weld's concentrated heat causes little distortion. Small fender flares can be welded with oxyacetylene, but more metal-finshing is required at the joints. In either case, avoid concentrating too much heat in one place. Numerous tack-welds will help limit distortion by creating a path for conducting heat.

After you've finished forming the side panel and it fits closely to the two others, tack-weld the three pieces together to make them a single piece. *Do not weld the pieces to the car yet.* After tack-welding, final weld the pieces while the "add-on fender" is Clecoed to the car. This will let you remove the part quickly so any warpage from welding can be bumped out. You can get to the *inside* of a part easily with it off of the car and on the workbench. The result will be a higher-quality final part. Little or no filler should be required. Metal-finsh at the bench whenever you can.

Now is a good time to fabricate the parts for the opposite fender. Any problems that were encountered in forming the initial parts can now be compensated for in the patterns. Remember: Parts for the opposite side are exactly the same in the flat, but they must form in the opposite direction.

Finishing—If you're satisfied with the finish of your add-on fenders, Cleco them back onto the car. Weld to the fender with heliarc, if possible.

Fill all Cleco holes with plug welds. You may need some filler material—like plastic or lead—where the fenders join the car. Some areas can't be reached to bump them out from behind. Grind everything very smooth. Send your car to the painter. You're done.

The Capri project involved more than just front fender flares. It also included construction of rear fender flares with side air scoops. The result was a unique body design. It had a racy stance and a smooth look. The rear flares were constructed similar to the fronts. However, they included functional air scoops for rear-brake cooling ducts. This kind of fender flare is a project that requires even more experience.

Final product in primer. Soft in appearance, the fender flare looks wider than it really is. Door-handle access hole can be seen in scoop.

Final Word—Regardless of the add-on part, it must have smooth clean lines that conform to and blend with the body it is attached to. The success of a design rests on its simplicity. It must be similar to the original vehicle. Add-on parts can't be abrupt—they must have continuity with the car body. Following this rule will keep a part you're adding on from looking like it's been slapped on. Keep working to make it look subtle and natural. The raciest, most dramatic effects are a combination of many small gradual changes. It is a mistake to try an overblown effect. It will end up looking like purple eyeshadow on a 10-year-old girl.

ROLL BARS & ROLL CAGES

Indycar roll bar hoops, such as the one on Rick Mear's PC-19, must extend at least 4 in. above driver's helmet. Hoop is manufactured of 4130 chrome-moly steel, bolted to carbon fiber tub and supported with steel gussets. Hoop and gussets are sheathed with a molded carbon fiber cover to improve aerodynamics. Photo by Michael Lutfy.

The main component of a *roll bar* is a steel tube bent in a U-shape. The bar, or *hoop,* is positioned directly behind the driver so it projects above his head. The bar, attached to the car's structure by welding, bolting or riveting, is usually braced and reinforced in some manner.

An expansion on the roll-bar concept is the *roll cage.* It incorporates an additional hoop positioned forward of the driver. The *front* and rear hoops are braced and tied together at the top, front, back and sides, usually with tubes of the same diameter and wall thickness. This encloses the driver in a "cage," thus the term.

All race cars need a roll bar or cage. Each type of race car requires its own special type. And the looks of a custom car may be improved with the addition of a roll bar or cage.

The primary function of a roll bar or cage is simple: to protect the driver from injury in an accident. However, protection isn't the *only* function. You may be surprised to learn a roll bar, particularly the *full roll cage,* can increase a car's performance. It does this by increasing the car's rigidity by reinforcing its structure over its entire length. A car will bend and twist significantly less with a correctly designed roll cage. A roll bar adds little, if any, rigidity to a car's structure.

This chapter covers the kinds of roll bars and cages, and the materials used for them. I tell you where and how to mount them. I tell very clearly what *not* to do. A good roll-bar or cage system adds safety, strength and performance without adding clutter.

PURPOSE OF ROLL BARS & CAGES

A roll bar or cage provides reasonable protection for the driver in a dangerous situation. It is the driver's "first line of defense" against injury. A properly designed and built roll bar or cage will prevent a roof or other part of a car from crushing onto and injuring the driver.

Early attempts at constructing roll bars and cages were laughable. The bars provided little protection. They rarely tied into car structures. A roll bar and cage *must* tie into as many places as possible to be effective. Attaching to the main frame at two points is not adequate. You must mount to many other points for maximum strength.

Welding—I hate to think of the bad roll bars and cages I've seen over the years. Some welds looked like bunches of grapes. I've even seen roll bars made from water pipe! Water pipe breaks on impact—as in a crash. Pipe has *no* business in a car. Pipe belongs in plumbing.

Thin-wall electrically welded tubing—like the kind commonly used for exhaust pipes—has been used in roll bars and cages. The idea was to save weight by cheating on the rules—two bad ideas. You cheat only yourself and whoever is in the car during a crash. Thin-wall electrically welded tubing splits or bends easily on impact. *It does not protect the driver.*

A bad welding job or poor tubing can result in roll-bar or cage failure. When one fails, the driver is usually hurt. Following a few sensible rules will greatly reduce the chance of roll-bar or roll-cage failure.

There are two major elements in fabricating a safe and effective roll bar or roll cage: proper design and high-quality welds. These factors are absolutely essential. Each contributes to the total result. Compromise one and you compromise the *integrity* of the roll bar or cage.

No Brazing—*Roll bars and roll cages must not be joined by brazing.* Braze simply is not strong enough. There are three acceptable ways to weld a roll bar or cage: MIG, TIG and arc-welding. Use the services of an expert weldor if you have the slightest doubt about your welding ability. The joints must have high-quality, maximum-penetration welds if they are to survive and protect the driver. A roll bar is worse than useless if it breaks like a pretzel or folds up like a sandwich. For instance, a bar that breaks loose can become a spear on which the driver can be impaled.

Weld Penetration—Weld penetration is crucial. Welds must completely unite the bars. Each weld must be as strong as the bars themselves. Otherwise a joint may break at the weld when the roll bar or cage is impacted. This type of failure can be fatal. So if you're not a proficient weldor, find someone who is. Have him weld your roll bar or cage.

Even though I strongly advise against anyone but an expert using TIG welding, I prefer TIG-welded roll bars. Properly done, TIG welds are

Drag-car chassis had to be stiff to work. I stiffened it by putting *every* bar to the best possible use. Bars on the passenger side haven't been installed.

Road-racing chassis must be light and strong. Efficient use of tubing is the key. Triangulation is used to reinforce cage, frame and important chassis points.

Rear view of the frame and cage shown at the bottom of page 125. Note that rear braces do not extend past the rear suspension shock mounts.

NASCAR Modified roll cage and door bars: main reason stock-car drivers walk away from seemingly unsurvivable crashes. Photo by Tom Monroe.

A roll bar is a serious piece of equipment. By the looks of how the bolt at the headrest collapsed this main hoop, I'd guess the material is exhaust tubing: very dangerous.

Don't build a "throwaway" 2000-lb car if it must weigh 2700 lb. Put the weight in the strength of the frame and cage, not in 700 lb of ballast. Tom Monroe designed this frame to bend behind rear wheels in a crash: Front is designed similarly. Photo by Tom Monroe.

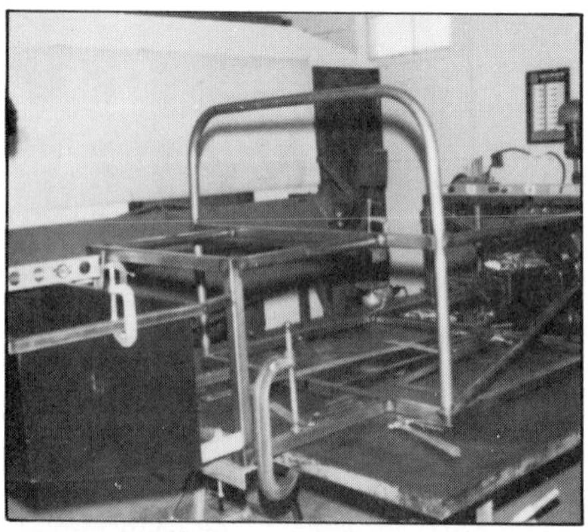

Early construction stage of this road-racing chassis shows beginning of fuel-container support. A good frame and roll cage includes this type of feature.

very strong and reliable. I use this method all of the time. The problem is that most people have trouble getting good penetration when TIG welding: Penetration is essential with a roll bar. So if you insist on doing your own welding—unless you've become an expert TIG weldor—MIG or arc-weld the roll bar or cage.

DESIGN ELEMENTS

Roll-bar and roll-cage design is determined by two major factors: the rules book and the design of your car. The sanctioning body your car will be competing in specifies details of roll-bar or cage construction in its rules book. These are rules your car must comply with: study them. This is the basic structure from which you must *begin.*

After you've determined what the rules require, adapt this to *your* car to make it most effective. For instance, a rule book specifies a *minimum number* of roll-bar or cage-to-frame attachments. The opportunity here is to add attachments or braces to points you think may need to be reinforced. Doing this often makes the difference between a car that handles well and one that doesn't. By using extra attaching points properly, you can minimize chassis twist or bending under high loading. This translates into better handling as the car's suspension works as intended. The frame is prevented from acting as a spring or *anti-roll* bar.

A good example of how a chassis structure affects a car's handling is one that occurs too frequently. Imagine an oval-track or road racer that needs more stiffening at the front to increase *understeer,* or *push.* The front anti-roll bar is adjusted for more front-end roll stiffness: Nothing happens. Why? The stiffer front suspension simply twists, or deflects the car's structure without adding additional front-suspension roll stiffness. The result: no appreciable change in handling. Properly designed braces that tie into the front roll-cage hoop would have avoided the problem. Read HPBook's *How to Make Your Car Handle* to learn about chassis stiffness and car handling in general.

Uniting a car's frame and/or body with a roll bar, particularly a full roll cage, results in a very strong vehicle. Professional builders strive for this. The idea is to build a roll cage so it results in the strongest, safest, most rigid car possible without adding undue weight. The driver will be protected and the car will go faster. That's the name of the game.

Plan carefully before you begin constructing any roll bar or cage. Although the general design of a bar or cage is usually based on a set of racing rules, you may not be concerned with any rules if you are building a custom car or street rod. It may be looks and/or your own safety you are interested in. The basic design of a roll bar or cage can be used to determine some important things: What kind and how much tubing you need and how many bends will be made? It also determines how much welding will be needed.

Measurements—If you came to me and asked me to build a roll cage, I would ask some questions: First, is the car for racing? If so, what class or what sanctioning body? The next thing I'd do would be to study the rules book, looking up "Roll Bars and Roll Cages." Then, whether your car were for racing or not, I'd jump inside it and make some specific measurements: distance from floor to roof; length of the driver's compartment; and width of the driver's compartment. These measurements would give me the basics for starting the design of the cage. I'd have a good general idea of the overall size of the cage needed.

Mounting Points—I'd then look for ideal mounting points. I would also want to have a good seat location and mounting.

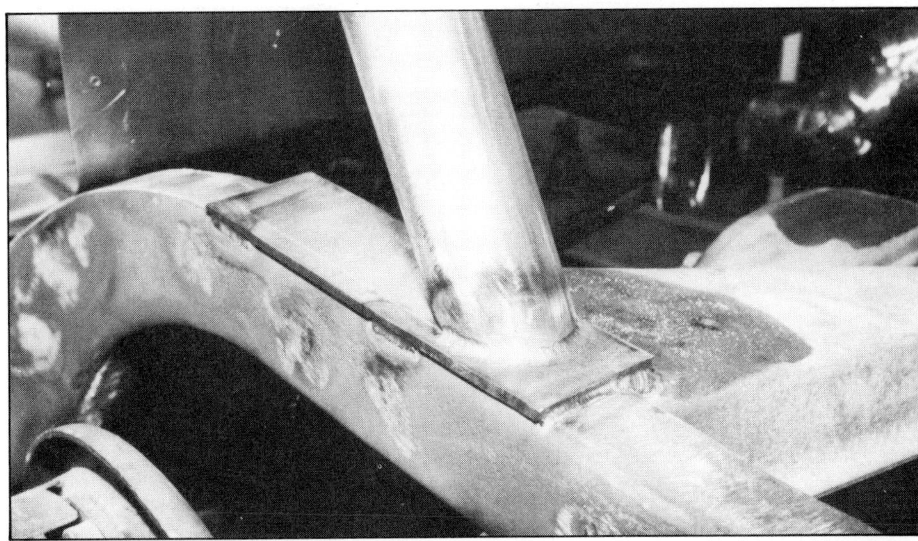

Just as a house is only as sturdy as its foundation, the same is true of roll bars. Mounts are critical. Mounting plate distributes bar-to-frame loads. Mount was completely welded before chassis was painted.

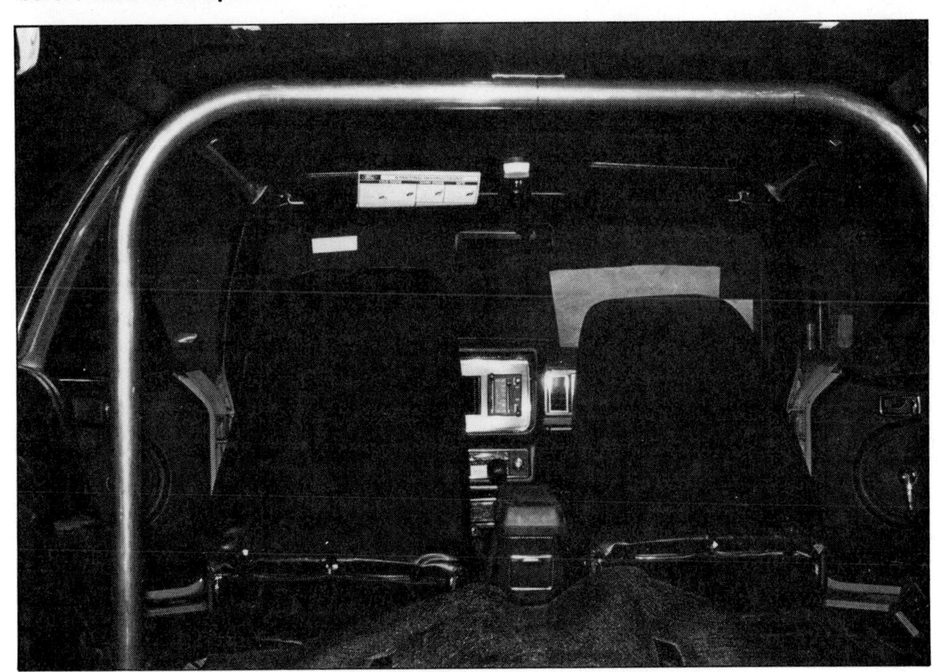

Main hoop comes first. When it's firmly in place, other bars can be fitted.

It's important at this point to consider the driver. His comfort is important and race-car seats are not easily adjustable once installed. I avoid crowding the driver with the bars. The driver must also reach the controls with ease. The idea is to put the driver where *he wants to be.* Some drivers want to be farther from the steering wheel than others. Some like more leg room than others. *You must consider the person when designing a roll bar or cage and mounting a seat.* Installing seats is covered beginning on page 134.

Determine where you can best attach the bars and braces for strength and simplicity of design. Anyone can make a complex, cluttered roll cage. The pro builder tries for a safe and *efficient* roll cage. An efficient roll cage is strong, takes up minimum interior room and requires the least work and material to install.

To ensure your roll bar or cage will be efficient, sketch several different roll-bar or cage designs. Recheck the rules. Remeasure and settle for the design you think best. Remember: It *must* be safe and it *should* be efficient.

Main Hoop—After measuring and sketching, I'd have a good workable

Roof hoop is held in place by square tube for welding and fitting the side bars. Brace is removed once welding and side-bar fitting is done.

Bar at the top of the windshield is bent down at the ends to raise it above the driver's line of sight.

Door bars are tied into the tube frame. Exceptionally nice-looking, well-built chassis, car is the same as that pictured on page 126 during later construction stage. Photo by Tom Monroe.

Alternative to the roof hoop: front hoop with roof-rail tubes connecting it to the main hoop. Photo at top center shows a third alternative: side bars with tube in between at the top of the windshield. Photo by Tom Monroe.

design. I would know how much material is needed, and how many bends are required. Construction is started with the bending of the main hoop. The main hoop is the U-shaped bar that goes over or behind the driver's head: It is the roll bar or the heart of a roll cage. The main hoop protects the driver's head. It also supports a roll cage's roof and windshield bars or front hoop. Rear support bars run from the main hoop to the rear body or frame rails.

Try to keep the number of bends in the main hoop to a minimum—two, if possible. You may need to use more bends to keep the hoop close to the *B-pillars*—the doors' lock pillars that sometimes support the roof—and the roof. *Fewer bends mean a stronger hoop.*

After bending the main hoop, I position it in the car *exactly* and tack-weld it in place. I can then fit the roof and windshield bars accurately with the main hoop firmly in place. If I were building a roll bar, all I'd need at this point would be to make some braces and fully weld the roll-bar assembly in place.

Roof Hoop—Sometimes I use another U-shaped bar that fits up into the roof. This roof hoop runs forward off the main hoop. It extends along the roof lines up to the windshield. The roof hoop is supported at the front by bars at each corner. These bars follow the windshield *A-pillars,* then drop straight down to the frame.

Side Bars and Front Hoops—A *side bar* is actually two long bars, each bent to fit inside the roof. Each runs forward along the roof rail from the main hoop, angles down the windshield A-pillar, and runs straight down to the frame or body structure. The side bars are connected by a cross bar at the bend at the top of the windshield A-pillar. Keep these bars *close* to the A-pillars so they don't inhibit driver vision.

A *front hoop* is similar to the main hoop: A bar goes up one side, across above the windshield and down the other side. The main difference is the front hoop must be bent back to follow the windshield A-pillar. A *roof-rail* tube ties the front and rear hoops together at each side of the car.

Door Bars—The roll cage may need door bars on one or both sides: The number of bars depends on the rules. Bend and fish-mouth these bars and fit them between the main hoop and windshield-pillar bar or front hoop. It is best to weld these bars in last. You need to be able to get in and out of the car while building a cage. Otherwise it makes getting in and out very awkward.

Now that you have a general idea of what it takes to fabricate a roll cage, let's look at the specifics.

BENDING

Bend *locations,* particularly in hoops and roof bars, are very critical. Bends largely determine fit. So be careful when deciding where to bend a bar. A full-scale drawing of the main hoop is

Slight bend in a roof bar gives driver more head room in each of these cars. Bars tie into common point for maximum strength.

Greenlee tubing bender. Tube and bending die are lined up with bend tangent point (arrow).

a good idea. You should also have a full-scale drawing of the side bars, a roof bar or a front hoop.

There are two popular brands of tubing benders: The Greenlee and the Hossfeld tubing benders both work well. Each costs about \$1,000—\$1,500. Both are hydraulic and are available with a wide selection of tube-bending dies. They can be operated manually or with an electric-powered hydraulic pump. If you will be doing lots of tube bending, I highly recommend the hydraulic pump. The hand pump is fine for occasional bending.

These benders make beautiful, wrinkle-free bends in tubing with 0.090—0.1875-in. wall thickness. Bend quality deteriorates as wall thickness gets thinner. This happens because these benders have two dies, both on the outside of the bend. There is no inner *mandrel,* or die, to keep the wall from collapsing.

Minimum Bend Radius—Bend radius is measured to the inside radius of the bent tube. Bending a tube in too tight a radius will weaken the tube significantly at the bend. Limit bend radius to *no less than three and one-half times tube OD,* or: minimum bend radius = 3-1/2 x OD. For example, if you're bending a 1-3/4-in. OD tube, minimum bend radius = 3-1/2 x 1-3/4 in. = 6-1/8 in. It's OK to make bends an even 6-in. radius.

MATERIALS

Roll-bar or -cage design includes chosing the correct materials. Rules books usually specify tube OD and wall thickness. Some rules books even specify the grade of steel to use.

Greenlee bender has an electric-hydraulic assist. It makes short work of forming bends.

Professional builders very frequently use *cold-drawn seamless steel tubing,* or *seamless mild steel.* This steel tubing is generally strong enough to do the job. It is also available in many OD's and wall thicknesses. Remember: The final roll bar or roll cage will be only as good as its design and workmanship—materials are a major part of the design. Begin with quality tubing of the correct outside diameter and wall thickness.

If you don't have a rules book to use as a guide for choosing roll-bar or cage material, use 1-3/4-in.-OD, 1/8-in.-(0.125 in.) wall, cold-drawn seamless-steel tubing. For a custom car or street rod, consider the *absolute minumum* to be 1-1/2-in., 0.083-in. tubing. Anything smaller on either dimension won't give the occupants any protection.

Fresh out of the bender, a high-quality, wrinkle-free bend.

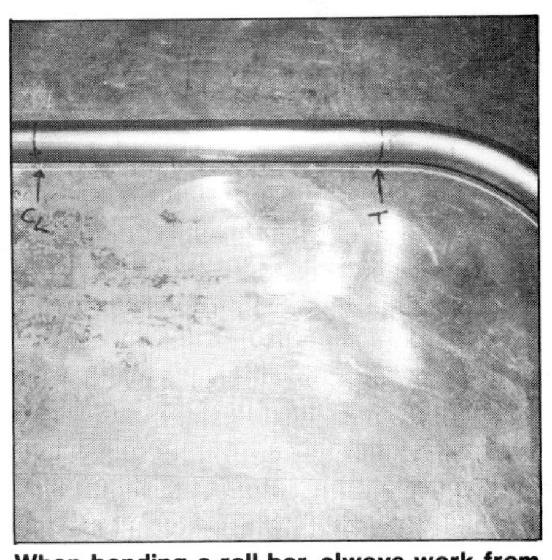

When bending a roll bar, always work from a full-scale layout with a center line: CL indicates center line and T indicates bend tangent. Use a felt-tip marker to mark the center line and bend tangent before each bend is made. Bar is checked to chalk-line layout after bending.

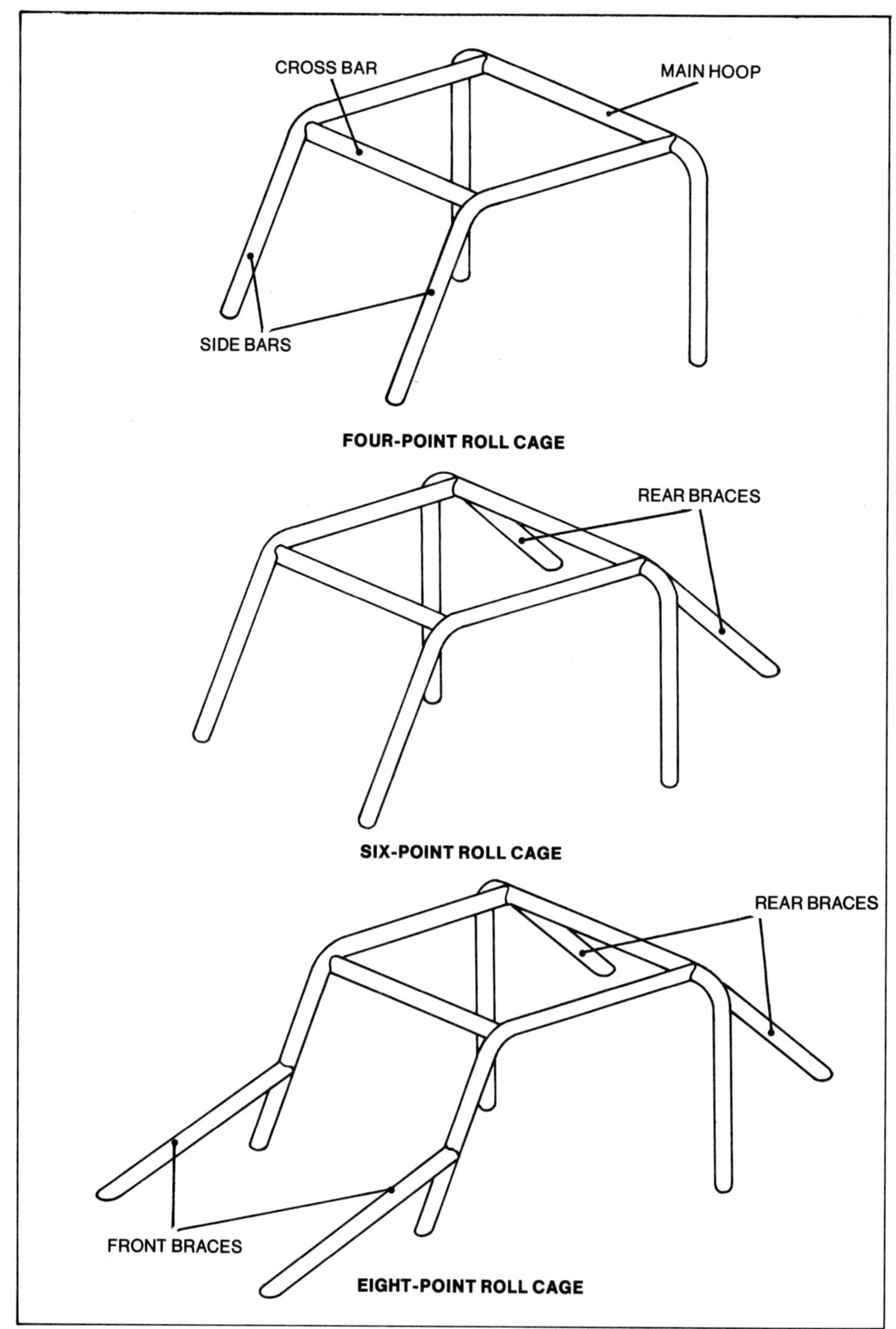

Basic roll cages: four, six and eight point. Additional bars—such as diagonal stiffeners and door bars—are added to stiffen the chassis and comply with rules.

Steel isn't the only material used for roll bars or cages. Stainless-steel and aluminum roll cages are sometimes used on exotic race cars. These materials present a completely new set of problems. The first problem is basic; cost is extremely high. Stainless steel and aluminum *must* be heliarc welded. Then there are rules: They are not allowed in stock cars—who cares about saving weight when the car must weigh 3500 lbs? I've only seen them in road-racing cars with high-buck sponsors. Of course, stainless and aluminum bars have their advantages. They are lightweight and look good.

Then of course there is *chrome moly*. Some builders prefer chrome-moly tubing. It is stronger than mild steel. It saves weight on the car by allowing you to use a thinner-wall material without sacrificing strength—but not *toughness*. Toughness is extremely important when you consider the primary function of a roll cage—to protect the occupants. Turn back to page 50 to read about the properties of

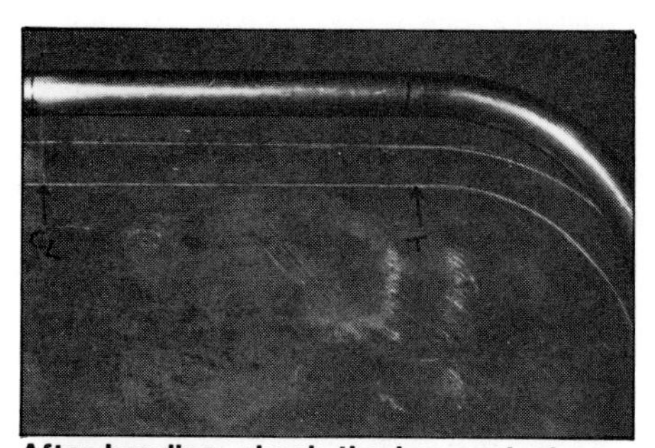

After bending, check the bar against your layout. You may want to add a few degrees bend—it's easier than taking a few degrees out.

chrome-moly steel. The second problem with chrome moly is cost. It easily costs three or four times more than mild steel.

Attachments—Other important aspects of roll-bar design are where and how the bars are attached. Rules books usually specify some basic attachment points. If the car has a true frame—not a unit body—the bars should be attached *directly* to the top of the frame. Naturally, a simple roll bar is easier to attach than a full roll cage simply because of fewer mounting points. A complex roll cage requires even more anchor points and is much harder to install.

ROLL-CAGE TERMS

Roll cages have unique terms. For instance, a *four-point roll cage* has four frame mounting points: Two mounting points are for the main roll-bar hoop; two are forward of the driver, near the front fire wall on opposite sides of the car for the front hoop.

A *six-point* cage is the same, but with two added supports. Each bar runs from the top of the main hoop—usually butting into the side opposite the roof-rail bars—back and down to each rear frame rail. A six-point roll cage adds to the safety and strength of a car. An *eight-point* roll cage takes the six-point cage a step further. It is the six-point cage, plus two more bars. These bars go forward from the front hoop or side hoops, immediately below the windshield, through the front fire wall and tie into the front frame rails.

The four, six and eight-point cages are standard structures. Depending upon the car and class, you may want to add bars. These additions may be door bars or bars you think will add stiffness: diagonals, X-braces and gussets for instance. In most cases, a dash bar is needed for mounting the steering columns and foot controls—if

There's enough head room, but bars in the head and leg areas should be padded. Padding can be found at your local heating-and-air-conditioning supply store. It's used to insulate pipes.

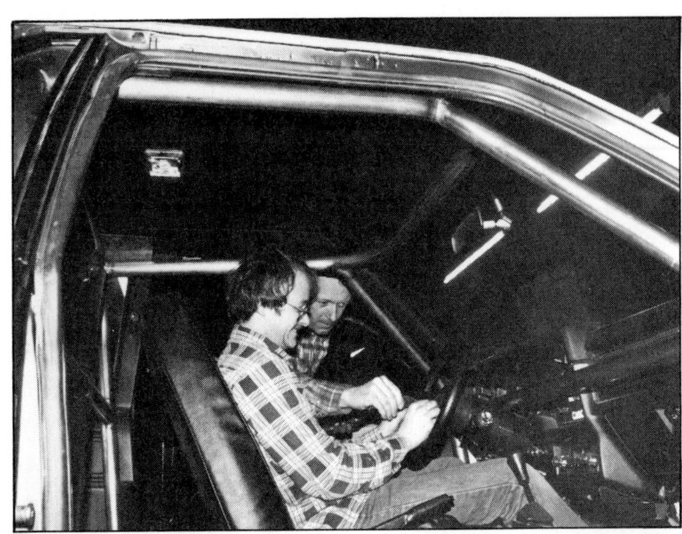

Dash bar—immediately above steering column—is well located. It's high and ties in with A-pillar bars at the bends. It will offer no danger to the driver's knees.

hanging pedals are used. The three cages I've described are basic. You must decide what extra features you want to use.

Precautions—For a roll cage to protect a driver, some things must be kept in mind: Always provide as much clearance as possible between the driver and the bars. A driver moves around during a crash—his body and belts stretch, even with a seat belt harness that is cinched tight. *A bar should never be close to the driver's head.* Clearance is needed to prevent his head—helmet or not—from striking a bar in a crash. The same applies to the rest of his body: legs, arms and shoulders, for instance. The more distance allowed, the safer the cage. *Never crowd the driver with bars intended to protect him.*

Bars that are closest to the driver—main hoop and dash bars—must be covered with thick foam-rubber padding. This padding is available at air conditioning-supply stores or some speed-equipment dealers.

Sharp brackets are dangerous. In an accident, the driver's arms and legs can be tossed around. A sharp bracket will cause serious cuts or injuries if the driver contacts it. There's no reason why any bracket should be a hazard. I refer to sharp interior objects as "stitch makers." Unless you're fond of sutures, avoid or eliminate any sharp edge the driver could possibly hit. *Remember he or she will bounce around upon impact.*

FRAME MOUNTS

Mount the roll bar or cage to the car's frame carefully. The mounting points are the foundation for the structure. Unless these mounts are strong and correctly welded, the whole cage will not have *structural integrity.* Structural integrity means equally strong throughout.

The world's strongest roll cage is less than worthless if it breaks loose from its mountings. So don't build a roll bar or cage with any weak spots. For instance, a bad weld at a mount can be a weak spot. A poor location for a mount can be a weakness. Never let an existing obstruction—such as a section of floor, roof or fire wall—keep you from making a really sound weld on a mount.

Rather than compromise a weld, remove sheet metal to reach a mounting point at the frame. After welding, rivet a small piece of sheet metal around the bar to seal the floor at that point. Do it right or don't do it at all.

UNITIZED-BODY MOUNTS

A *unitized body*—also called *unit body* or *Unibody*—has no separate and distinct frame and body. The frame is combined with the body structure: They are one *unit.*

Mounting a roll bar or cage to a unit body is different than mounting one to a frame. Because there is no true frame, mounting involves tying into the main structure of the unit body.

This main structure runs lengthwise through the car, similar to the frame rails, but with thinner metal. Unit bodies are typically made up of the front rails, center rails—or rocker panels, or boxes—and the rear rails. These three sections are tied together with *torque boxes*—one at the base of

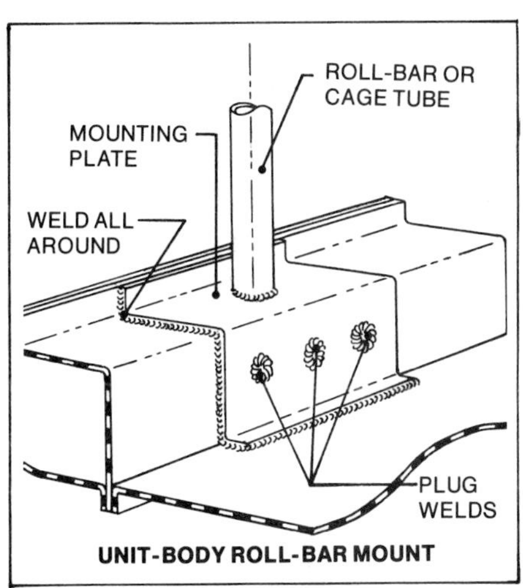

Although unitized bodies are strong, baseplates must be used to spread the loads of a roll bar or cage over a large area. Otherwise, the bar or cage mountings will tear out of the thin body sheet metal.

Drill a small vent hole in each tube before welding it. This lets hot, expanding air escape, preventing it from blowing the weld puddle out onto you and the torch.

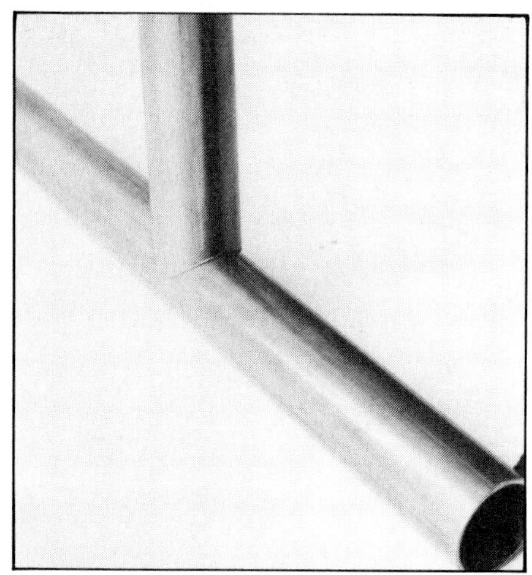

Properly fish-mouthed tube has no gaps when fitted to another. A no-gap fit will ensure the best possible weld, all other things being right.

Save time building a roll cage by getting a tubing notcher. This one is from Williams Low Buck Tools. It speeds fitting by eliminating a lot of grinding and deburring. Photo by Tom Monroe.

Plate welded between roof hoop and inner roof panel adds considerable stiffness.

Corner gussets add strength where it's needed. If a roll cage is going to twist or fail, it will do it at the corners. Photo by Jeff Hibbard.

each A-pillar and one immediately ahead of each rear tire. Members run across the car between these rails in about four locations, depending on the vehicle: at the front of the body, between the front suspension, at the transmission and at the rear of the body. These are the strong areas of a unitized body.

Other strong areas of a unit body include: upper cowl area, spring towers, and front and rear wheel houses.

Some people think unit-body cars are weak—wrong! The unit-body structure is similar to that of an airplane or the monocoque structure of an Indy or Formula-1 car. Lightweight-steel stampings are ribbed and spot-welded together to form a structural *shell.* This arrangement is considerably stronger per pound than a frame/ body setup.

Another incorrect notion is that a roll bar or frame can't be suitably mounted in a unit-body car: You *can* safely attach a roll cage to a unit-body car. But you must do it differently and with more care than if you were mounting it on a separate frame.

Mounting Plates—To install a roll bar or cage in a unit-body car, first locate the strongest points. Check the metal thickness at these points. You will probably find the metal is 1/16—3/32-in. thick. Common sense says a 1/8-in.-wall tube would easily pull out of such a thin base. Steel plates must be used between tubes and the unit body to get more weld area at the thin metal. I use 0.090—0.125-in.-thick mild steel for the mounting plates. It is close to the wall thickness of the tubing, yet it's not so thick that there's much difficulty in welding it to the unit body.

Mounting plates must be big enough to cover a large area of the unit body. This helps prevent the bars or plates from pulling loose at the weld or punching through body sheet metal when loaded. The proper size and shape of a mounting plate depends on two main factors: body shape and metal thickness at the mounting area. However, a plate with about a 25 sq-in. area and a 20-in. periphery should be OK. A 5-in.-square or 4 X 6-in. plate will work for most situations. Mounting plates must be welded 100% around all of the edges to the unit body. *Plug welds* can also be used.

Mounting-plate periphery is important so there will be enough weld to hold the mounting plate to the thin body sheet metal. To get additional mounting-plate-to-body weld without using a larger plate, use plug welds. This will make the mounting-plate-to-unit-body attachment stronger. Drill some 3/8—1/2-in.-diameter holes in the plate similar to that in the photo, page 131. Weld into the holes, filling them and penetrating into the unit-body metal. Now, when the roll bar is welded to the plate, the roll-bar and car structure will have more integrity.

It is impossible to attach a mounting plate too securely. Remember, the steel plates are the foundation for the roll bar or cage. And a roll bar or cage must have strong mountings before it can be strong itself.

GUSSETS

Gussets—triangular pieces of metal welded in the corner of a joint—are one way to reinforce roll bars or cages. The addition of gussets gives extra strength and stiffness to a welded joint. USAC and NASCAR rules books specify 0.090-in.-steel gussets at *all* of the welds on roll bars. Each book details where the gussets should be placed. There are two advantages to gussets. One is generally known, the other is sometimes overlooked.

First, gussets are an insurance policy. Gussets provide back-up strength to a welded joint—particularly one of questionable strength. I believe this is why the rule makers require so many gussets. It is a good idea.

Second, steel gussets stiffen the entire roll-cage structure by strengthening each joint. This strength helps the car resist twist, increasing the car's *torsional rigidity.* This makes a safer and potentially better-handling car.

Extra strength was gained by the addition of the two front braces. V-ing out from the center of the front frame rail, one goes to the front bulkhead and the other to the A-pillar.

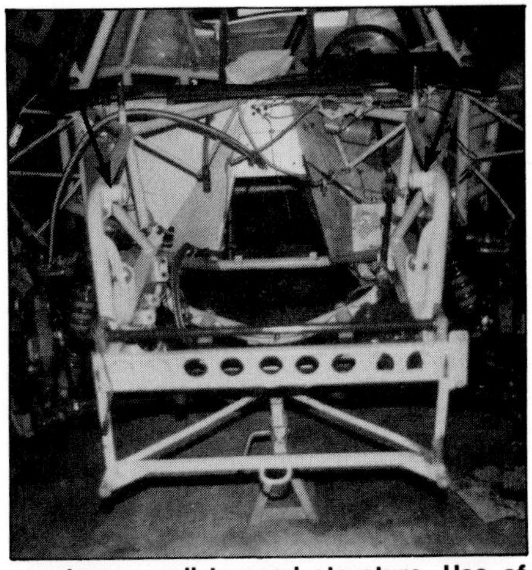

Well-thought-out frame and roll cage creates a strong, well-trussed structure. Use of lightening holes, like those in the front crossmember, is effective. Stubs off the front roll-cage tubes (arrows) are for an engine-bay brace. This is Mo Carter's car being prepared for the 24-hour race at Daytona.

A-pillar was recapped after fire wall was cut away to make room for a foot box. A strong A-pillar provides important support for the roll bar and door.

Because gussets work so well for both safety and torsional rigidity, use them wherever you can. Even though a rule book specifies where they must go, add extra gussets to strengthen any joint you think needs it. And don't forget: A roll bar or cage must protect the driver. Make the cage as bulletproof as you can.

EXTRA STRENGTH

There is a way to gain extra strength and safety with a roll bar or roll cage, whether in a frame or unit-body car. It can make a big difference in the overall effectiveness of the bars.

You already know that a roll bar or cage should be mounted securely at the bottom. You may not know that if you *also* attach the roll bar or cage to the car in other locations, you increase the strength of the roll bar or cage and car a great deal. Not many fabricators know this, so very few do it.

The main hoop usually is placed near the B-pillars. Because the hoop and pillars run parallel, it is often easy to weld them together. You can create a sort of bulkhead by attaching the main hoop to the B-pillars at the tops and bottoms. Use 0.090—0.125-in.-thick plates 3—5-in. long, cut to fit between the bar and the B-pillar. The car will be stronger with very little increase in weight.

Another area where you can tie a roll cage into a car is at the windshield. It can be attached high, above the windshield. Tie in with plates at these points. The cage can also be tied in about halfway down at instrument-panel height. Bars going through either the front or rear fire wall should be welded to the fire wall to pick up extra strength. All of these areas provide free extra strength. Take advantage of them. You will build a very secure, strong vehicle. All you need to do is tie these areas together.

FIRE WALLS AND ROLL BARS

A caution about roll bars and fire walls: *When a bar goes through a fire wall, the gap around the bar must be sealed.* Either weld completely around the bar, or attach a metal plate to seal the fire wall. An open hole in a fire wall is a hazard. Fumes can enter through the hole, or if there's an accident, fire may enter. *Seal all of the openings in a fire wall.*

BOLT-IN ROLL BARS

Some racing-car classes allow only bolt-in roll bars. Showroom Stock, for example, allows only bolt-in roll bars or cages. The important thing here is the same with a welded-in bar—a strong *foundation.* You *must* look carefully for the ideal mounting places for this kind of installation. Be sure before you begin that every mounting area is easily accessible from above and below.

Baseplates—Although you cannot weld the bars, you can secure them effectively. Do this by using extra-large baseplates—somewhere between 6 X 6 and 8 X 8 in. Make two identical plates for each mounting point. Use 0.25-in. or thicker mild steel if there's no weight problem. A plate welded to the bottom of each bar goes inside the car. This plate is then bolted to the floor and to its matching plate mounted under the car. The plates sandwich the metal mounting area between them.

You must be sure the plates fit exactly to the surface over and under which you are working. If you have to bend the plates to make them match the contour, do it. Make the surfaces identical.

It is wise to form the plate and drill the bolt holes *before* any welding is done. This way you avoid having to drill the heavy plate inside the car—impossible or very hard work. It is also extremely awkward. Drill the plates. After you've checked the bar's location, use the plate as a drill guide

Bolt-in roll-bar baseplate should be large and fit the mounting surface. A similar piece is installed underneath to sandwich the wheel house. Sandwiching prevents the attaching fasteners from pulling through the sheet metal.

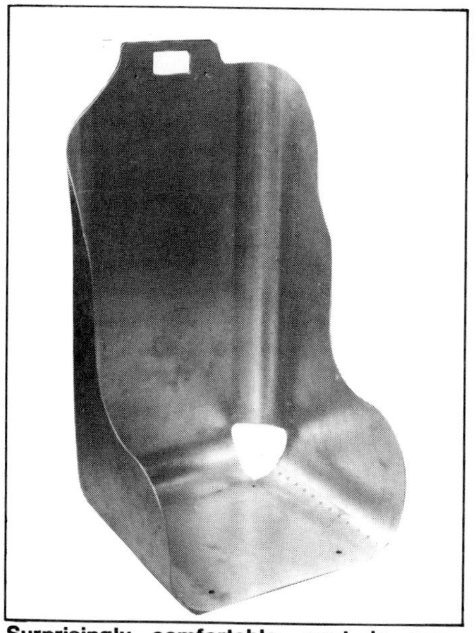

Surprisingly comfortable, seat is constructed from one piece of 0.063-in., 3003-H14 aluminum. Note holes for seat belt and shoulder harness. Photo by Tom Monroe.

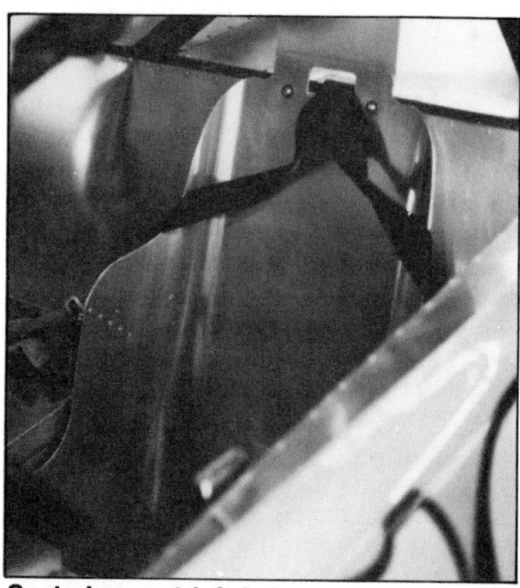

Seat shown at left bolted into car. Seat is designed to give driver good lateral support under high cornering loads. Photo by Tom Monroe.

to drill the mounting area. Following this order saves work and makes a more accurate installation.

Bolts—Each mounting plate should have at least four bolts for strength. Use at least Grade-5 or stronger 3/8-in. bolts. Grip is crucial to the strength of the sandwich-type mounted roll bar.

Otherwise, bolt-in roll bars or cages follow the same basic rules I've previously mentioned. Use quality materials. Use care to determine where to mount them. Design the bars or cage carefully. Don't weld the bars yourself unless you are very proficient. Follow the rules—they were written to protect the driver.

SUMMARY

Whatever type of roll bar or roll cage you are building, remember its primary purpose: It must protect the driver. To do this well, it must be made according to rules. You need quality materials. The mounting points must be strong. Welding has to be *extremely good.* And the design should be as simple and uncluttered as possible.

These requirements will produce a safe, strong roll cage. Such a cage will increase the car's overall strength. Don't skimp on the roll bar or roll cage to save weight. Chances are it will have to pass a technical inspection someday—stick closely to the rules of your sanctioning body. More than tech inspection, the roll cage may have to survive a crash. Survival is often a matter of building the cage *right* the first time. Many a driver has been saved by a well-built roll bar or cage. Drivers with poor cages have rarely offered comments.

SEAT MOUNTING AND BELT ANCHORS

Seat and seat-belt mountings are big safety items. Some rules books make a big deal about them. This makes sense. Other rules books overlook seat mounting completely. I place great importance on seat mounting in particular because I've seen what kind of injuries can result if a seat isn't mounted well.

Seats—Seat belts and a roll bar or cage are not all that's needed to keep a driver safe in an accident. *Where* and *how* the seat is mounted are extremely important. If a seat breaks loose upon impact, the driver loses the protection of the seat belt. He becomes a flying object. The roll bars intended to protect him then become lethal weapons. I know of four drivers who lost their lives because of poorly installed seats—they broke loose from their mountings. Two were stock-car drivers. Two were drag racers. In any kind of racing, seat mounting is equally important. This applies equally to custom cars.

Four things are required for good seat mounting. And a seat-mounting system is more complicated than you might think. It relies first on a strong seat. It also relies on a seat-back brace and head rest. Also, seat-belt anchors must be attached properly and to the right place. The fourth element is seat mounting—a seat must be mounted to the car itself. All of these elements work together to create a good, safe seat. Each one needs careful planning.

Start with a strong seat. If the seat is to be installed in an oval-track or road-race car, it must have good lateral strength and support. It must withstand *high side loading.* Side loading is created by *centrifugal force* while cornering—similar to the force in a string attached to a ball you're spinning over your head. Centrifugal force pushes the driver's body sideways—the faster the cornering speeds, the higher the force. A seat must be comfortable and hold up under these loads and the higher loads that would be encountered in an accident. So consider a good strong seat as a necessary part of the driver's safety equipment.

The seat-back brace has an important job. It supports the top of the seat. Loads placed on a seat top are tremendous and bracing is needed to support these loads. The driver is forced back against the top of the seat back by the shoulder harness. This force requires that the seat-back brace must be fastened securely to a roll bar. Do this by welding a brace to the roll bar. Bolt the seat to the brace. This method will allow you to adjust the seat location without compromising brace strength.

A head rest is needed to protect the driver's head and neck in case of an impact from the rear. The seat-back brace adds significantly to this protection.

Seat Installation—Just as the rules vary in other areas, race-organization rules vary as to how seats should be installed. Although you should not ignore the rules on installation, seats should be mounted as I'll describe whenever possible. It has taken me years of experience to learn what is safe and reliable.

Assuming you satisfied the first requirement by starting with a strong seat, you must determine *exactly* where to mount the seat. The strongest seat in the world is no good if the driver can't operate the controls comfortably. This means you have to set the seat in the car on blocks or whatever and have the driver check for steering-wheel and foot-pedal reach. Move the seat up and down, back and forth and tilt it one way or the other until he is perfectly satisfied. Also, make sure he has enough clearance to interior components such as the roll cage. Measure, make sketches, take pictures, whatever. Don't lose track of the where the seat must be mounted.

The safest and most secure way to mount a seat is to tie it in with the main roll-bar structure. Make a mount for the seat with holes for bolts. Add gussets to the mount to stiffen it. The mount is welded to the main hoop. It's also welded to the floor at the right side. One or two legs go down to the floor. These legs are welded to 0.090-in. steel plates about 4 X 4 in. on top of the floor. The plates are welded to the floor and will distribute impact loads. If all of the other factors are satisfied, this seat should be safe and secure during an accident.

You cannot mount to a roll cage if the car doesn't have one. Then you must mount directly to the floor. To make it work right, you have to reinforce the floor mount area. Find the strongest floor area. Mount the seat directly over this area, or as close as possible to it. Reinforce all of the points where the seat bolts to the floor. Weld 4 X 4-in. mild-steel plates wherever you will bolt in the seat. This is not an ideal way to mount a seat, but it works if done right. Weld the plates carefully and completely.

Seat-Belt Anchors—Seat-belt and shoulder-harness anchors are the base for a restraint system. Unless they are located and installed carefully, the system can fail. Like roll bars, roll cages and seats, each racing body has slightly different requirements for seat-belt anchors. So start by reading the rules carefully. Examine your car. Follow the rules book so you'll have a safe restraint system and one that will pass tech inspection.

Once you know where to put the seat-belt and shoulder-harness anchors, you need to install them correctly. Check the areas. See if they need to be reinforced. For instance, a seat belt anchored directly to a 0.030-in.-thick sheet-metal floor will fail on impact: It will rip out.

Mount a seat belt and shoulder harness to something substantial—roll bars or frame rails, for instance. Don't even think about anchoring a restraint system to a seat. If you *must* mount to the floor, reinforce it. Add metal plates that will distribute the load over a large area. A 6 X 8-in., 0.090-in.-thick mild-steel plate mounted beneath the floor for each anchor will do. Weld all around it. Add plug-welds if you want. Then bolt the anchors through the floor and the plate.

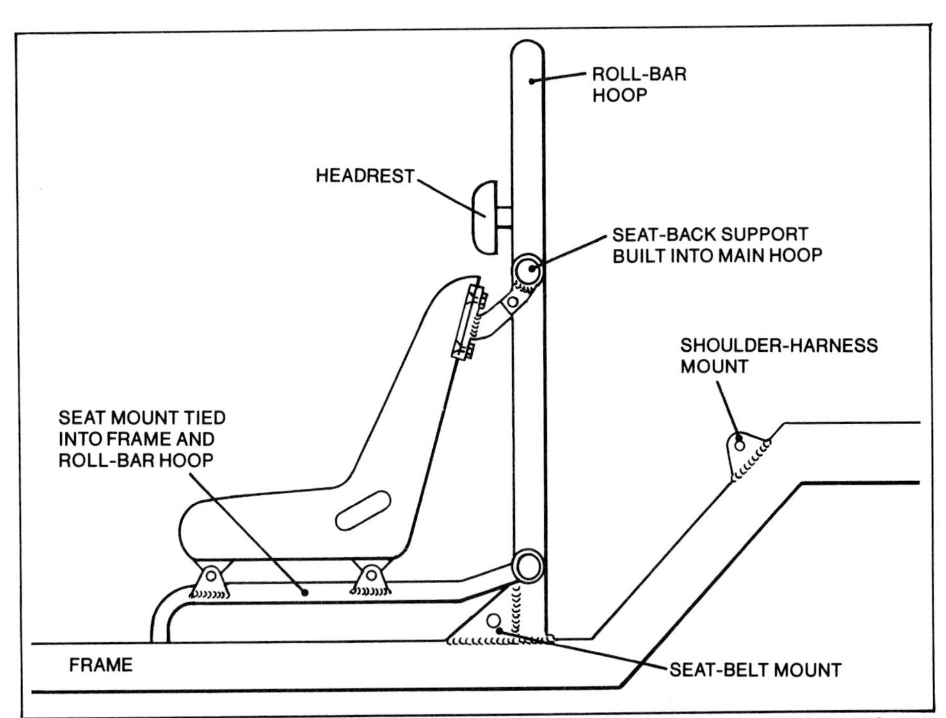

Good seat and restraint mounts are crucial. It's best if they are designed as an integral part of the chassis. Note how seat mount is built into the frame and roll cage. Seat-belt and shoulder-harness mounts are welded to the frame.

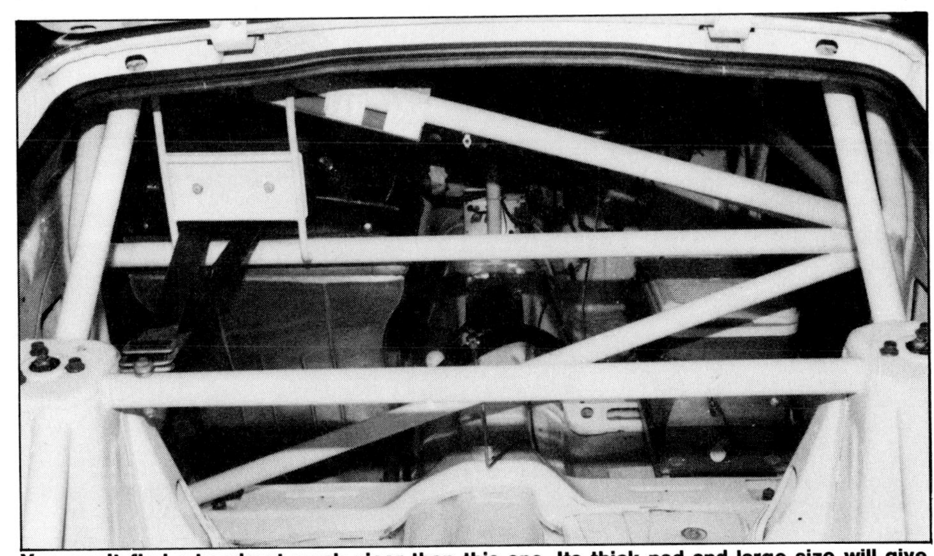

You won't find a headrest much nicer than this one. Its thick pad and large size will give good support. Shoulder harness is anchored to rear spring-tower support tube.

TANKS

Fuel cell, container, pumps and filler necks. This setup is well-designed and constructed. No wonder, it belongs to Bob Leitzinger's IMSA GTU championship-winning Nissan 240SX. Photo by Michael Lutfy.

Three kinds of metal tanks are commonly found in race or custom cars: fuel, coolant and oil. All cars have fuel tanks. Most have coolant, or water tanks. Cars with dry-sump lubrication systems need an oil tank, or reservoir. All of these tanks are fairly good projects to exercise your metalworking skills. Generally they are neither too big, nor too complicated.

Having a tank custom built can be expensive. Save this expense by doing your own fabricating. Follow my suggestions for materials, welding techniques and design features. I explain how to determine the shape of a tank and how to calculate its volume. I also tell where to get the special components you will need for your tanks.

All tanks have some elements which must be carefully considered. Tank location must be carefully planned so it will fit with the car's design. Sometimes the location must also conform to specific sanctioning-body rules. For instance, some organizations do not allow an oil tank in the driver's compartment for safety reasons. The irony here is that another organization specifies that the oil tank must go in the driver's compartment for the same reason—safety! So read the rules book.

Rules may also specify tank materials. Frequently you can choose the materials you think are best. The decision may be swayed by appearance or weight of a given metal. Most custom- or drag-car enthusiasts prefer aluminum tanks because they are light and they look good. Steel is preferred for fuel-cell containers because of its superior strength.

Volume—how much liquid a tank holds—is a prime consideration. A tank, regardless of its purpose, must hold a certain quantity. The fuel tank I recently built for a street rod had to have enough capacity for a 200-mile driving range. Race cars usually have a maximum-capacity limit—22 gallons

At left is a dry-sump oil tank and at right is an overflow/breather tank. Vent line is routed from the oil tank across the fire wall to the overflow tank.

Moroso aluminum oil tank has a removable top for easy cleaning. Aluminum brackets bolted to roll cage and straps make for a secure mount. Photo by Tom Monroe.

is typical. In any case, tank size determines the amount of liquid it can carry.

An oil tank must be large enough to supply adequate oil to the engine. How much oil will be circulating through an engine's lubrication system is basically proportional to engine size. Consequently, a larger engine requires a larger tank. Engines used for long-distance racing—as opposed to drag racing—will also require a larger tank. The dry-sump-system manufacturer will recommend the size of tank to be used with his system.

An oil tank should be designed to hold the required amount of oil, plus *at least* one third of its capacity for air space. This air space prevents the oil from blowing out of the vent.

Many racers use a surge tank in the engine's cooling system. A surge tank helps to bleed off air pockets in the engine water jackets. Such air pockets could prevent proper cooling. A racing engine needs all of the cooling it can get. The cooling system is filled at the surge tank rather than at the radiator.

FUEL TANKS

Aluminum Fuel Tanks—Aluminum fuel tanks are extremely popular because of their light weight and good looks. Aluminum fuel tanks are most popular in street rods, drag cars and racing boats.

Many interesting features can be included in an aluminum fuel tank. Aluminum can be heliarc-welded or gas-welded. It can be anodized clear or with a bright color. You can add beading for strength or appearance. There's almost no end to the variety of possible designs. Use cardboard to experiment. Come up with something special as well as functional. Impress your friends. The time and pride you put into your work will show. Every piece of work is a self-portrait of the person who did it.

The *shape* of a fuel tank is usually determined by where it will be located. *Size* is generally determined by driving-range requirements, rules or the number of runs needed from one fuel fill.

Location—Decide *where* you want the fuel tank. Most tanks are placed under the car, behind the rear axle. This is not always the best place. For instance, it may be hard to route a filler neck through existing body work. Check for this and other problems *before* deciding where to install the tank.

Once you've decided on fuel-tank location, you need to know three things: fuel-tank size, filler-neck location and fuel-outlet location. After you have answered these questions, make a full-size template that describes the area for the tank. This template will be one reference for determining tank volume and shape. Even though all of this space may not

Fuel reserve tank on Bob Leitzinger's GTU 240SX demonstrates fine welding technique and design. Note placement and weld of fittings. Photo by Michael Lutfy.

be occupied by the tank, it is important to see exactly how much space is available for the tank.

Patterns—Develop a tank cross-section template *within* the area described by the template. Trial-fit the template *across* the space where the tank will go. When the template fits, calculate its area in square inches. First measure the sides. Assuming the template cross section is rectangular, area (A) in square inches (sq in.) is calculated by multiplying the height (H in inches) by the width (W in inches): H X W = A. Example: a 10

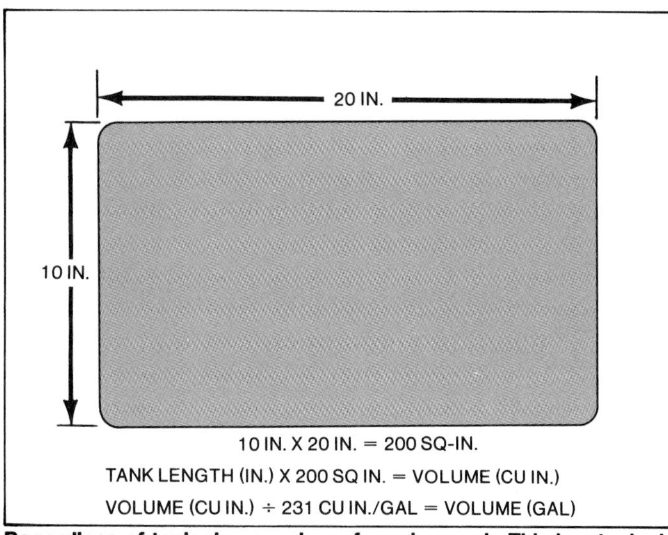

Regardless of tank shape, volume formulas work. This is a typical example. Formulas are on this page. Calculate tank capacity *exactly* using a formula or combination of formulas.

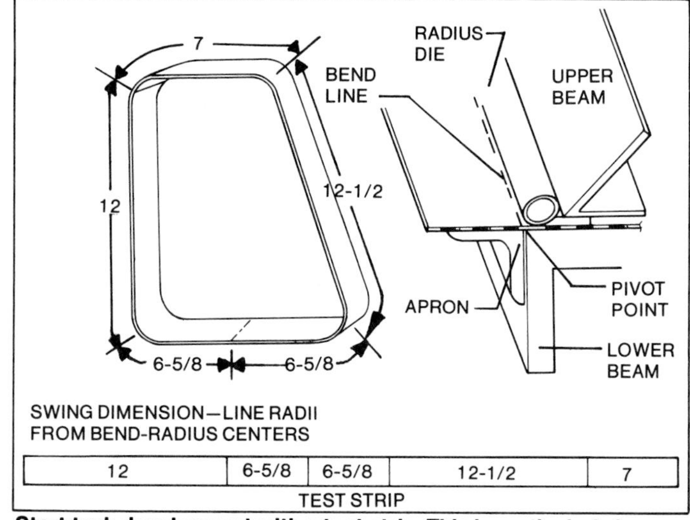

Start tank development with a test strip. This is particularly important if large-radius bends will be used because they reduce blank length. Final blank length and bending sequence are also determined. Lay out strip as shown indicating its center line, tangent points and distances between points.

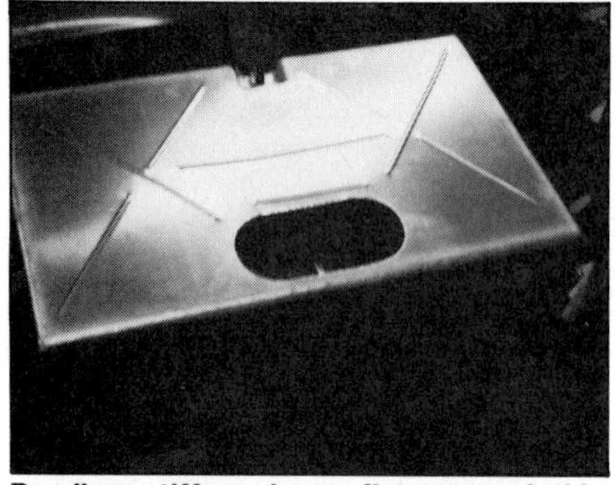

Beading stiffens large flat area of this 0.050-in.-thick aluminum fuel-cell container. Beading must be done before bending.

X 17-in. cross-section template has an area of: A = 10 in. X 17 in. = 170 sq in. If your cross-section template has an irregular shape, break it into a series of simple rectangles and triangles and calculate their different areas. Add or subtract the areas of these different shapes to arrive at the total area of the template. If your tank has a circular cross section, the area of a circle in inches is: $A = 3.14 \times R^2$ where R is the radius of the circle in inches. Write the area *on the template.*

Volume—There are 231 cu in. for every gallon of liquid. Multiply 231 by the number of gallons (G) to get required tank volume (V) in cubic inches. To get tank length, divide your answer by the cross-section-template area A in square inches or: $L = (231 \times G) \div A$.

The volume of a tank with a constant cross section is found by multiplying the area of cross-section template A by tank length L.

In the above example, if a 10-gallon tank is needed, $L = (231 \times 10) \div 170 = 13.6$-in. long.

Test Strip—Now that you have a cross-section template and you know how long the tank must be, make a *test strip.* You'll need a strip of sheet aluminum about 1-in. wide and at least as long as the periphery of the cross-section template. Use the same kind of aluminum for the test strip as you'll use to fabricate the tank. I recommend 3003-H14, 0.063-in.-thick aluminum. The length of the test strip in the example should be (2 X 10 in.) + (2 X 17 in.) = 54 in.

Bending Test—The first test is to bend the test strip the same way you intend to bend the tank. This exercise can save time and material. You'll first learn the order of bending. And if you round the corners of the tank, its peripheral distance will be less than with square corners. So the thing you'll learn is exactly how much material you'll need. With all of the bends made, cut the excess material off one of the overlapping ends so the ends will butt. The strip now represents the length of the metal blank for the tank.

A test strip also helps find bend *tangent points*—the points where each bend begins and ends. Mark those points on the test strip with a pencil or felt-tip pen.

Placing the Strip—After you've bent the test strip, set it into the tank location. Check for fit and clearances. A fuel tank should not rub against any sharp edge. Move the test strip back and forth to make sure the tank will have clearance over its full length. If it doesn't clear, revise the tank cross section and recalculate the capacity. It's better to make changes now than to scrap a tank that you've invested time and material in. Take time to make sure it will fit.

Layout—Once you've worked up a satisfactory test strip, you can proceed with the sheet-metal layout. Measure the test strip to get the blank length for the fuel-tank body. You calculated its width. The cross-section template will serve as a template for making the tank ends.

The *first* metal piece to cut out is the blank for the tank body. It forms the bulk of the tank: front, top, back and bottom. Lay out the blank and cut it out. Transfer the bend tangent points from the test strip to the blank. Remember: *Do not use a scriber.* A bend line might end up being a *tear line.*

If you want to bead the tank, now is the time to do it—before you bend the blank.

Bends—Bend and fold the blank according to your tangent marks. Although wider, the blank should bend the same as the test strip. If it doesn't, something is wrong. Follow the same bending order you found to be right with the test strip.

Recheck the tank profile using the cross-section template. It should be exactly the same. Tack-weld the seam formed where the two ends butt.

If you've changed the bend radius, make the same change to the cross-

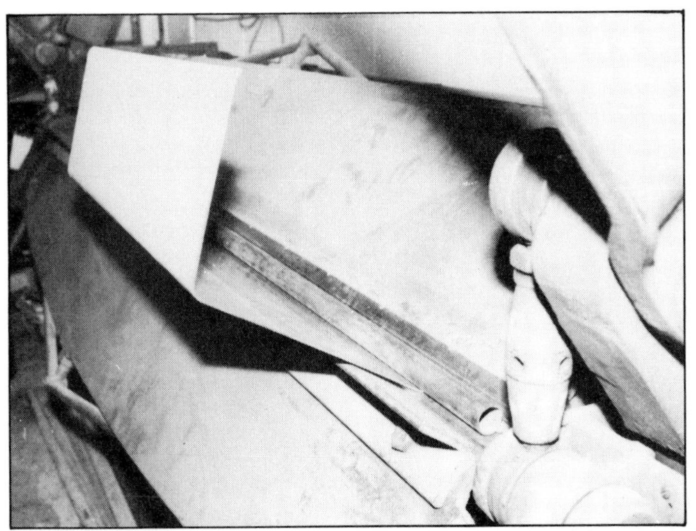

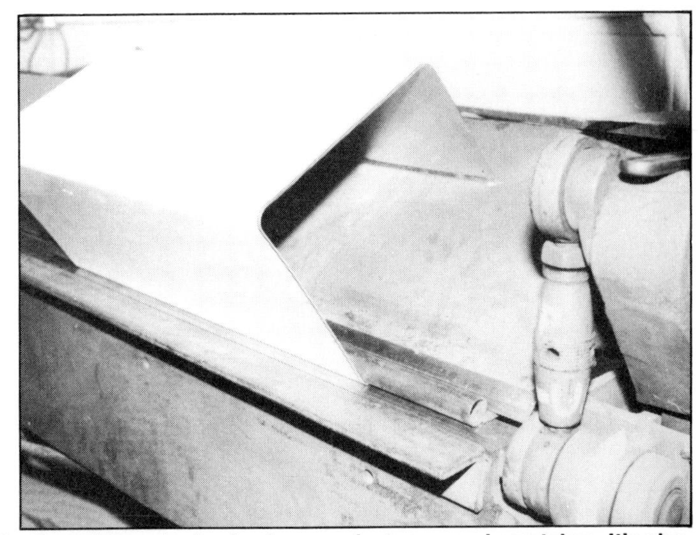

Bending tank with radius dies following sequence found from test strip. Covering protects aluminum against mars and scratches. It's also handy for marking cut and bend lines.

section template. Transfer the shape of the template to metal *twice* and cut out the tank-end blanks. If you want to bead the ends, do this now before you weld them on. Otherwise, forget beading them.

Holes—The fuel tank will need two or three holes. This is best done *before* the ends are welded on. Without the ends, you can easily remove bits of metal that fall into the tank. It's nearly impossible to make holes in a sealed tank without leaving some metal fragments or particles. Debris in a fuel tank will restrict or clog a fuel line and cause engine problems.

Fittings—For the same reason, weld the fittings—such as those for a fuel-gage plate, filler and fuel outlet—to the tank *before* welding the ends on. There are other advantages to leaving the ends off: You can clamp onto the tank-body surface to hold it. And there will be less heat build up, lessening the chance of warping the tank when welding.

Fuel-Gage Sending Unit—If you want a fuel gage in the instrument panel, you'll need a fuel-gage sending unit in the tank. Stewart-Warner has a very reliable—and inexpensive—sending unit, Stewart-Warner part 385-B. This sending unit must be used with their fuel gage for proper calibration. I used this sender in the fuel tank shown on page 140.

Mounting Plate—The Stewart-Warner 385-B sending unit comes with a steel mounting plate: No problem with a steel tank. Because steel cannot be welded to aluminum, you must make an aluminum mounting plate. As you did with the mounting

Radius bends on tank end being made with leather-faced slapper and T-dolly.

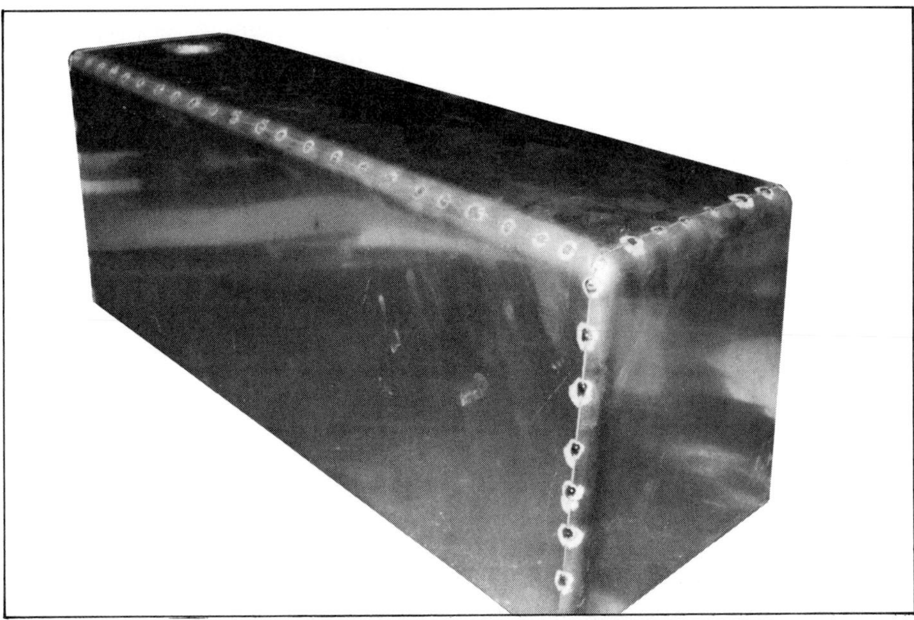

Tank has softer appearance with large radiuses than it would with sharp corners. Note weld seam in center of half bends. Warpage would be a problem if weld seam were on a flat surface. Tank is ready for final welding.

Strong, large-area tank mounts can withstand force of the fuel shifting in the tank during a crash. Short filler neck—about 3-1/2 in.—is securely fastened.

Typical 22-gal steel fuel-cell container. Fuel-cell manufacturers supply containers for their cells. Photo by Tom Monroe.

brackets, use the same alloy for the sending-unit plate as you used for the tank. The steel plate can be used as a template for making the aluminum plate.

The sending-unit plate should be at least 1/4-in. thick for two reasons: The thickness is needed for drilling and tapping mounting-bolt holes. Otherwise, there won't be enough thread engagement and the threads will strip. The thickness also helps prevent warpage as you weld the plate to the tank.

Filler Neck—The tank may need a filler neck rather than a directly mounted cap. If a long filler neck is used—more than 8-in. long—brace between the tank and neck. Such a brace will prevent a failure at the neck-to-tank weld joint.

Welding the Ends—Make the required holes and weld on the fittings. Remove any burrs or slag that may drop off once you seal the tank. Tack-weld the ends onto the tank. It is very important to space tack-welds closely—one about every inch. This will help prevent the tank walls from distorting during final welding. Final-weld the ends to seal the tank.

After the final weld, *the tank must be tested for leaks. Under no circumstance install an untested fuel tank.* Fill the tank with water, cap off all openings and turn the tank in all positions while checking for leaks. Look carefully at all weld seams for leaks.

Mounting—How a fuel tank is mounted is critical. Fuel is heavy. Gasoline weighs about 6.5 lb/gal. So 22 gallons of fuel weighs about 143 lb. This figure is small compared to inertia loads generated under racing conditions.

A fuel tank and its mountings must be strong to be safe. The tank must first be mounted to a structurally sound part of a car, such as a reinforced rear floorpan. This area must support a *full* tank while cornering, accelerating, braking *and* possibly crashing.

Brackets—The brackets welded onto the fuel tank must also be strong enough to withstand these loads. Weld 1/8—3/16-in.-thick-aluminum brackets to your aluminum tank. Use the same alloy for the brackets as you used for the tank: 3003-H14.

Bracket thickness and mounting points aren't the final answers. As you know from previous discussions, loads must be distributed. The brackets must be spread over a large area to keep them from pulling out of the tank. Two 2-in.-wide, 3/16-in.-thick strips of aluminum as long as the ends of the tank should be sufficient. Bend the strips into 1 X 1-in. angles and fully weld them to the tank ends. These brackets should be positioned so they'll fit flush against the mounting surface in the car.

Bolt through the mounting brackets with at least two 3/8-in., Grade-5 or better bolts at each end. Use locking nuts and a large flat washer under the head of each bolt and nut. Bolt strength must match bracket and mount-area strength.

Steel Fuel Tanks—A relatively good-looking steel fuel tank can be made by following the same basic construction rules as with an aluminum tank. There are differences: Material thickness can be less by about 0.050-in. You can also use heliarc or gas welding, and you'll have to paint the exterior of the tank.

FUEL-CELL CONTAINERS

The majority of competition cars *must use fuel cells.* A fuel cell is a rubber bladder made of a very tough space-age synthetic-rubber. It is crash-resistant. A fuel cell is unlikely to rupture even in a severe crash. Otherwise fuel would spill and probably cause a fire.

As strong and well designed as fuel cells are, a metal container—sometimes called a *can*—is needed to hold a cell safely in place. The container also protects the cell from being ripped by jagged metal in a crash, or from tears and abrasions encountered while racing: stones, metal fragments, dirt and dust.

Specifications—Racing rules specify when a fuel cell is needed. When one is required, container material and thickness are usually specified. Rules also specify braces or straps when they are required. The number of bolts securing the fuel-cell-container lid may also be specified. *Follow these rules to the letter—don't bend them at all.* They are written and enforced by specialists who are trying to save lives and make racing safe.

Let's look at some examples: USAC's stock-car division requires that fuel cells that hang below the

Construction of this fuel-cell container is simple. A few folds with extra-long flanges did the job. Flanges are for mounting. Long flange will become part of the floor.

Fuel-cell container mounted in the car. Container is very strong because it's part of the body structure. Rear fire wall is in place just forward of container—a requirement of most race organizations.

Front-to-rear view of container showing fuel cell in place. Beads stiffen floor-mounting flange. Container top bolts on to seal *bladder*, or cell, in place.

Drop-in fuel-cell container does not form a part of the car. Several strong mounting brackets are fastened to a square-tube structure that's built into the car.

frame rails have a 20-gage (0.036-in.) steel container. This container must be *part* of the car, not just an add-on item. In some cases the type of metal trunk floor that extends downward in a U-shape will hold a fuel cell. A metal lid or cover is placed over the fuel cell and bolted at its periphery to seal the container. In other cars, a separate fuel-cell container is bolted to the frame or body structure. *In all cases, a fuel-cell container must be very well constructed of quality materials. It is a major safety component.*

Sometimes rules allow a fuel cell to be mounted at the extreme rear of a car. If this is the location, an additional bar is needed behind the cell for protection in case of a rear-end impact.

Extras—As with a fuel tank, a fuel-cell container can incorporate some special touches. You may bead the

Structure being fabricated from 1 X 1 X 0.063-in. mild-steel tubing will support the ATL fuel cell. A strong structure need not be heavy if properly designed and constructed with well-chosen materials.

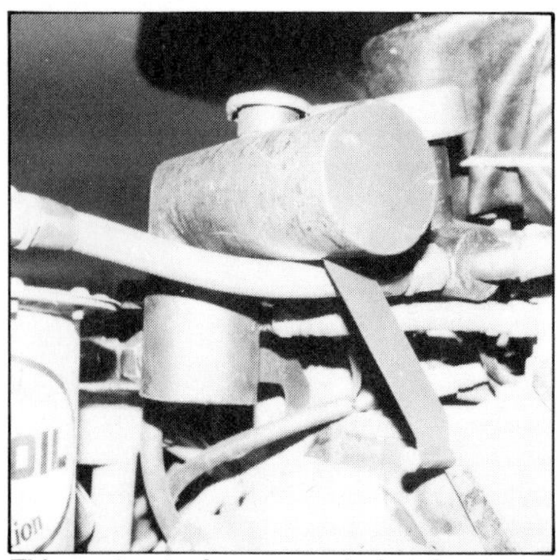

This water tank may work, but I suspect engine vibration will eventually break its mount. Don't hang any tank on an engine.

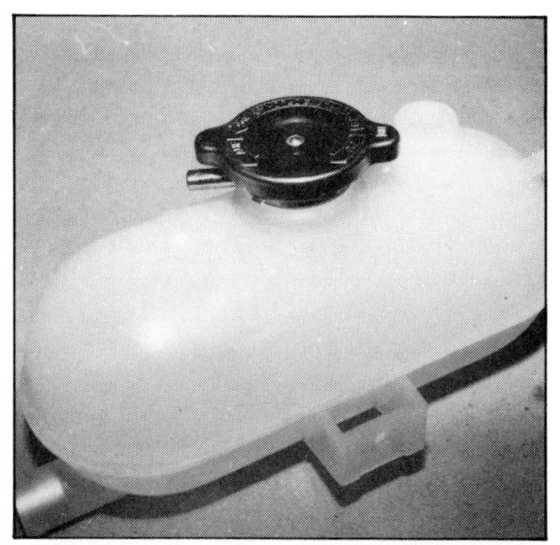

BMW makes this light, sturdy plastic coolant surge tank. It even has built-in mounts. Tank measures about 4-in. diameter X 9-in. long.

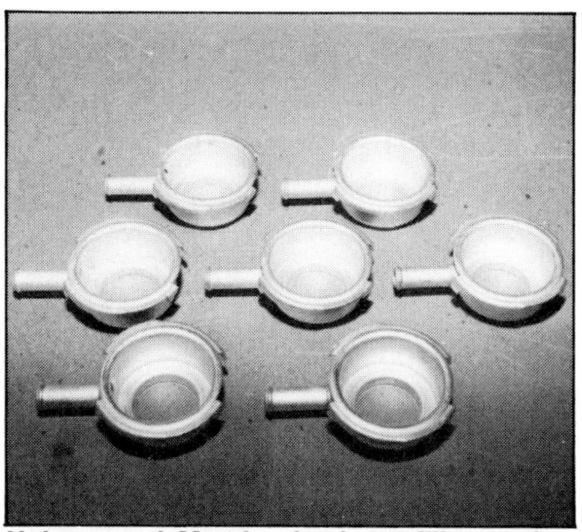

Holman and Moody aluminum filler necks. Buy one of these necks and you won't have to cannibalize an aluminum radiator or tank.

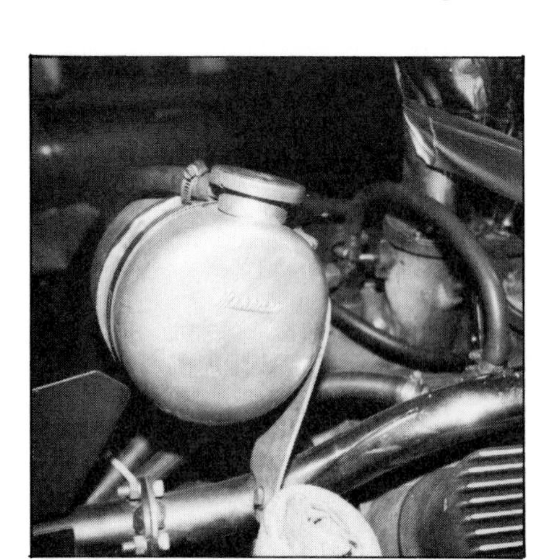

Harrison aluminum surge tank made for the Corvette is not mounted very well. It will probably fail during a race, putting the car out of competition.

Harrison surge tank is well mounted. Tank is retained to saddle mount with screw-type stainless-steel hose clamps. Photo courtesy of Bob Sharp Racing.

container blank before bending and welding it. It can have radius bends to soften the lines. Be artistic.

Aftermarket Containers—Fuel-cell containers are frequently available from fuel-cell manufacturers. These commercially manufactured containers are a good buy for several reasons: They are usually certified or approved by most race organizations. The container is made to accept the specific fuel cell. The proper materials are used in their manufacture. They are painted. Considering the cost of materials and the time required for construction, manufactured fuel-cell containers are a good buy.

SURGE TANKS

Surge tanks do two important jobs: provide a convenient way of filling a cooling system and bleed off air pockets in the cooling system. This will prevent air pockets which can make an engine overheat.

How It Works—To work properly, *a surge tank must be mounted higher than the radiator and the engine.* A cooling system can only be filled to the level of the fill neck.

Location—Unless you can install a surge tank and attach its hoses and can fill the cooling system easily, neither the tank nor its location will work. *Never mount a surge tank directly to an engine.* Engine vibration and the weight of the surge tank and its contents will combine to break the tank's mounting bracket/s. If a surge-tank mount breaks loose, the cooling system fails. Such a failure will put the car out of a race.

Basic Construction—Although size and shape vary, a surge tank is relatively simple. It uses uses two hoses, a filler cap and a mounting bracket. A surge tank can be any shape, just so it has at least a 1-quart capacity. Stainless steel, mild steel or aluminum can be used in its construction.

Surge-Tank Sources—You can also buy surge tanks that are made of these materials or plastic. They work well. Take your choice. I recommend three surge tanks. BMW makes a plastic one. Harrison aluminum surge tanks for Corvettes are very popular. If these don't satisfy you or you have a special shape in mind, you have a third alternative—make your own.

Surge-Tank Rules—Like other custom tanks, building and installing a surge tank has its rules. Start by deciding on a location. It must be above the radiator and engine.

Two hoses are attached to a surge tank. A large hose—about 1-in. diameter—is connected to the bottom of the tank. This hose goes directly to the engine. It's the main source of water from the tank to the cooling system. The second hose is smaller—about 1/4-in. diameter. It goes from the highest point on the radiator tank to a location high on the surge tank. This hose allows the cooling system to be *purged*—bleed air from the cooling system—as the cooling system is filled. The engine block and radiator can be *completely* filled with coolant.

Make a full-size cardboard pattern. Convert this pattern into metal pieces and weld them together to make the tank. *Include the two hose fittings and filler cap.* Weld a filler neck—some are pictured on this page—into the top of the tank. Weld on the mounting bracket/s and you're ready to install the tank.

DRY-SUMP OIL TANKS

Dry-sump oil systems are used in many race cars. This type of engine-lubrication system carries a large volume of oil, allows the engine to be

Although not good for air/oil separation, straight-down oil returns are popular in sprints and modifieds. Tall, round tank could be used to better advantage if return entered tangent to curved tank wall.

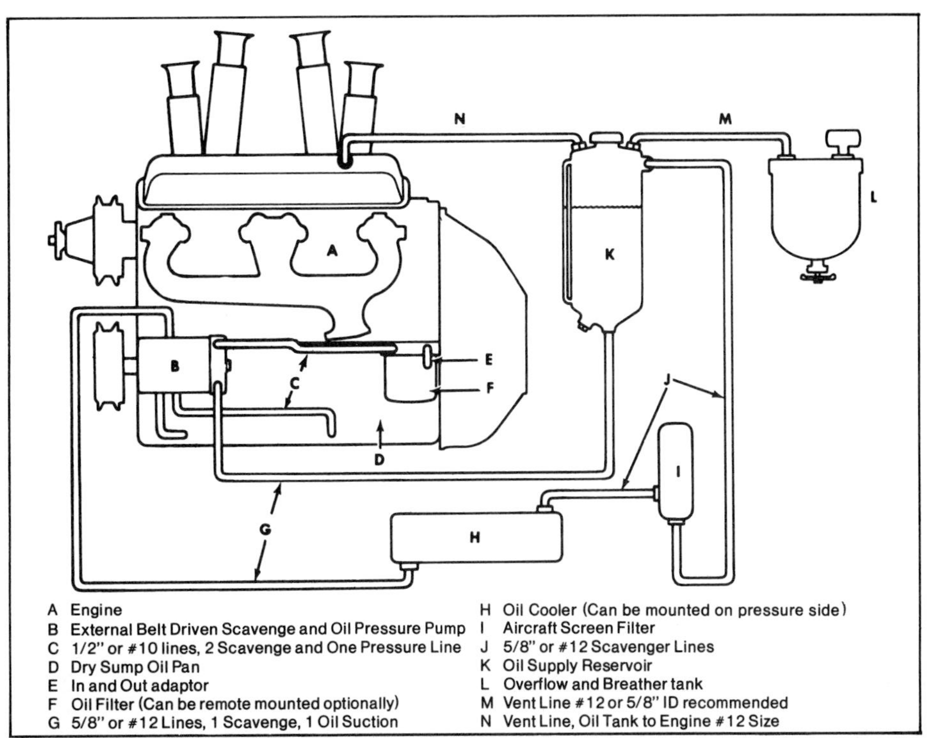

Typical dry-sump lubrication system. Vapor is vented from oil reservoir K **into overflow/breather tank** L**. Drawing courtesy of Chevrolet Product Promotion.**

Oil tank would get better air/oil separation if it were round rather than square. Height of tank is good. Steel straps provide good mounting. Photo by Tom Monroe.

positioned lower in the chassis and supplies a constant source of oil to the oil pump. It must do this under extreme conditions—in high-banked turns at 200 mph or bouncing across the desert at 50 mph.

The name *dry-sump* explains where the oil supply is relative to the engine oil pan. The conventional *wet-sump* oil pan has a deep section that contains the engine-lubricating oil—so the sump in the engine is *wet.* Inside the sump is the engine's oil-pump pickup. The oil pump draws oil from the sump, through the pickup and feeds it to the engine-lubrication system under pressure.

Unfortunately, a wet-sump system does not work well under racing conditions. Even with a modified oil pan, oil will slosh or flow away from the pump pickup under high cornering, accelerating or braking loads. When this happens the pump draws in air instead of oil. No oil, goodby engine! Goodby engine, goodby race—and many bucks, too.

The dry-sump lubrication system eliminates this problem. It has a much-smaller oil pan because it is "dry"—the engine oil pan is not the reservoir for the engine's *pressure pump.* Its main job is to collect oil which is pumped to a remote reservoir, or *tank,* by a separate pump.

The dry-sump system uses two or more pickups to draw, or *scavenge,* oil from the engine. Rather than storing the oil in the oil pan, the main volume of oil is now held in reserve in the dry-sump tank. This tank can have a much greater capacity than the standard oil-pan/wet-sump system. This is possible because the tank design is not dictated by its need to fit to the bottom of the engine, over chassis components and between the engine and the ground. Instead, a dry-sump tank can be designed to perform one function without these restrictions: provide a constant supply of air-free oil to the engine's lubrication system.

Because a greater volume of oil is available from a remote tank, the dry-sump system is much less likely to pick up air. This means better engine lubrication at high engine rpm under all racing conditions.

There are other advantages: The oil remains cleaner and cooler simply because there's more of it. It can be cooled further by adding a cooler. And one little-known fact: An engine with dry-sump lubrication makes more power!

This 0.063-in. aluminum oil tank has two large side-mount brackets. Cone-shaped bottom makes for an efficient oil pickup.

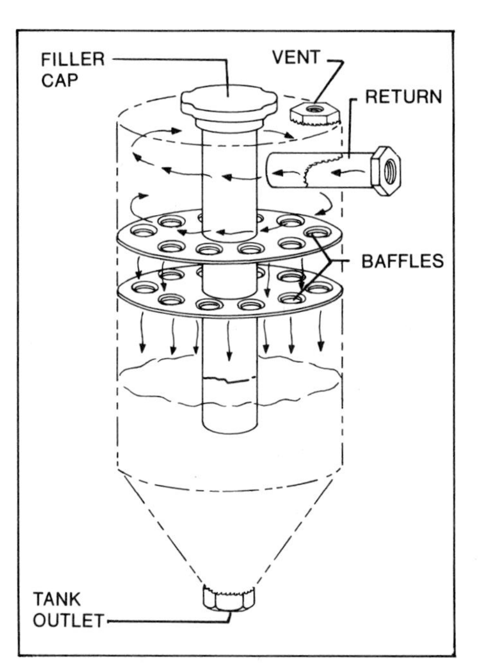

Properly designed oil-reservoir tank: fill tube at the center, two or more baffles, tangent oil return, and cone-shaped bottom. Vent is at the very top of the tank.

Oil tank in this off-road car is securely mounted to passenger-side door bars. Photo by Jeff Hibbard.

The reason is basic: A wet-sump system allows oil to be whipped by the crankshaft, robbing engine power in the process by creating drag on the crank. This is greatly reduced by a dry-sump system because the bulk of the oil is removed from the crankcase. And the dry-sump-scavenging system not only draws oil from the engine, it removes air and engine blowby. This reduces crankcase pressure, further reducing drag on the crankshaft. So dry-sump lubrication not only does a better job of lubricating, it lets the engine deliver more power.

Sources—Dry-sump oil tanks are available commercially. Moon, Moroso and ATL are popular brands. However, why buy one when you can make your own? That's why you bought this book—right?

Besides providing another project you can exercise your metal-working skills on, custom fabricating your own tank can save money. It's also possible that none of the commercially made tanks will fit where you need or want it. If an unusual size or shape is needed, build your own tank. Follow the procedure I describe and you should have no problems.

Material—Oil tanks made of mild steel or aluminum work equally well. Most commonly, oil tanks are made of 3003-H14, 0.063-in. aluminum. I love that stuff! This alloy and thickness is easily formed and welded. It also holds up well under racing conditions. Finally, it's lighter than the frequently used 0.050-in.-thick mild steel.

Oil-Tank Design—Three elements govern the design of an oil tank: location, shape and size. These have to be worked out. But before tackling the design of the tank, let's look at how a dry-sump oil tank works. You'll then be able to decide on its location, shape and size.

There are two basic kinds of oil tanks: round ones and all of the rest. If the intended location is tight, that decides which tank shape will fit best. Shape and how a tank is proportioned are very important. They affect capacity and how well air and vapor bubbles are separated from the oil as it flows back into the tank. *Air separation is crucial to a lubrication system.*

Round Tank—Let's take a moment to look at how a dry-sump tank works. Oil is pumped from the engine to the tank in a great volume. Because oil, blowby vapor and air are pumped in together, the oil is *emulsified,* or loaded with bubbles, similar to a giant milkshake. To remove these bubbles, the returning oil must flow into the tank tangent to the tank wall. The oil then spirals downward against the interior wall of the tank at high velocity in a *centrifuge* action, compressing the bubbles and separating them from the oil. The free vapors and air then rise and are vented through the top of the tank. Bubble-free oil flows to the bottom of the tank and is drawn back back into the engine.

Not-Round Tank—It is important that air and vapor bubbles are separated from the oil, regardless of the shape of the tank. Not that it can't be done, it's just more difficult to accomplish with a cubical or rectangular-shaped tank. The problem with a flat-sided tank is there's no curved surface to centrifuge the oil against. To solve this dilemma, you must install a curved surface *inside* the tank to spin the incoming oil against.

Oil/Air Separator—The term for such a curved inner piece is *oil/air separator.* Logical name, right? This separator must be cylindrical in shape. Its inside diameter (ID) should be at least 6 in. and it should be as tall as you can make it. The oil return *must* be located so the oil enters the tank tangent to the wall or separator, regardless of the tank shape.

Oil-Tank Size?—Contact the manufacturer of your dry-sump system to see what capacity oil tank you should use. The manufacturer will also specify the size lines needed between the pump and the tank. These will determine the fittings you'll use for the return and pickup at the tank. Your engine builder should also be able to help with what size tank should be used. The manufacturer and engine builder tank recommendations should be similar. If there is a disagreement, build the larger tank. A few extra quarts of oil won't hurt anything. Not enough will.

Capacity—An important factor to consider when calculating oil-tank volume is *air space*. An oil tank should not be filled to the top with oil. *You must allow for air space.* How much?

Say, for instance, you need a 4-gallon-capacity oil tank for your system. The oil-tank-volume formula is simple: 2/3 oil, 1/3 air space. Put another way, there must be half as much air space as there is oil. This is a *minimum* requirement for air. Total tank capacity must be at least 6 gallons. If more space is allowed for air, it will be a better tank. More air allows for better *breathing*—or venting—and *separation*. And the tank is less likely to spill oil out the vent with additional air space above the oil.

Each gallon of liquid—oil in this case—requires 231 cu in. of volume. Total oil-tank volume, including air space, is calculated as follows: 1-1/2 X (gallons needed X 231 cu in.) = total capacity in cubic inches. Do your arithmetic. Recheck. Remember that this formula provides for *minimum* air space. So include a few more cubic inches if the space allows. Then begin making templates.

MAKING AN OIL TANK

Here are some pointers on how to make an oil tank. *Tall and skinny* oil tanks work best. Plan your tank to be taller than it is wide, if at all possible. This height allows separation and venting to work best. A vertically mounted cylindrical tank is the most efficient oil tank. A cube-type tank can gain efficiency with the use of an *oil/air separator* that protrudes out of the top of the tank—sort of like a smoke stack. The smoke-stack arrangement is shown in the nearby drawing. This protrusion may also make it easier to mount the oil return.

Use the photos and drawings as a guide to plan and construct your tank. Make a cardboard mockup and check it for fit. Convert the cardboard pieces to metal and form them. Before welding the pieces together, finish reading this chapter. There are more pieces to be made before you're finished with your tank.

Vent and Overflow—Just as an engine must be vented to release crankcase vapors, an oil tank needs venting. Weld a 1-in. diameter tube into the top of the tank. A hose connects to the fitting and runs to a remote *overflow tank*—another tank you must provide.

An oil-overflow tank, or *catch can,* is required by most race-sanctioning bodies. This tank collects oil or moisture that is vented from the oil tank, keeping it off of the track. Overflow capacity should be either 1 qt or 1 gal, depending on the rules.

Wedge-shaped for "aerodynamics." Breather line from valve cover is routed to top of tank.

Before the end went on. Air/oil separator in this non-round tank is completely enclosed inside the tank.

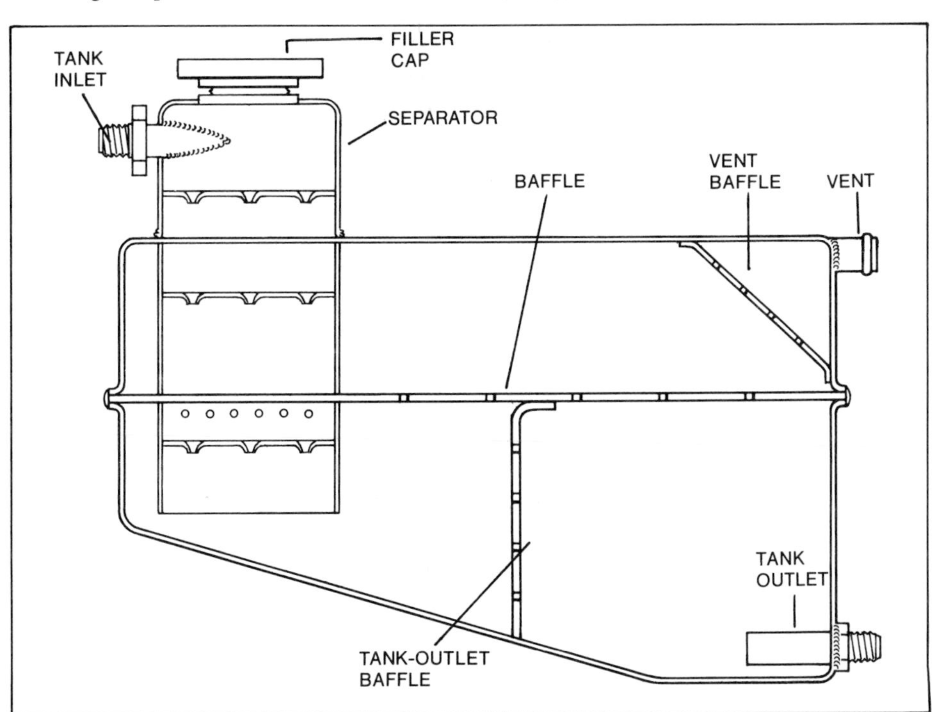

Non-round, low-profile oil tanks require extra work to make them efficient. Smoke-stack-type air/oil separator is at left. A baffle is placed near the vent, another baffle controls oil at the pickup and a horizontal mid-tank baffle separates the breathing-chamber portion of the tank from the oil in the bottom.

Assuming it has sufficient capacity, a plastic windshield-washer-fluid container can be used for an oil-overflow tank. It is transparent, so you can see how much oil is venting from the oil tank. It is inexpensive and readily available.

Oil Pickup—The oil *pickup*—tank outlet—must be located as far away as possible from the *oil return,* or tank inlet. The pickup should be on the very bottom of the tank, and the oil return should be on the top. The pickup and oil return should also be on opposite sides of the tank. This arrangement decreases the chances of the oil pickup drawing in air.

Baffles—Bubble separation improves when baffles are used. A baffle is *another surface* over which the oil must flow, causing more bubbles to

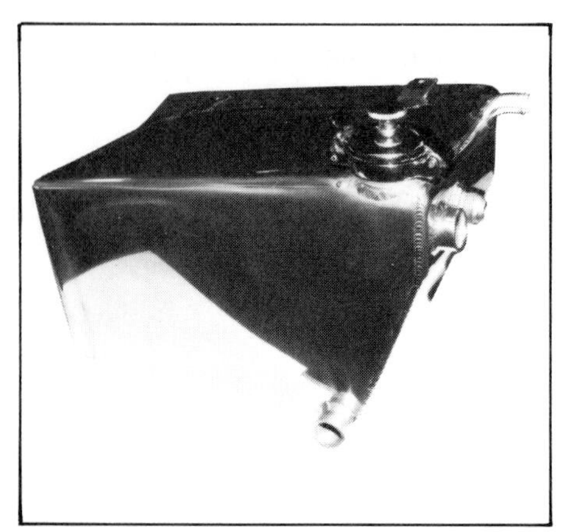

There wasn't space for a tall, protruding separator. Narrow tank bottom channels oil to the pickup.

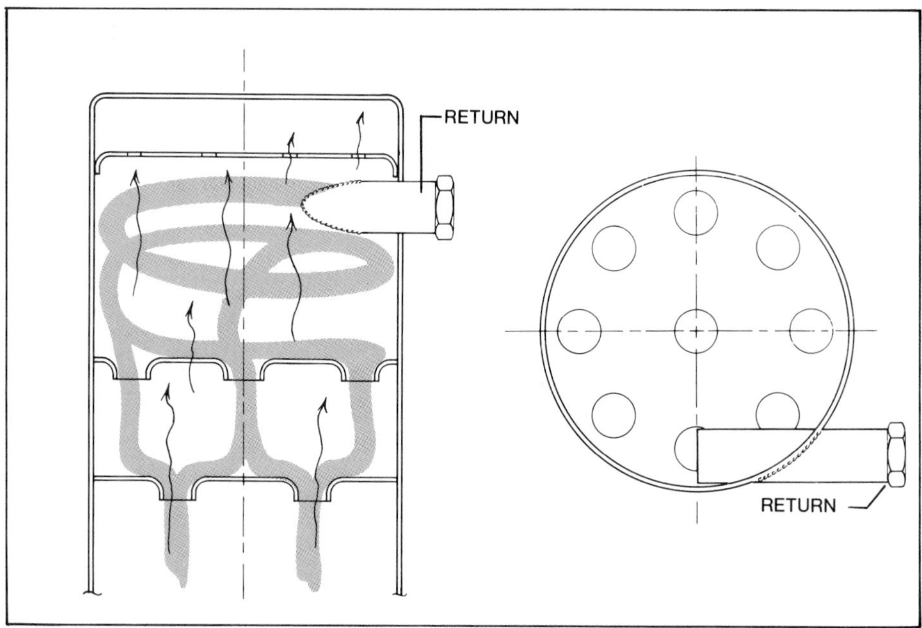

As oil swirls inside the tank and spills over the baffles, air and vapor separate from the oil. Air and vapor rise to the top to be vented as oil flows to the pickup.

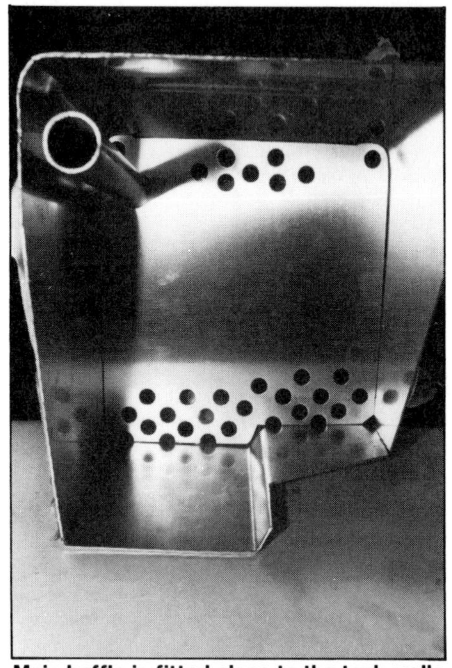

Main baffle is fitted close to the tank walls for welding in place. Holes at the bottom let oil flow to pickup. Holes at the top are for breathing. Long tube is oil return.

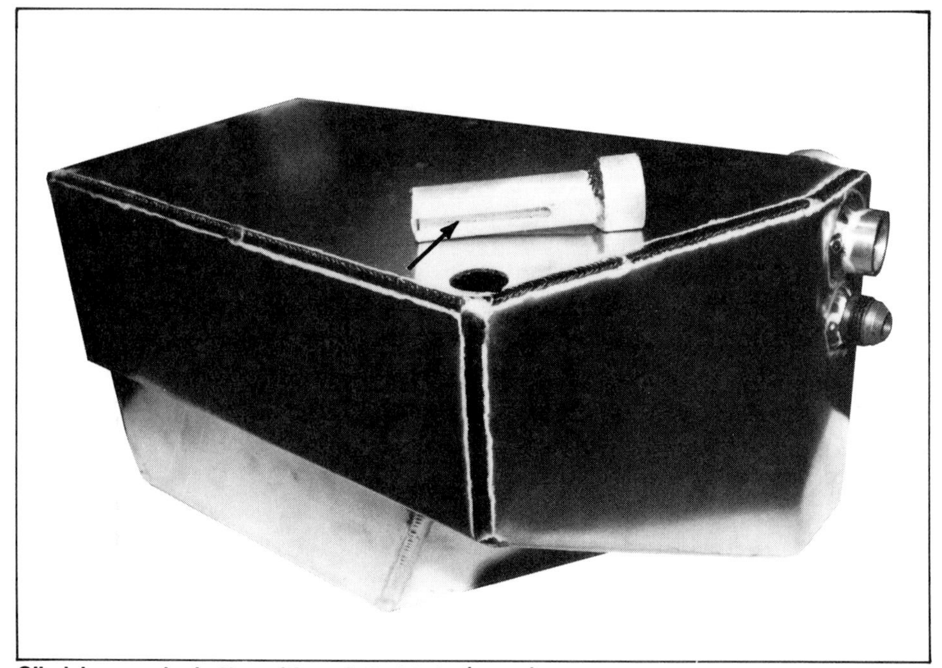

Oil pickup works better with a slotted hole (arrow). Plug tube end with a cap. Oil pickup should be mounted very close to bottom.

burst and release their air and vapor. Each disc-shaped baffle should have many 1/2-in. holes punched in it. These holes should be formed, or *belled,* like those shown.

As oil flows across the baffle and bubbles break, air and vapor is vented out the top. Oil flows down through the holes and to the bottom of the tank. By the time oil spins around the curved tank or separator wall and flows over a couple of baffles, it should be bubble-free. This will increase the life of your engine. Bubble-free oil is pumped to the engine for lubrication.

Additional Baffles—Two additional baffles are needed. One must shield the vent and one shields the tank inlet. The vent baffle should cover the vent opening to keep oil from splashing out. Oil vented from a dry-sump tank can be very dangerous. It may end up on the tires of your car as well as those on other cars, causing an accident. Oil loss won't do an engine any favors either. Close off the vent end of the tank with the baffle, or plate, drilled with at least ten 1/4-in. holes to allow oil flow to the pickup.

If the tank doesn't have a cone-shaped bottom like that pictured on page 144, place another large baffle at the bottom to stop large amounts of oil from sloshing or flowing away from the oil-pump pickup. If this happens, one of the main advantages of the dry-sump system is lost. The oil pump draws in air rather than oil, damaging the engine.

Mount a large baffle across the inside of the oil tank using the drawings as a guide. The oil will then be contained so it cannot slosh completely to the opposite side of the tank, away from the pickup. It will be blocked by the baffle. Even under high cornering loads such as those encountered in banked turns, the oil will flow smoothly from tank to oil pump. Although a mid-tank baffle is absolutely necessary in a low-profile tank, I recommend that you also use one in a tall, skinny, round or rectangular tank.

Mounting—An oil tank must be *securely* mounted. Mounting brackets must survive high forces from vehicle movement and weight shifting as oil moves in the tank. Tank mounting must also endure engine vibrations.

Design the oil-tank brackets to cover a large area of the tank. Secure them with strong welds. Use *double-thickness* brackets. If the oil tank is made of 0.063-in. aluminum, use 0.125-in. aluminum for the brackets. The same rule applies to a steel tank and its brackets.

Fittings—Any hardware needed for the oil tank—such as filler caps, *bungs* and Aeroquip-type, or AN, fittings—can be ordered from Earl's Supply, Russell's, Nelson Dunn or Moon Equipment Company. Your local speed-equipment store may carry some fittings.

A bung is a machined fitting that is designed to be welded to a tank. Oil-tank bungs are threaded to accept the hose fittings which are used in dry-sump systems. Bungs are used for the return and pickup openings.

Final Word About Tanks—Fabricating a tank is not a simple project. Because of the safety and engine-durability implications, care and planning must be used when designing and building any tank. Use your skill and brains. Produce an efficient and good-looking tank. You may save some bucks, you'll definitely sharpen your metal-working skills, and let's hope you pick up a few compliments.

BUILDING AN OIL TANK

Constructing an oil tank using available space most efficiently can be a big problem. This situation is best handled by developing a cardboard pattern and using it for a mock-up to trial fit and calculate tank volume. Placing the cardboard mock-up into the car will show how the actual tank will fit. Make your adjustments now, so you won't be sorry later.

Taping the pieces together to check fit and volume before making the metal blanks.

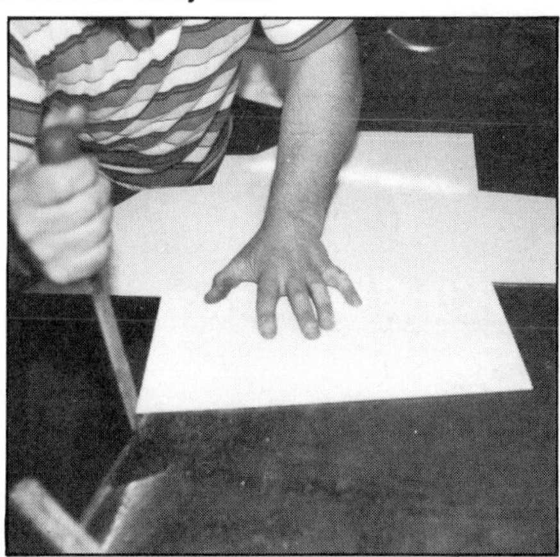

When the pattern is developed, transfer it to metal and cut out the blank. File all edges smooth.

Bend blank *after* filing edges and beading. Round tube welded to square tube makes a simple radius-bend die. Clamp die to a bench with the blank in between to make bends.

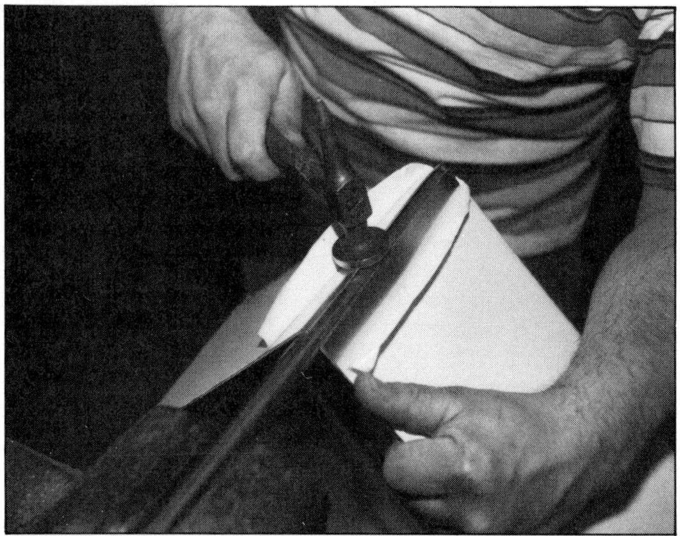

Rounding corners at the weld seams. Tap metal gently over a steel bar or T-dolly until edges touch. You may have to trim some metal along an edge to get the desired radius.

After folding and welding the main oil-tank body, I develop the back-side pattern. Working to metal rather than cardboard gives a more accurate pattern.

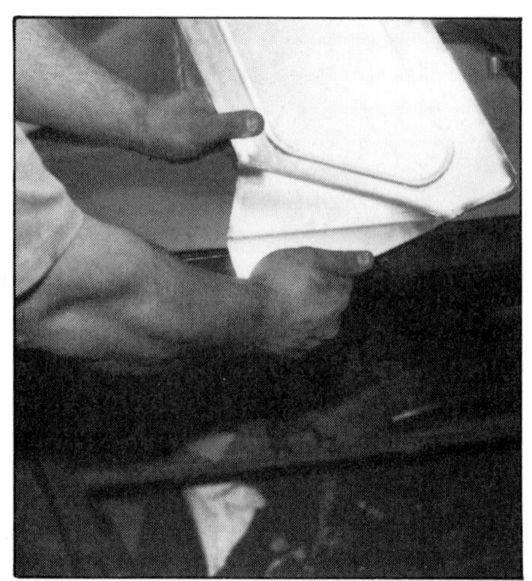

Remember, *bead before bending.* I also radiused the edges with hammer and T-dolly before bending.

A couple more degrees of bend and this part will fit just right. With the protective cover in place, hand work on the tank back can be done without scratching the surface.

Welding the air/oil-separator cylinder. It's about 6-in. diameter X 14-in. tall.

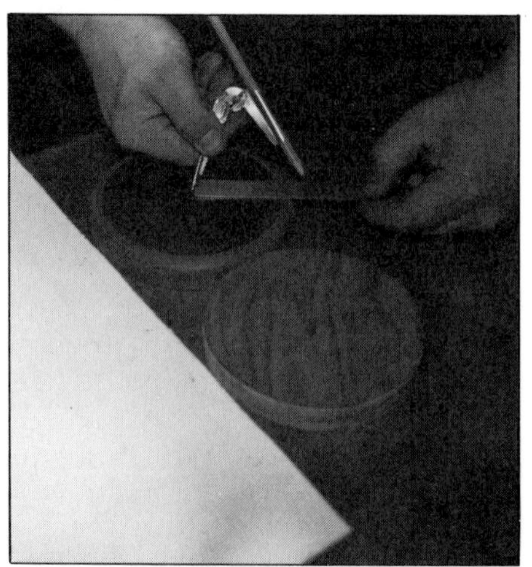

When building a protruding air/oil separator, I cap it with a hammerformed top. Using a hammerform that's equal to separator ID, measure the blank radius. Clamp blank between the base and top and gradually work the edge down over the base.

Using a protective cover leaves an unmarred surface, even after hammerforming.

Hammerformed top welded in place. Large tube at right is the oil return: A -14—read "dash-fourteen"—(14/16-in. hose ID) AN fitting will be welded to it.

File the edge where the air/oil separator fits into the tank. Clean out all filing chips. Masking tape on top of the tank was for marking the separator opening. It also protects the surface.

Air/oil separator was beveled for clearance to the inside of the tank. Large baffle to the right of the separator prevents oil from sloshing away from pickup.

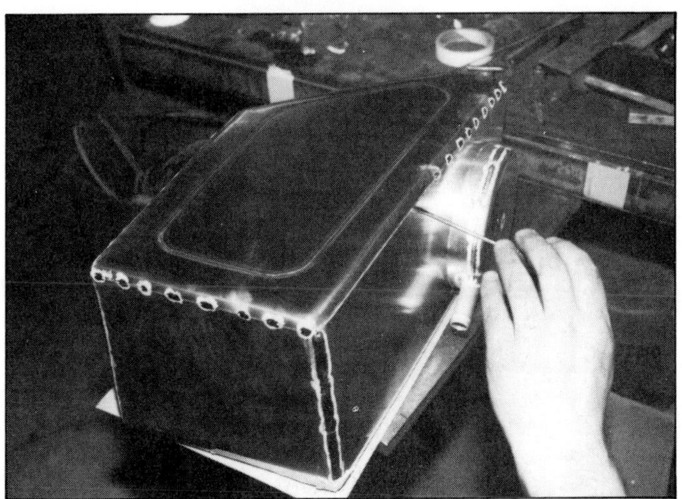

Small screwdriver or knife blade can be used to align edges *exactly* for tack-welding. Take time to make it right. Then finish welding the seams.

After oil tank is 100% tack-welded and tack-welds are filed smooth, final-weld the seams.

Completed oil tank installed in the car. Four mounting brackets secure tank.

EXHAUST HEADERS

Carefully designed and properly built headers play a very important role in engine power. Photo by Michael Lutfy.

Properly designed, custom-built exhaust headers increase the power of an engine. They also look good. Most header manufacturers make headers for specific stock-engine/vehicle combinations. A few manufacturers also make headers for popular engine swaps. So you can usually buy headers if you are equipping a common engine-swap car—a small-block Chevy in a Model-A Ford. If, however, your plans for a street rod involve a rare engine/vehicle combination—a Boss 429 Ford in a Willys—you'll need custom-built headers. If you are building a race car, custom headers are usually a must.

KINDS OF HEADERS

Custom headers can be divided into two general groups: the cosmetic header and the race header. Cosmetic headers are designed primarily for appearance and "easy" installation. A slight power improvement can be expected. How much and at what rpm—who knows? Street vehicles often use this type of header. The cosmetic header is easier to build than a set of race headers because the pipes aren't held to an exact length. Naturally, it would be better if they were close to the same length—it just isn't a live-or-die situation.

Racing exhaust headers are a different kind of custom-built headers. If they look good, OK. The purpose of a racing header is pure function: to improve engine performance. Certain rules *must* be followed closely to achieve this. Tube diameter, tube length and good exhaust flow are of the utmost importance.

Building custom headers is a project requiring *lots of time,* planning, thought and patience. And it can be frustrating because your first attempt may not be right. The dynamometer may be the final judge. With street headers, if they fit and look good, who is going to notice a couple of horsepower? But race headers: What

Stahl headers for a Datsun 240Z are beautiful! Primary-pipe lengths are equal and they flow nicely. Few welds are used. It is obvious that a lot of planning went into them: Planning is a must when building custom headers.

Custom-built headers installed on a big-block Chevy in an East-coast modified. Equal-length primaries are critical for maximum efficiency. The fabricator who built these headers knew what he was doing. Note unusually short secondary pipe.

a racer won't do for one more horse! That's why race headers are more complex and exacting. Therefore, if you've never built headers before, begin with street headers. You'll learn a lot in the process.

If at all possible, begin by building headers for a four-cylinder engine. This will introduce you to the basics of header building without forcing you to dive into a complex exhaust system. Total frustration will set in if you tackle more than you can handle on your first attempt.

Components—Exhaust headers are made up of four basic and easily identifiable components. First comes the *header flange,* or *header plate.* It bolts to the cylinder head. Next are the *primary pipes,* or *tubes.* These pipes, one for each engine cylinder or exhaust port, are welded into the header plate. Each of the primaries enters a *collector.* Unless the header is the *crossover* type, all of the primaries from one engine bank enter the same collector.

The collector does what the name suggests. It collects the exhaust, channeling it from the primaries into the *secondary pipes.* The secondary pipes funnel the exhaust from the car. Some exhaust systems incorporate a *balance tube*—a tube that connects the secondary pipes. Whether or not you should use a balance tube must be determined on an engine dynamometer. This has to relate to the rpm range in which power is required. The balance tube is usually used to "flatten" the power curve.

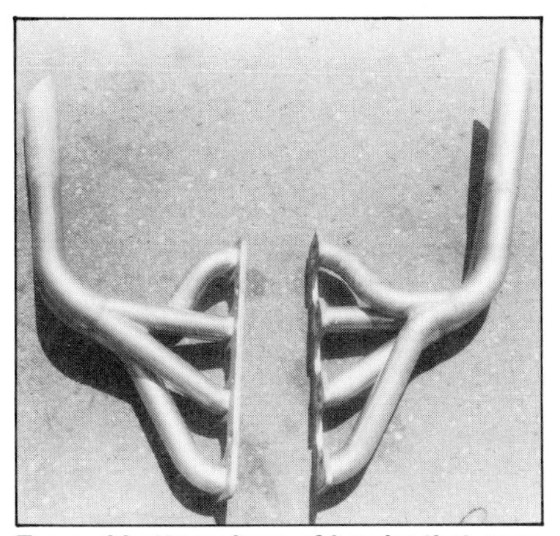

Top and bottom views of header that goes on the right side of a four-cylinder race-boat engine. George DeLorean built this efficient header with a simple design that *works.*

Materials—Use the same materials for custom headers, whether they are for a race car or a street car. If the system is for a turbocharged engine, use stainless steel. Otherwise use *prebent* 90° and 180° mild-steel tubing. See the photo of the series of tube bends. Tube diameter is determined by the requirements of the engine. Prebent tube is available in a wide range of bend radii. Major sources of prebent tube for headers are Cyclone, Hooker, Mercury and Stahl. These suppliers also provide a line of bolts and *header flanges*—what the tubes weld into for attaching the headers to the cylinder head.

HOW TO DO IT

I'll describe how to build race-car headers for an eight-cylinder engine.

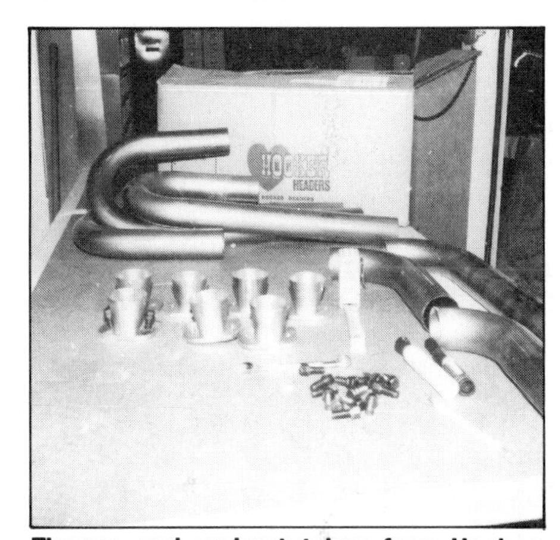

Flanges and prebent tubes from Hooker. Buying these pieces rather than making them yourself saves a lot of work.

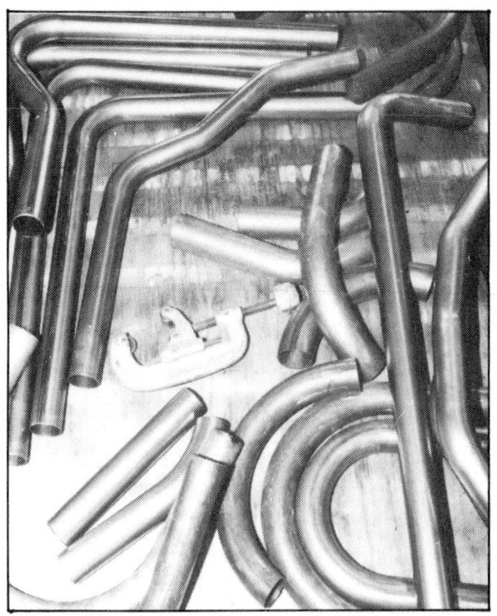

Stahl, like other header manufacturers, has a big market supplying prebent tubes to guys who build their own headers.

Headers on this Pennzoil Chaparral Indy Car present the most difficult requirements: maximum performance, high heat, tight package and high durability. Photo by Dick Johnson.

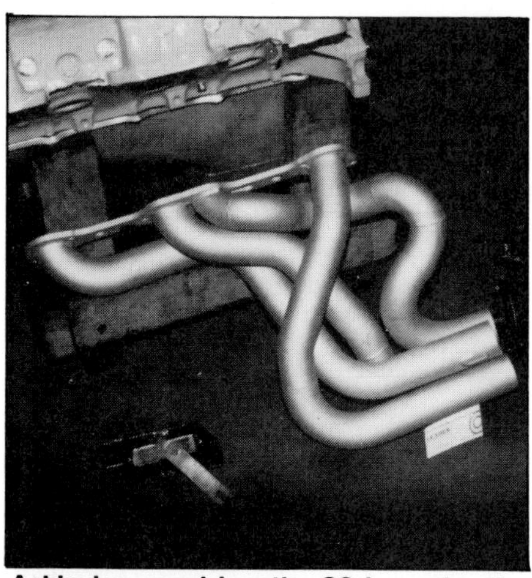

Achieving equal-length—30-in.—primaries with these Pontiac drag-car headers was not easy. Equal length was achieved through careful development. Photo courtesy of George DeLorean.

If yours are for a street car, there will be few differences. Header-to-body and chassis clearances should be the same. The headers should be equally well designed and constructed for maximum exhaust-gas flow. If you understand race-header construction, you will probably be able to build cosmetic headers without a hitch.

Maintaining equal-length primary pipes is less critical with cosmetic headers. Bend radii are also less critical. They can be "tighter" to clear an obstruction. These are the differences that make cosmetic headers more for looks than for horsepower. Power is compromised when a primary-pipe bend radius is reduced. So keep bend radii as large as possible.

Basics—You must follow precise rules when building headers for a race engine. Begin by consulting the engine builder. He should be able to tell you what diameter and exactly how long the primary pipes must be. In addition, he should give you specifications for collector length and diameter, and the length of the secondary pipes. With this information you can go to work.

Building equal-length primaries is difficult. After you've done it, no one will be able to convince you otherwise. Although each pipe must end at the same place—at the collector—each begins several inches apart on the cylinder head. In addition to clearing engine-compartment obstructions, bends are used to compensate for the different beginning points.

The job of building headers is complicated by engine-compartment obstructions—wheel wells or spring towers, steering components and frame rails. The trick is to weave from the cylinder head—through the engine compartment—and end up with the same-length primaries at the collector: a real puzzle! A tape measure and flexible exhaust tubing—available from most exhaust-system suppliers—can be used as "patterns" for getting through the tight areas. More on these later.

CHROMING HEADERS

If you know the headers you are building are to be chromed later, be extra careful about matching tube sections perfectly. And be extra-careful to have excellent penetration at each and every weld. Chromed headers can look great, but they have to be strong enough for the polishing preparation they go through before chroming.

A final note about chrome. If your headers have tight slip fits, chrome will cause a real problem. Chrome adds to the effective thickness of the tubes. What this means is slip-fit joints will end up being interference-fit joints. You won't be able to get your headers together!

COLLECTOR

Determining where the collector will go is the next step. Primary-tube length is the most critical factor in determining collector location. It must allow for chassis, body and ground clearance. Be sure to provide room behind the collector for the *secondary*, or outlet, pipe. The collector *must not* be close to the car's floorpan. Otherwise it will make the interior hot and it could burn the driver's feet. So allow lots of clearance—try for 4 in. or more—between collector and the floor. If you can't get this much clearance, you'll need to add a *heat shield*—a protective baffle— between the floor pan and the collector. Heat shields are discussed on page 159.

Sources—Either make or buy a collector. The companies that produce tube bends—Cyclone, Hooker, Mercury and Stahl—also produce high-quality collectors. Numerous collector sizes are available. So unless you need an unusual size or design, making a collector is unnecessary. Buying ready-made collectors will save a lot of your time and effort.

Making a Collector—You'll need at least seven separate pieces to make a *four-into-one* collector. Start with four rings that will *accept* the primary tubes. Because they must fit over the primary tubes, the ID of these rings must equal the OD of the primary tubes. The rings can be short sections of exhaust tubing—the same type of tubing as the primaries, only with a 1/8-in.-larger ID. The fit between the rings and the primaries should be close, but it should allow for an easy

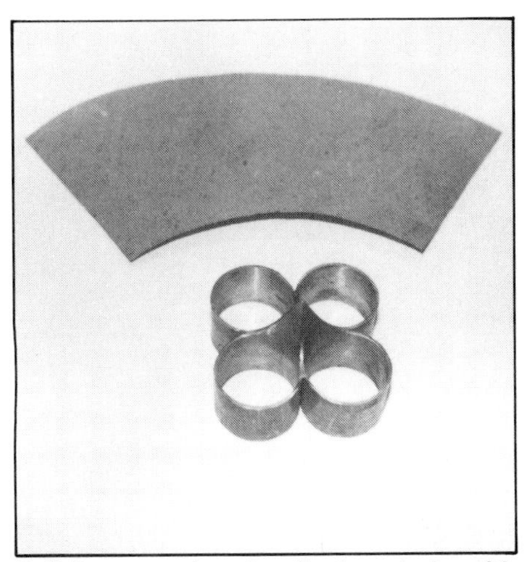

Building an exhaust collector starts with this: four short lengths of tubing welded together and a pyramid at the center. Pattern for cone is at top. After cone blank is rolled and welded, tack-weld the cone to the rings.

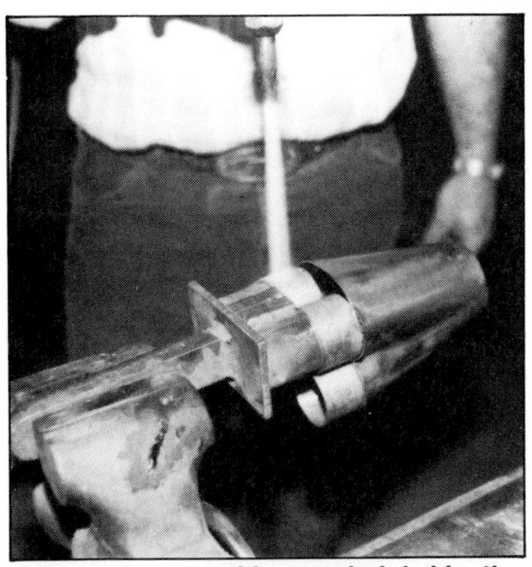

Homemade assembly mandrel holds the rings as the collector is constructed. It also supports the rings so they'll be round after you form and weld the cone.

After heating, cone is formed one side at a time. Cone is rotated so I'm always working on a top surface. Handmade corking tool forces the cone tightly around the rings.

installation. Make the rings about 2-in. long.

The four rings are joined by tack-welding them together in a square pattern, see photo. A diamond- or pyramid-shaped sheet-metal assembly is welded in the center of the four rings to seal the opening and provide for a smooth primary-to-collector exhaust flow.

The diamond-shaped center-piece is fabricated from 0.042—0.050-in. sheet steel. Make a cardboard pattern using the photo as a guide. After welding the "diamond" together and smoothing it, weld it to the four rings. I find heliarc- or gas-welding equally good for constructing headers.

The rings feed into a cone which tapers down from an opening large enough to slip over the four rings. The cone-shaped tube is made from 0.042—0.050-in.-thick sheet steel. Make the cone about 8-in. long and large enough at the big end that it will overlap the four tubes about 1/8 in. The OD at the small end of the cone should match the secondary-tube ID.

Turn to page 36 to refresh yourself on how to develop a cone and form one using rollers. After you have it cut out, formed and tack-welded, weld the cone lengthwise at its seam. Slip the cone over the four tubes and tack-weld it in place. Now comes the fun part. The cone must be formed so its section will change from the group of four rings to the round section at the end of the cone. This process is shown above.

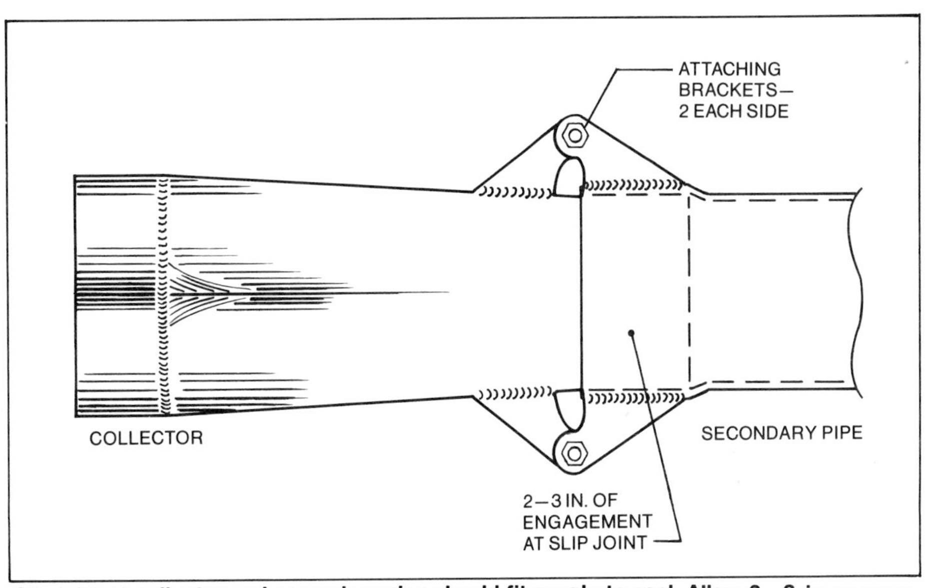

Slip joint at collector and secondary pipe should fit snugly to seal. Allow 2—3-in. engagement at the joint. Two tab-type brackets attach the parts securely.

Slip two of the four tubes over a fixture made using solid-steel rounds or thick-wall (1/8-in. minimum) tubes of the same OD as the tubes' ID. The two rounds or tubes are welded to a steel plate to maintain their center-distance. See the photo for reference. Heat the large end of the cone lengthwise at the "crotch" of these tubes. Using a hammer and a tool similar to the *corking tool* shown in the photo, push the cone down between the tubes. This will form a smooth transition from the tubes to the small end of the cone. Do this three more times and you can finish-weld this end of the collector.

The 3—4-in.-long tube that joins the small end of the cone can either be a rolled piece of sheet steel or a straight section of tubing. Either way, it must be small enough so the secondary pipe can slip over it. Butt-weld the section of tube to the cone and you've got a collector. You are ready to fit it to the car.

Placing the Collector—Now that you have a collector, you can place it in the car. Eventually the primary tubes will support the collector. In the meantime you'll have to make provisions to hold it in position. It is the "target" for the primary tubes. The object is to route equal-length tubes from each cylinder-head exhaust port to the collector.

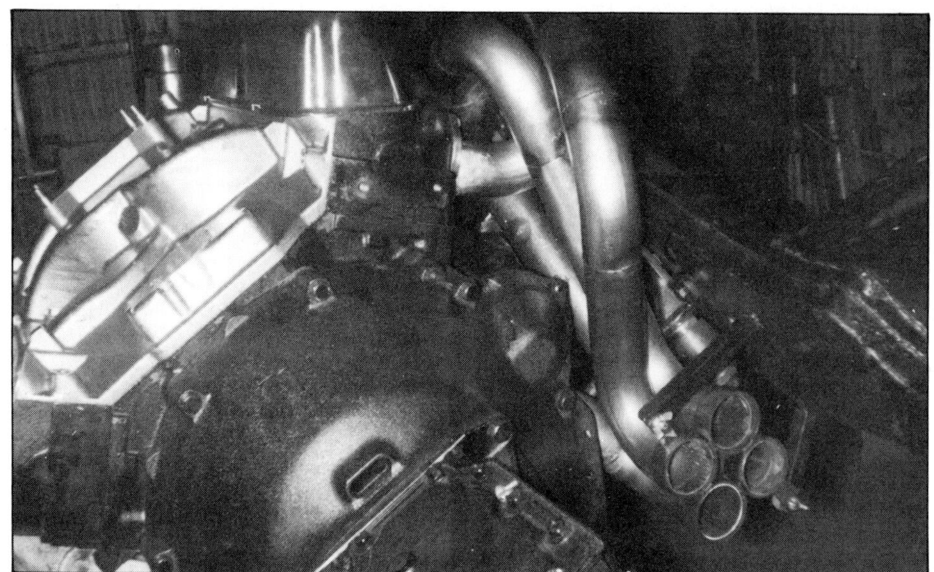

It took a lot of work to hold clearances and determine a good collector location for these stock-car headers. Note angle-iron jig tack-welded to the frame and collector rings. Wing nuts allow speedy removal of the pipes from the jig.

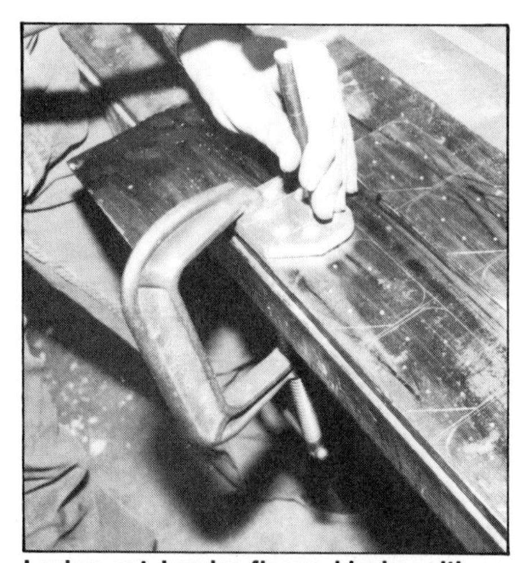

Laying out header-flange blanks with an aluminum template. Holes in the template allow centerpunching holes before drilling. Close nesting of the flange blanks saves time and material.

Some header flanges need a *starter tube* to make the transition between the shape of the port and the shape of the pipe. This starter tube begins as a cone.

Starter tube is hammered over steel mandrel to blend shape of large, round primary pipe to that of the smaller rectangular-shaped exhaust port.

Hold the collector in place by clamping it to nearby car parts, or by making a holding device for it. *Make sure the collector will stay put.* It must be stable.

Usually it is harder to route headers on the driver's side of an engine. The steering column and linkage are on that side. This can be further complicated by a starter motor. Because I like to get problems behind me, I start with the left side. Things will get easier.

Exactly where the collector should go is a "chicken-or-egg" situation. It not only depends on where it will fit best, it also depends on the primary pipe that has to reach the farthest. Use your tape measure to help determine about where the collector should go. Extend the tape to the length the primaries should be and measure from the farthest exhaust port to the trial-location of the collector.

HEADER FLANGES

Although you can buy or make your own header flanges, I recommend that you buy them. Suppliers who carry bends and collectors also sell these flanges. Although you can make header flanges from 1/4-in. mild-steel plate, buy them. They are well worth the price. Making header flanges is not easy and it takes lots of time and equipment. Study the series of pictures if you are skeptical. If you still insist on making your own header flanges, or you have no other choice, follow the illustrations.

Attaching Flanges—Bolt the header flange/s to the cylinder heads. Tighten it only enough to hold it on. Finger tight is enough. You now have a starting point for each primary pipe. All you have to do now is connect each header flange to its collector with equal-length pipes.

PRIMARY PIPES

Study the area you'll be routing the primary pipes through. Look for potential problems: wheel wells, roll bars, shock towers or any other obstructions. The primary tubes must be routed around them.

Clearances—An engine that is supported by rubber mounts rotates from side to side as it runs. Because headers are bolted to the engine, they also

Starter tubes fit inside flange opening. Gaps must be completely closed before welding.

Once the fit is perfect, completely weld the flanges to the tubes.

Tight-fitting headers on this Can-Am car illustrate use of heat shielding. One heat shield is mounted to headers to protect plug wires; another is attached to valve cover at back of engine to shield injectors. Photo courtesy of Ford Motor Company.

move. The farther the collectors are from the side of the engine, the greater this movement becomes. A tube that moves 1/16 in. at the cylinder head may move 1/2 in. or more at its farthest point from the side of the engine. Allow room for this movement. Otherwise the pipe will hit whatever it's too close to and bend or break something. Clearance is crucial.

Here are some specific clearances to be aware of: Headers must clear the tires and steering linkage. These components move as the suspension travels up and down and the wheels are steered back and forth. So be sure the tires and steering linkage clear, even when the tires are fully rotated as they would be for a turn. Another critical clearance area is at the spark plugs. You have to be able to remove and install the plugs easily. Check with a standard sparkplug socket to be sure the headers aren't a problem. Avoid building a nightmare.

Exhaust headers can create a heat problem for the driver. If there isn't enough clearance between the exhaust system and the fire wall or the floorpan, the driver can become very uncomfortable. You have two options: Allow at least 4-in. (6-in. with a turbocharger) to passenger-compartment sheet metal, or shield the exhaust system.

Measuring—Use a tape measure or flexible exhaust tubing to get accurate pipe lengths. Both methods work. Generally, though, the tape method is best for tubes that will require few bends. Flexible exhaust tubing is best to work out a problem routing. First,

When exhaust pipes are routed near suspension parts, move the suspension through its full travel during the design stage. You must be sure the suspension will not strike the headers, causing damage to either the headers or the suspension. Photo by Michael Lutfy.

Six-cylinder headers used with a turbocharger begin the same: after header plates and collector are in place. Collector is being positioned. Round tube taped to steering gear represents an obstruction that will be present later.

You can't build primary tubes from flex tubes, but they are great for establishing lengths and routings.

Fabricating headers involves experimenting with bends to get clearances and lengths right. Header plates and collector must be in place to do this.

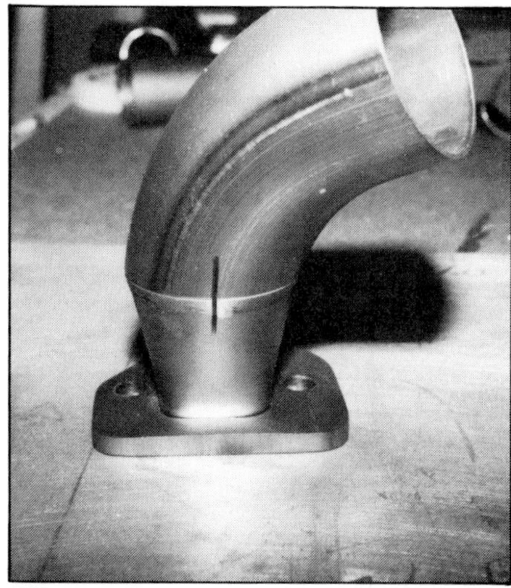

Witness line made with felt-tip marker shows where pipes should line up for fitting and welding. Another line on the opposite side ensures accurate positioning.

let's take a look at the tape method.

Extend the tape to the length of the primaries. Generally, exhaust ports farthest from and the closest to the collector are the most difficult ones to route primary tubes from. Attack the hardest ones first. Hold one end of the tape against the flange farthest from the collector. Weave the tape around in the accessible space to the collector opening. Get a three-dimensional idea of the best pipe routing from the tape. Do the same act with the closest pipe, followed by the other two. You'll quickly find out which area, or which pipe, will present the toughest problem.

Flex Tube—If you encounter a routing problem, get some flexible exhaust tubing. Use it as a "pathfinder." Cut the flex tube to the length needed for the primaries. Put one end against the header flange. Hold it *firmly* in place at the cylinder head with one hand. Bend the tubing with your other hand, trying to get the end to the collector.

You'll probably have to try several times. Before deciding the route and shape you think is best, be *sure* you can duplicate the shape of the flexible tube with the bends you have. The flexible-exhaust-tube route is useless unless you can reproduce it using steel exhaust tubing.

Bends—Study the curving tape or flex tubing to see what bends are needed. If the pipe has to make a 90° bend right out of the cylinder head, choose or cut a pipe using the prebent sections so it turns a full 90°. Hold it up to the flange. Rotate it so it will mate to the next tube: perhaps a 45° bend.

Primary pipe exiting rear port had to be routed forward to clear brake master cylinders. I allowed plenty of room so heat wouldn't affect the brakes. Note witness line: Witness lines are crucial when fitting pipe sections.

When you've turned the 90° bend to where it will meet the next tube, mark it. Using a black felt-tip pen, draw a *witness line across* the tubes where they join. The witness line "testifies" as to where the tubes must be positioned. For convenience, I often use two witness lines drawn on opposite sides of the tubes. Number- or letter-code each section to keep the pieces in order. Once you've marked two tubes, remove them from the car for tack-welding.

Remeasure—While you're piecing and assembling each primary tube, continually check to make sure it will end up the correct final length. You don't want to end up with a tube that's longer or shorter than is necessary after all of your work.

To get accurate tube-length measurements, lay each tube flat on a table and measure its length *at its center.* Mark the length on the tube section with felt-tip pen. It's easy to get confused if the sections aren't marked.

Tacking—Tack-weld the first bend to the second, using the witness marks

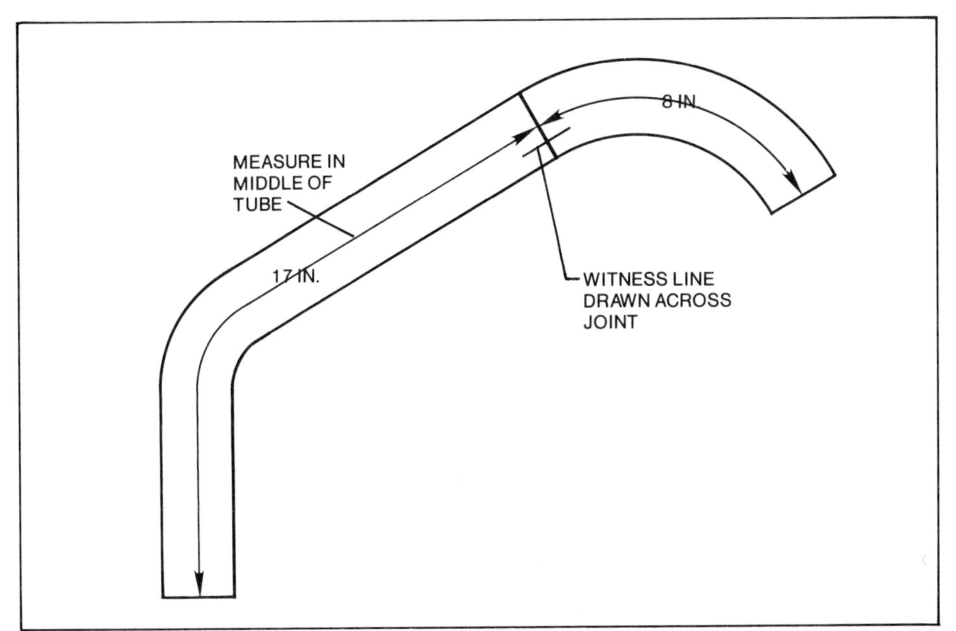

Measure length of bends at the tube center. Do this while the tube is lying flat. Mark the length of each section on the pipe. Be *sure* to revise this figure if you change pipe length by cutting or adding on.

With three pipes tacked together, the fourth pipe can be completed. There's plenty of room for removing and replacing sparkplugs.

to line them up. Each successive piece of tubing is tacked on. One after another, fit, mark, measure and tack-weld sections of exhaust tubing. When you finish at the collector the tube must be the correct length. Remeasure the final tube.

Adjusting—It's common to have a tube end up a little long or short. If it's longer, you may be able to rotate a bend up or down. After making some adjustments to get the tube aimed at the collector, trim off the excess tube-length. Remeasure and remark the tube section you trimmed. If the tube is short, add straight sections of tube after making the appropriate adjustments.

Remember: *Tube lengths must be accurate.* Otherwise engine power is compromised. Properly designed and fabricated exhaust headers are the cheapest horsepower there is. So take time to do them right. Whether you have to add more bends or not, be sure the tubes are exactly the right length. You can't recheck often enough. It's better to check and be right than to make them wrong.

When you are satisfied with the length and routing of the first primary tube, go to the next one. *Do not fully weld anything yet.* Each tube must be fitted, routed, remeasured at the center, tacked together and checked step by step. As you finish each tube, slip the tack-welded primary tube into its collector ring. This way each tube is fitted for its length, minimum gaps

Turn the steering to full lock in both directions to check tire clearance. Be sure to check it at both full jounce and full rebound. It won't lie. Just make sure you use the correct wheel and tire for checking.

for welding *and* for routing it *with* the other tubes. Don't forget to maintain clearances to body and chassis components.

WELDING

Once you're satisfied with tube length and fit, remove the headers from the engine for final welding. Loosen the bolts holding the header flange to each cylinder head. Remove the tack-welded headers.

Welding the Tubes—At this point you'll need a spare cylinder head to use as your welding fixture. Mount the headers to the cylinder head. The cylinder head holds the flange/s in the correct position/s and keeps the flange/s from warping during welding.

Welding the Flange—Weld *around* each tube at its flange. The best and strongest way to weld the tubes to the header flange is a combination of heliarc and brazing. You can create strong tube-to-flange joints by heliarc welding inside and low-temperature brazing on the outside. First, heliarc the inner joint. Then braze all around the outside of the joint as shown on page 63. Leave clearance for the attaching bolts. The braze forms a wide collar that supports the tube and flange.

At the Collector—When you finish welding the primary pipes to the flange, weld them to the collector. If you want to keep the slip joint at the collector rings for ease of installation or whatever, don't weld the tubes to the collector. Instead, make a pair of tabs for bolting each tube to the collector, shown in the photo. These will secure the collector to the primaries. Otherwise, weld the collector to the pipes. *It is important to seal this joint with weld.* No exhaust gases should escape from the exhaust system at the collector. Don't worry if you kept the

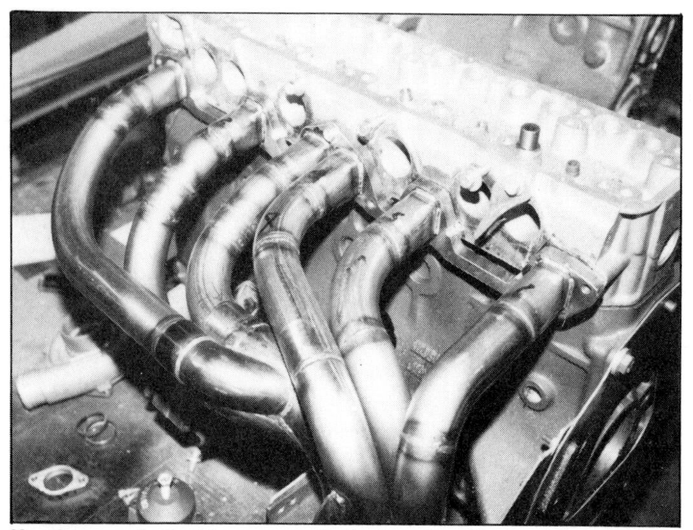

Headers are bolted to spare cylinder head for final welding. Long engine required one collector to be vertical and the other to be horizontal; unusual.

BMW collector and secondary pipe is ready for another coat of paint. Reinforcing gusset at collector-to-secondary-pipe joint is to prevent cracks from engine vibration.

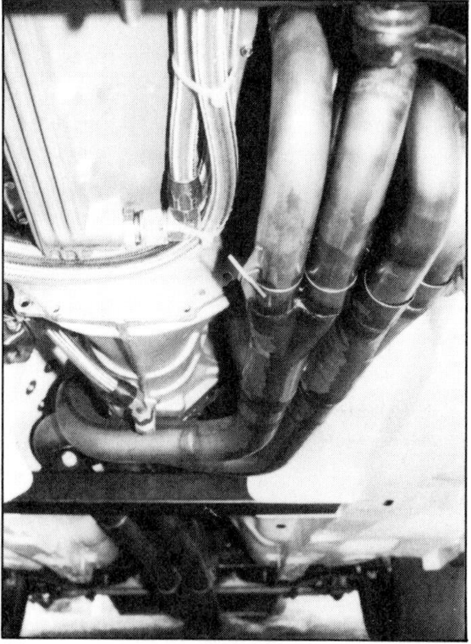

Pipes tucked up under the body for ground clearance. Builder ran the pipes above the transmission crossmember: the low point at the center of the chassis. It wasn't easy, but the right way rarely is.

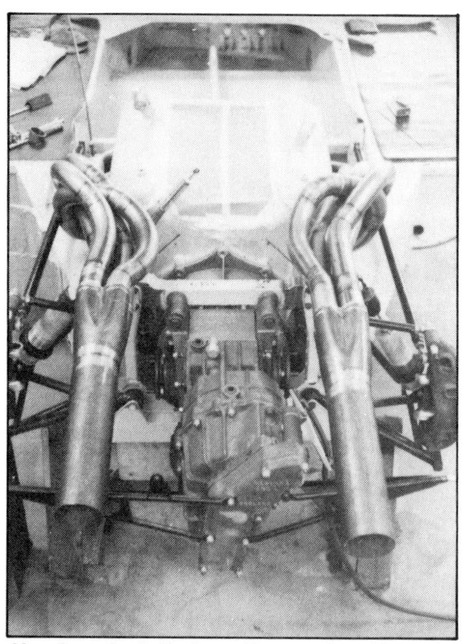

Headers installed on the engine. Now the system needs to be supported. This should be done with a flexible mount—like a saddle under the secondary pipe with a heavy spring hooked over the top.

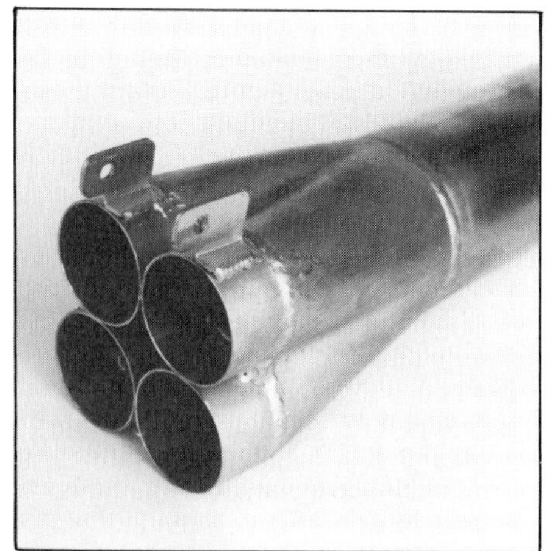

Finished collector with attaching tabs. Companion tabs are on the primary pipes.

slip joint. If the slip fit is snug, the inside pipe will expand from the exhaust heat and seal as effectively as a welded joint.

SECONDARY PIPES

Your welding job is nearly finished, assuming you used a slip-fit collector-to-secondary-pipe joint. All you'll need to weld are two pairs of mounting tabs to hold the pipe to the collector.

FINISH

After the headers are fully welded, remove them from the cylinder head. Inspect your welds closely. Be *sure* they are sound and reliable. Exhaust gases should never escape at the seams.

The headers are almost done. Steel tubing is reasonably attractive immediately after construction: but it rusts. If you want your headers to look good and last a long time, protect their surfaces before installing them.

Paint—Heat-resistant paints that can withstand up to 1200F (649C) are readily available. Black, silver and white are the most popular colors. If heat may be a problem, here's something you should know: Light paint acts similar to a shiny heat shield. Radiated heat is reflected, keeping the surfaces of the headers cooler

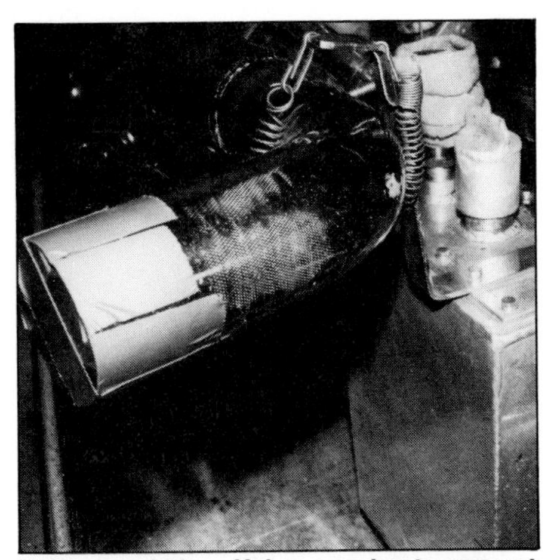

Half a good idea: Using a spring to support the secondary pipe is good, but mounting the bracket on top of the oil tank is not. Unless the tank is made from 1/4-in.-thick aluminum, I think he'd need a miracle to last four laps before the top of the tank rips off.

Extensive heat shielding is required on this GTX turbo Mustang. The most effective shield is the stand-off type (arrow) attached to the fire wall. It allows air to flow between the shield and fire wall. Shiny surfaces reflect radiated heated. Photo by Tom Monroe.

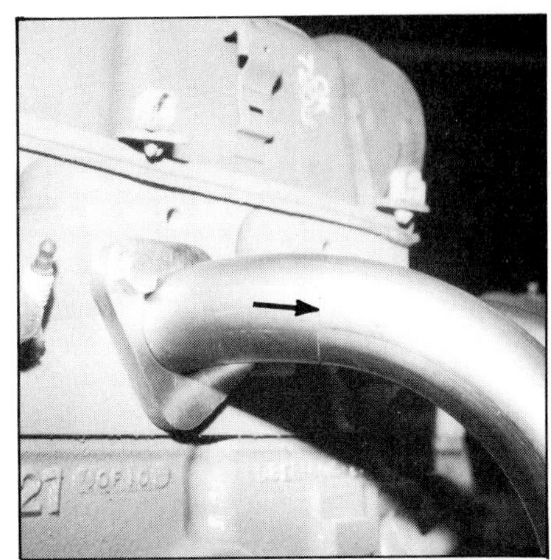

Fit on this weld joint (arrow) is so *good* you can barely see it. Fit must be good if you plan to grind the welds smooth for chroming.

Polished headers on the engine side of the flange are ready for the chrome shop. Not only are they pretty, there's plenty of clearance around them.

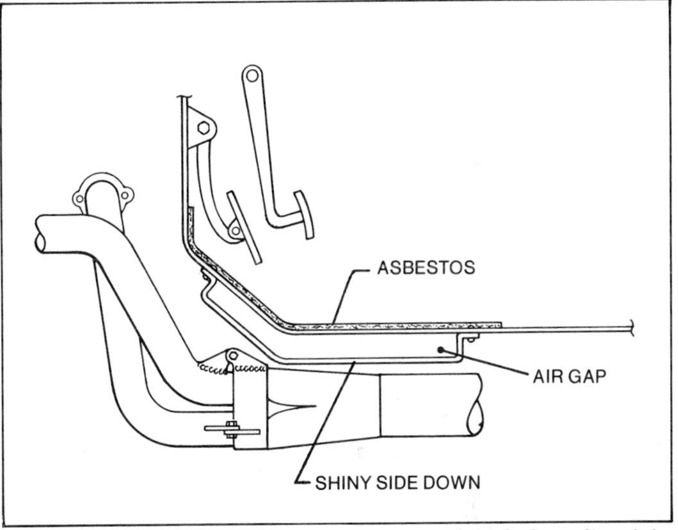

There's no set rule about the size of a heat shield. It doesn't weigh much, so make one as large as you need to keep the driver cool. A shield works best with an air gap and an asbestos carpet in the driver's compartment.

than what they would be with a dark color. So instead of using black, paint them white.

Most race cars have painted headers. Paint holds better than chrome under racing conditions. But if you're building headers for a street rod or custom car, you have a choice. Chroming headers is common practice when a good appearance is important. Before toting your headers off to the chrome shop, you need some special information.

Chrome—Many street rodders prefer chrome headers. There are problems associated with chroming headers. To understand them you need to know *how* chroming is done. The first thing a chrome shop does, after taking your deposit and your headers, is to grind and polish all of the welds very smooth. This is a real problem if those welds didn't have good penetration. Grinding off much of the bead significantly weakens a weld. And if the tubes don't match perfectly at the seams, the chrome-shop grinder will thin or grind *through* the mismatched surface while smoothing the seams. So preparation to chrome headers *starts* with good welds.

SUMMARY

Headers are *not* an easy project. They can be tricky. Your skill and your planning will show in the final product. But if you do a good job, you'll gain some horsepower and have something specially custom made for your car. That isn't a bad reward for the investment in money, time and patience.

HEAT SHIELD

If the exhaust system ends up too close to the floorpan or fire wall, install a heat shield. Fabricate the shield from shiny aluminum or stainless steel. Bend or attach mounting legs with flanges to the shield to allow a 1-in. air space between it and the body. Drill 1-in. holes in the legs for additional air flow. Mount the heat shield between the body and the exhaust with the shiny side facing the exhaust. The shield will reflect radiated heat. Asbestos cloth can be attached to the interior-side of the firewall or floor with large-diameter-head rivets to provide additional shielding.

SHEET-METAL INTERIORS

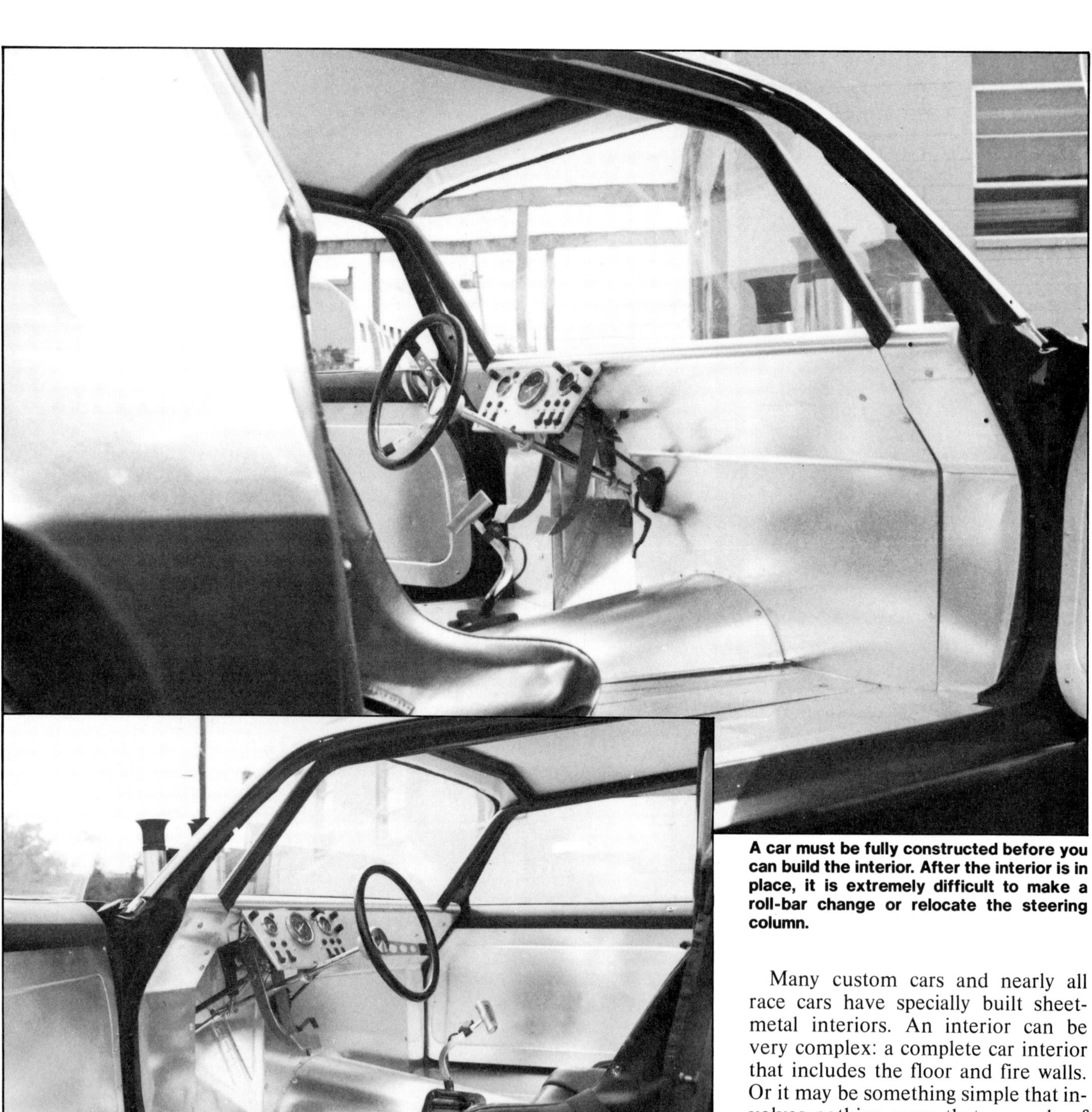

A car must be fully constructed before you can build the interior. After the interior is in place, it is extremely difficult to make a roll-bar change or relocate the steering column.

Small, efficient instrument panel. Measure all gages and switches before building panel.

Many custom cars and nearly all race cars have specially built sheet-metal interiors. An interior can be very complex: a complete car interior that includes the floor and fire walls. Or it may be something simple that involves nothing more that a couple of door panels.

DRAG CARS

Many drag cars—such as those in pro-stock and gas classes—use *full* aluminum interiors: I use 0.050-in., 3003-H14 sheet. It is easy to bend and shape. Also, 3003-H14 is lightweight,

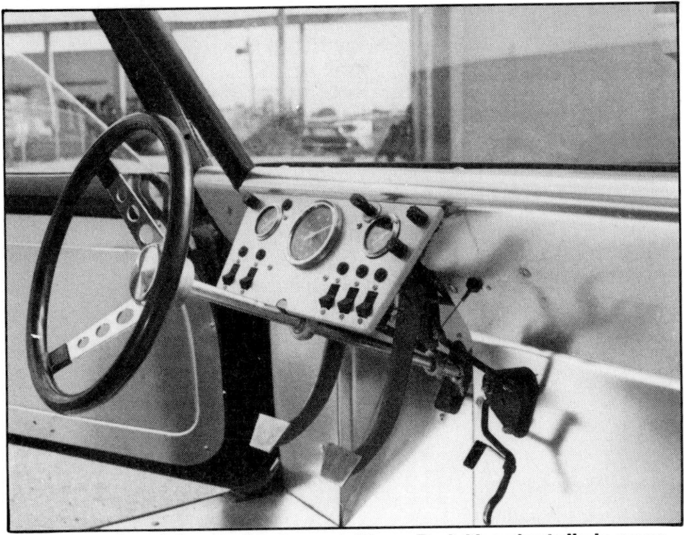
Location of pedals required a pedal box. Pedal box installs in opening in fire wall.

Rear compartment is made up of several large panels. Put your beader to use stiffening those panels.

easy to weld and it anodizes well, readily taking color or clear finish. And 3003-H14 aluminum sheet is durable and resists cracking reasonably well.

These interiors include every metal or trim panel normally found inside a passenger compartment or engine compartment. From front to back, the interior is totally hand-fabricated—a custom package. Custom aluminum dashboards, or instrument panels, are even found in altered classes.

ROAD-RACING CARS

Road-racing cars—such as grand-touring cars or sedans—also use aluminum interiors. These interiors are generally not as extensive as the ones used in drag cars. IMSA and SCCA rules books spell out which stock panels must be retained. Stock floorpans are mandatory in some classes. Other road-racing classes have more flexible or lenient rules. In these classes—Grand Touring Experimental (GTX) or Grand Touring Prototype (GTP)—you can use a *complete* aluminum interior.

STOCK CARS

Stock-car interiors are usually a combination of steel and aluminum. Steel components are used for critical areas such as floorpans, and front and rear fire walls. Because it resists heat better, and is generally stronger and more crash-resistant, it is best to use steel in those areas: 20-gage (0.0359 in.) minimum. The rest of the interior panels are better in aluminum. Door panels and the dashboard are almost always aluminum. Lighter weight and good looks make it the best choice for less-critical areas.

GT car retains stock floorpan as required by rules. Custom-fabricated aluminum instrument panel is anodized blue.

Rules—Many different stock-car-racing groups exist. Each one sets its own requirements. Each may specify what types and thicknesses of metals can be used for specific interior parts. It is vitally important to obtain the rules book from your sanctioning body. Study it. Follow the rules. If you are in doubt, be conservative. Always use the material that will make the safest interior.

Before building any interior part, carefully study the rules that apply to

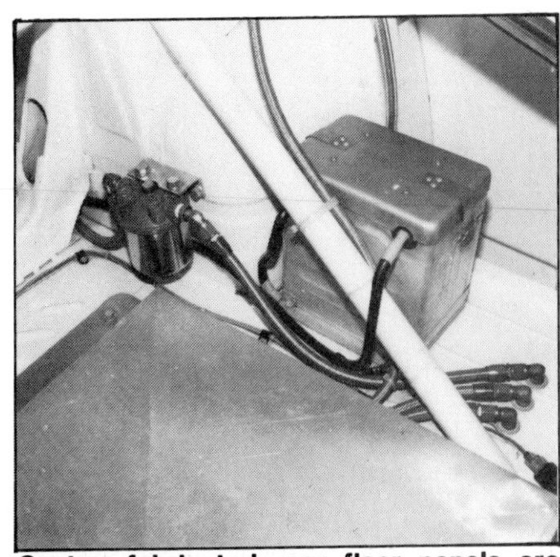
Custom-fabricated rear floor panels are used in GT car. Note tidy custom battery box, fuel-cell cover and fuel-filter bracket.

Steel interiors can be nice too. Interior in this late-model stock car was well planned: It had to be because of its complexity. Most panels are riveted, but a few are retained with quick-release fasteners—Dzus in this case.

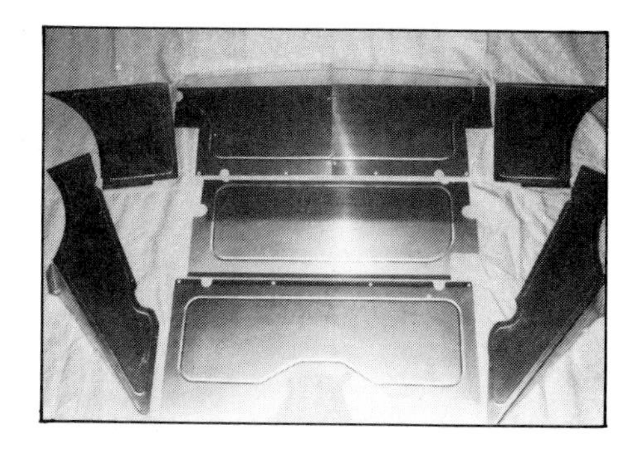

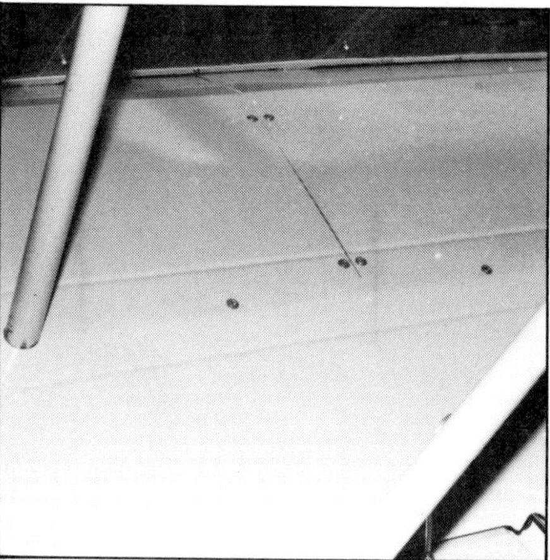

Interior panels in and out of GT car. Display of interior panels shows simple, but effective beading and mirror-image right and left parts. Panels fit tightly to each other and around roll bars. Dzus fasteners hold them tightly in place.

NASCAR modified under construction. Cutout in fire wall is for transmission tunnel and opening at right for pedal box. Rear fire wall is very simple with a flange at top and bottom for mounting other panels. Photo by Tom Monroe.

your car. These rules will specify how extensive the metalwork should be. You will also get an idea *why* the interior is regulated. It may be for safety reasons, or to ensure that you don't reduce weight illegally. Frequently, certain stock sheet-metal panels must be retained to give the "illusion" of a stock-based car. The idea is to have safe and competitive cars racing against one another.

BASICS

Whatever type of car you're working on, some general rules will apply to its interior. Let's start with the basics: A sheet-metal interior must protect the driver and should be attractive. An interior doesn't have to be ugly to be safe. Nor does it have to sacrifice protection to be pretty. It can serve both functions if you have a good design and you build and install the panels with care. Let's look at

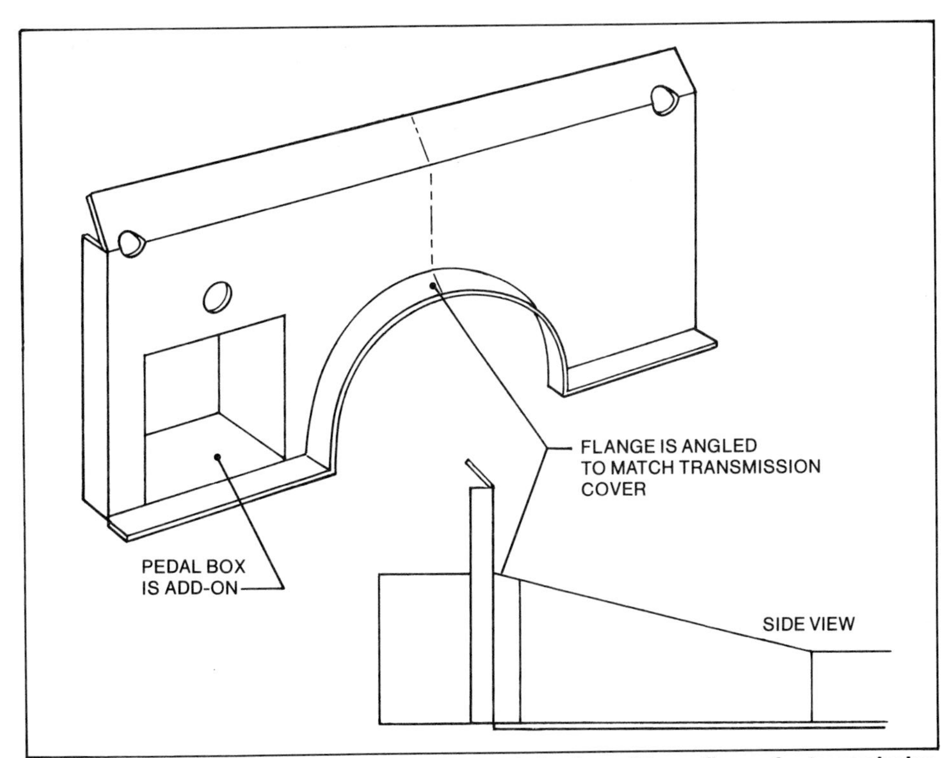

Typical fire wall has some or all of these items: holes for roll bars, flange for transmission cover, floor-mounting flanges, steering-column hole and a pedal box. Make a sketch and take measurements when starting development of a fire wall.

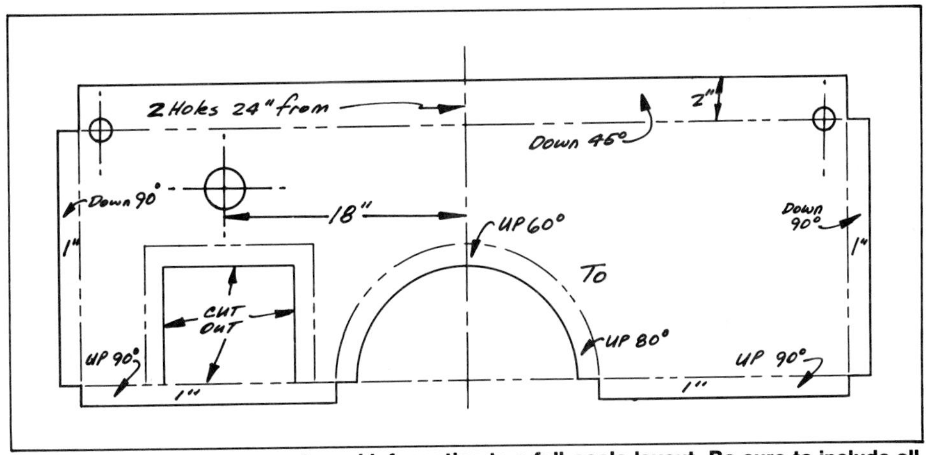

Transfer all the measurements and information to a full-scale layout. Be sure to include all information: flange to be added, bend lines and amount or direction of bends.

some basics on how to accomplish *both* goals.

Driver Protection—The first goal of a sheet-metal interior is to protect the driver. What *kind* of protection can an interior give a driver? The interior shields the driver from three potentially dangerous elements: fire, fumes and debris.

A protective metal interior must be of the right materials, fit well and be durable. Rules books usually specify the metals for certain areas, and how thick it must be. For an interior to be durable, it must be carefully designed, fabricated and installed. Panels must be fitted tightly to each other and the car. Take great care to prevent gaps and misfits. Similar to a leak in a boat, a gap or misfit will keep an interior from sealing the way it should. Remember: *One small gap can let in enough fumes or fire to endanger the driver's life.*

Installing interior panels can make or break the safety of an interior. Not only must they fit well, they must be *tightly* attached with the right type and number of fasteners. A loose panel is as bad or worse than no panel at all. Usually a loose panel will not come off completely. If a corner pulls loose in an accident, it may become a giant knife blade. It can cut the driver. It can also become an open door for burning oil or fuel. Any way you put it, an poorly installed interior is hazardous. A well-installed interior protects the driver. Let's take a look at specific installation techniques.

HINTS

Fit each panel to the car and to the next panel as you go. This allows you to get a gap-free interior more easily.

The order for making and fitting interior panels is very important. Following a certain order works best. Because most panels must fit to other panels, it is necessary to make some first, others second, and still others later. Save yourself some grief. Build according to the order I describe.

Start with the front fire wall. It is often the most complicated panel. Take time to develop this panel and make sure it is right. In many cases the front fire wall includes a wide lower flange where the front edge of the floor rests. It also includes a flange around the transmission opening to secure the transmission cover.

When fabricating an interior, don't overlook the fact that you're building a *complete interior*, not a bunch of unrelated panels. So keep the "big picture" clearly in mind as you go. An interior must have continuity; all panels must combine to be functional, simple and attractive. Use large-radius bends wherever possible to soften the visual impact and create a beautiful interior. Avoid sharp edges and bends when you can.

FRONT FIRE WALL

A front fire wall is usually the most difficult interior panel because of the many components it must accommodate: engine and transmission (with a front-engine car), steering, driver and driver controls, frame and roll cage, and exterior body structure and panels.

The procedure used to develop a pattern for a front fire wall is identical to that used to develop *any* interior-panel pattern. Follow the same procedure used for pattern development, metal cutting and forming, fitting, and on to installation when making the remaining interior panels.

The engine, bellhousing, and transmission must be in the car before you can fit the front fire wall—if yours is a front-engine car. The same applies to a rear-engine car when doing the rear fire wall. Start with the basic dimensions: Measure across the front firewall area from door pillar to door pillar. Measure up from the floor, or where you think it will be, to the underside of the windshield cowl. Construct a rectangular-shaped pattern from a big piece of cardboard using these two dimensions. If you don't have a piece that big, tape smaller pieces together. Draw a vertical center line on the pattern.

Locate the bellhousing opening. Measure from the top of the cowl to the top of the bellhousing. Transfer this point to the pattern. Determine the bellhousing radius. Find its center and draw it on the pattern. Add 2-in. to the bellhousing radius to provide bellhousing clearance. Snip out the bellhousing opening from the pattern. This opening will have to be modified later to allow for a transmission-cover flange. Fit the pattern over the bellhousing and check for accuracy. If there are any gaps or fit problems, correct the pattern.

Pedal Box—Depending on engine or driver location, your fire wall may need a *pedal box*—a protrusion of the fire wall that extends into the engine compartment. The pedal box accommodates the clutch, brake and accelerator pedals, and the driver's feet. It will be a five-sided box attached to the engine side of the fire wall. The pedal box must be large enough to clear the pedals and allow for comfortable foot and leg movement. The bottom surface of the pedal box should blend in with the floorpan.

After you've developed a pedal box, mark its location on the fire-wall pattern. Add 1-in. flanges to the firewall panel for attaching the pedal box. Indicate the direction that you'll want the flanges bent. *Make the marks on the side the flange will be bent toward.* Bend the pedal-box flanges toward the engine compartment.

Pattern Details—The front fire-wall pattern must include openings for components that must pass through it such as the steering column or roll-bar braces. Each opening must match the size of whatever part fits through it and it must be located *exactly* where the object will be.

Cut a slit from the edge of the pattern so you can fit it around parts you're cutting holes for. While you are pulling the pattern in and out of the car to mark and cut holes for the steering column or roll bars, fit its periphery. The pattern should be marked and trimmed to match the areas you'll mount it to.

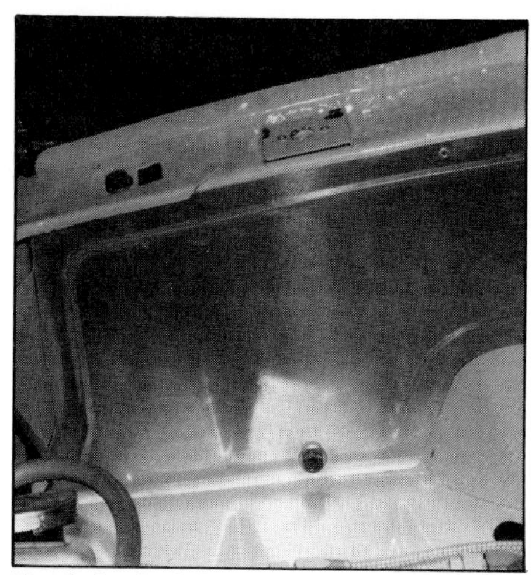

Bead simply follows the periphery of the panel. This type of bead is mainly for appearance.

Make the fire-wall pattern a map or set of directions. For instance, to add a 1-in. flange at the bellhousing opening, write ADD 1 IN. on the pattern. Point to where you want the flange. This flange will be bent toward the passenger compartment, so mark on the passenger-compartment side. A 1-in. flange is needed at the bottom of the fire wall to support the leading edge of the floorpan. Indicate this on your pattern too.

When you transfer the pattern to metal, add the flanges and transfer the instructions. When you cut out the metal, you'll have the extra 1 in. you need for the flanges.

This system will let you develop a simple, clear cardboard pattern and a set of instructions so you can construct a complicated metal panel. All you have to do is remember to make your notes *while* you're developing the pattern. Afterwards, be careful to read them before you cut or bend the metal.

Don't be surprised if you end up scrapping the pattern and have to start over. But first, follow standard practice by patching the openings with taped-on cardboard pieces until the pattern fits. You won't have any problems telling when it fits. The pattern will slide into place and fit tightly to all the mating surfaces and components without bulging. When you have the pattern fitting well, remove it and take it to your workbench.

CARE OF THE METAL

Metal Covering—If you want a scratch-free-aluminum interior, use aluminum with a stick-on cover. Although it costs more than bare aluminum, it is worth it. There is less chance of damaging the metal surface as you work with it. As a bonus, the plastic or paper stick-on cover is easy to mark. Keep this cover on as long as you can. You can do a lot of metalwork—such as bending, beading and drilling—without removing the cover. To weld, merely peel back the immediate area of the cover—say 6 in. from the weld line. When you've finished welding and the metal has cooled, lay the protective cover back in place. It may not fit perfectly, but it will protect the surface as you continue to work.

Anodizing is another reason to use sheet aluminum with a protective cover. If you intend to have the aluminum panels anodized it is important to keep them unmarred.

ANODIZING

Although anodizing can be done in a range of colors, colors tend to fade from exposure to sunlight. I recommend a clear-anodized finish for all interior metal panels. It protects the aluminum from oxidation, looks good and won't fade. Without clear anodizing, aluminum panels oxidize to a dull, dreary gray color. They lose their original shine. When you consider how much you'll have invested in those interior panels, it's only smart to have them anodized.

Never send an aluminum panel to an anodizer with a steel rivet or fastener in it. Steel makes the anodizing process impossible.

FROM BLANK TO PANEL

Marking the Blank—Lay the firewall pattern on the sheet metal. Mark the metal or its protective cover with a soft black pencil. Follow any instructions you have on the pattern. *If you want to bead the panels, this must be indicated on the pattern.* You should decide if and where you want to bead the fire wall. Remember: A beader's throat-depth limits how far a bead can be put from the edge of a panel. Mark the beading instructions on the metal blank.

Beading—Beading serves two purposes: It stiffens a panel so it flexes less easily. It also enhances the looks of a panel. Just remember not to get carried away. A little beading can go a long way. It should be simple and consistent on all interior panels. I've seen a lot of overdone beading jobs. They made me think a seven-year-old kid got hold of a beader and thought it was a neat toy.

Cut out the metal blank according to your instructions on the pattern. Once it's cut out, the blank is ready for beading or *stepping*. Stepping is a means of forming overlapping joints between panels without unsightly raw edges showing. It also keeps the surfaces of the two panels flush with one another.

Stepping—Stepping is done with a special set of beader dies. You either make them yourself or have them made according to how deep a step you need. Making special beader dies is discussed on page 38.

The step should be as deep as the thickness of the overlapping panel. So if the overlapping panel is 0.050-in. thick, the step must be the same: 0.050 in.

Stepping, like beading, must be done before any other forming operations. This order has to be maintained for a simple reason. Folds or bends will interfere with the beader shafts. You'll have unbeaded areas near any fold if you don't bead or step the blank *before* bending.

Flanging—Bend flanges after beading or stepping. Straight flanges can be made in a sheet-metal brake or over a straight edge. Curved flanges, such as the one at the bellhousing opening, can be made by hand. Bend the flange with glass pliers, hand seamers or a hammer and dolly.

Fitting the Panel—As you make any interior panel, do your final fitting. Custom tailor the panel until it fits perfectly. Each edge must match up to its mounting. If the panel doesn't fit, snip or file it until it does. If the panel is too small you can do one of two things. Scrap it, or add metal to it. Neither is desirable. Be sure you allow a little leeway in cutting the panel in the first place. Don't be like the guy who said, "I don't understand. I cut it twice and it's still too short."

Installation—When you have it right, Cleco the fire wall in place. It will stay there until you're ready to remove it. It will also be easy to remove for fitting neighboring panels, allowing you to fit as you go. Mate each panel to the adjacent panels. Only when all the panels are formed, fitted and anodized should you final-install anything. Once all the panels are in place, they should match perfectly. It is not uncommon to find an interior with dozens of pieces. Matching them right is no small accomplishment. *Take your time!*

Tube construction allowed me to make most interior panels before the body was installed. Inboard wheel-house panels are hammerformed. Photos courtesy Steve Allen.

WHEEL HOUSES

Rear *wheel houses,* or *wheel tubs,* are next. You can't make a rear fire wall or floorpan without them in place. Here are some important factors to take into account while developing wheel-house patterns:

Measurements—Clearance must be provided for the tires as they move up and in with the rear suspension. Drag cars need additional rear-tire clearance.

Rear drag-car tires *grow*—their outside diameter increases as the tires spin. Unless the rear-wheel housings allow for this growth, the tires will rub against the wheel houses. Don't let this happen as tire failure could result, causing a crash.

If your car is a drag car, find the *ex-*

Check all clearances before building a wheel-house tub. Tire is held in its full-up position. Added clearance is needed at the end of the rocker panel.

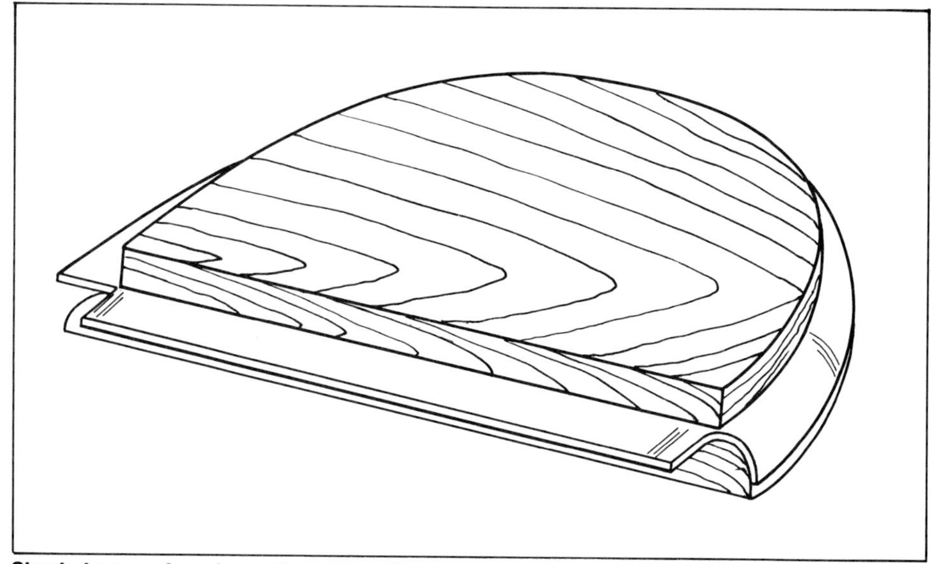

Simple hammerform is worth making. Use one to produce a beautiful, professional-looking wheel tub.

pansion factor, or how much the rear tires will expand. The tire manufacturer can supply this information. Use the expansion factor in combination with the other tire and suspension measurements to construct rear wheel houses that will provide sufficient tire clearance.

For example, say the tire diameter is 30 in.—it's 30-in. "tall." If its diameter expands 3-in., you must allow 33 in. for the tire. Allow a 2-in. clearance in front and in back of the tire. The width of the wheel house—front-to-back—should measure 33 in. + 2 in. + 2 in. = 37 in. Allow for suspension travel at the top of the wheel house. No two cars have the same suspension geometry or wheel travel.

To determine with accuracy what clearance you must allow at the top and sides of a rear tire, run the rear suspension through its travel with the wheels and tires installed. Remove the springs—unless it's a standard leaf-spring rear suspension—to make the job easier. Measurements must include the width of the tire plus clearance on the inboard side of the tire. Make a vertical section template and check it with the tire all the way up. Position the template to a "hard" point such as the frame rail. Match mark the template—use witness lines—so it can be accurately repositioned for checking the wheel house.

In some gas and pro-stock drag cars the width of the tires can be a problem. The frame itself must be designed and built very narrow. Otherwise you'll never be able to enclose the tires under the body. If the frame was not specially built or designed for drag racing, you may be faced with narrowing it before you can fabricate the rear-wheel tubs.

Wheel houses aren't hard to build. Just follow some basic guidelines. First of all, keep the wheel dimensions and clearances in mind and write them down. Make templates from these figures. Trim the cardboard pattern to fit against the inside of the quarter panel and frame, then tape it in place.

Shape—The shape of a wheel house is simple. A semi-circular side panel joined by a curved top: like a 55-gal oil drum cut lengthwise in equal halves, then the halves cut in equal lengths.

Although it's not the method I use, here is a simple way to make a wheel house. Make a semi-circular pattern for the side. Use this pattern to make a sheet-metal blank. Cut and roll a long rectangular strip of metal wide enough to enclose the tire, plus some clearance. This piece curves around the tire. Weld the curved piece to the side piece and the result is a functional wheel house. I think the method produces a useful and fairly attractive wheel tub. *There is a better way.*

Hammerforming makes a more professional-looking wheel tub. You still need a full-size pattern for the side and for the long strip. Make the pattern for the side according to your measurements. Measure the periphery of the arc of the curved side. The rectangular strip is as long as this arc, plus additional length for front and rear mounting flanges. The width of the strip is determined by tire clearance on the inside, plus tire width and the distance from the outside surface of the tire to the quarter panel.

Here comes the trick. Transfer the pattern of the side of wheel house to a large piece of 3/4-in.-thick plywood. Cut it out.

Hammerform—The plywood piece is your hammerform base. Grind a *smooth* 1/2-in. radius around *one* of the curved edges. The opposite edge should remain just as you cut it—square. Leave the straight edge at the bottom alone.

Make the upper half of the hammerform from another piece of 3/4-in. plywood. Trim it about 1 in. smaller along the curved edge. Use this piece to secure the metal blank to the hammerform base for forming the side panels.

This hammerform can be used to do two jobs. The radiused curved edge is used to form a metal edge on the wheel-tub side. The flat straight edge can be used to bend a straight flange at the bottom of the side panels.

Tub Side Blank—Cut out a metal blank using your pattern. Clamp it between the hammerform base and top. Make sure it is tight. Use as many clamps as you can. Use a rawhide or wooden mallet to hammerform the edge. Work all around the curve gradually. Do not flog away in one spot. Turn the metal down over the radiused edge with repeated soft blows. It should form over the hammerform base very evenly. Be careful not to leave mallet marks in the metal.

If the side piece is to include a

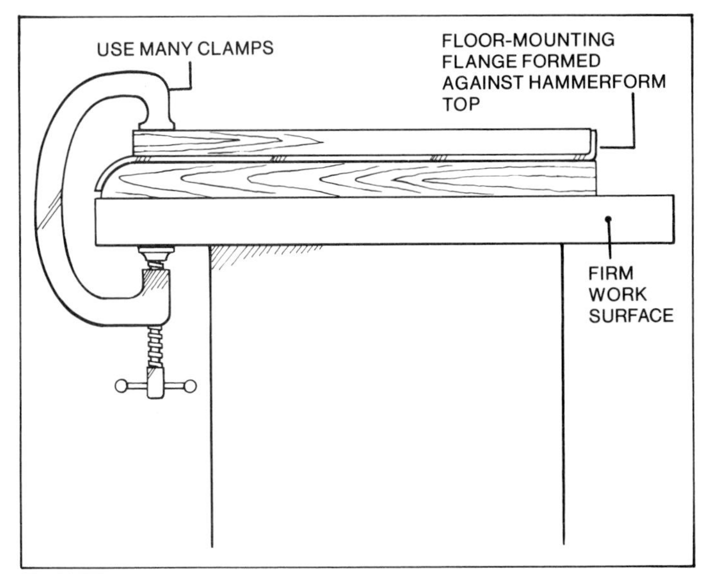

Although only one C-clamp is shown, use several. Clamp blank securely to prevent it from shifting during hammerforming.

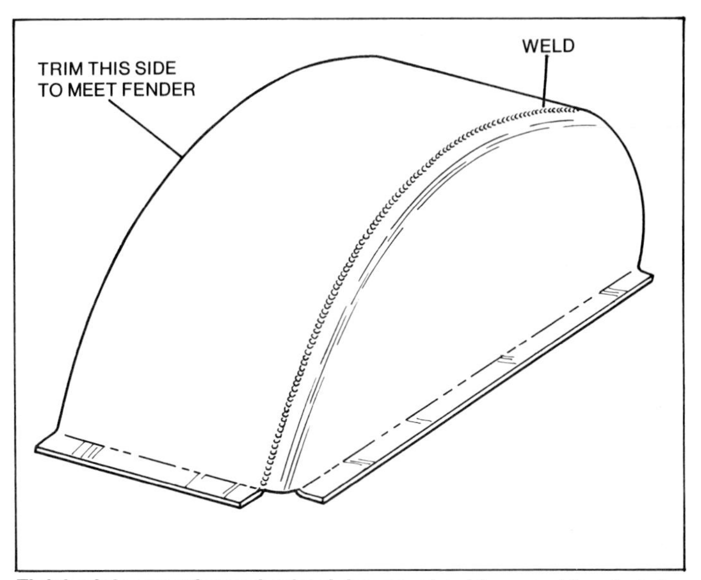

Finished hammerformed wheel house should resemble sketch. Mounting flanges may or may not be straight, depending on the floor panel they mate to.

flange at the bottom, bend it up against the hammerform top. Make sure the the hammerform top lines up with the intended bend line of the flange before you do any bending.

Remove the wheel-tub side from the form and fit it to the car at the frame. Clamp it in place and check for fit. If and when it fits properly, make the long strip that forms the wheel-tub top.

Tub-Top Pattern—The long rectangle which makes the tub-top pattern can be fitted against the hammerformed side to double-check it for length. If you want it to include mounting flanges, add cardboard to the end of the pattern. Trim the pattern if it's too long. Now's the time to make any changes.

Tub-Top Blank—Use the cardboard pattern to cut the metal blanks. If flanges are used, transfer their bend lines. Roll the blank to form a curve. Keep checking it against the hammerform base. When the strip is rolled close to the desired shape, check it in the car. Fit it over the side piece you clamped into place. Do any necessary adjusting to fit the pieces. The outer edge of the tub may require additional trimming to fit it to the inside of the quarter panel. When the pieces fit closely, remove them from the car.

Welding the Tub—Tack-weld the wheel-house pieces together at the workbench. Before final-welding, recheck their fit in the car. If the fit is correct, complete the welding. Cleco the finished tubs in the car. You can then build around and to them with the other panels. The finished tubs have a nice curved inner edge, giving them more of a professional appearance than those built without a hammer form. Although it took more time and effort, it was worth it—right?

REAR FIRE WALL

Next comes the rear fire wall. It attaches to the rear wheel housings and forms a wide flange at the bottom for the rear floor. Although considerably less complex than a front fire wall, the rear fire wall is constructed very much like it. But it is no less important. It must fit equally as well: no gaps or bulges.

Follow the same rules used to construct the front fire wall. *Be sure to seal any gaps or openings in the fire wall securely.* As with the front fire wall, add a 1-in. flange at the bottom edge of the rear fire wall. This flange will support the rear edge of the floorpan.

In many cases the rear fire wall is the main defense between the driver and the fuel tank. It has to be strong and well sealed—and remain sealed in the event of a crash. If it doesn't you may be very sorry. It should be as reliable as a pilot's parachute, and for the same reason: to save a life.

SIDE PANELS

You might think the floor panels should come next in the normal sequence of fabricating and fitting interior panels. Not so. Imagine having the floor in place. You would have to work on top of it or remove it to fit the side panels. Make the side panels next and you'll save the floor from the abuse of working on it. Or you will save the time to remove and reinstall it. *Make the side panels next.*

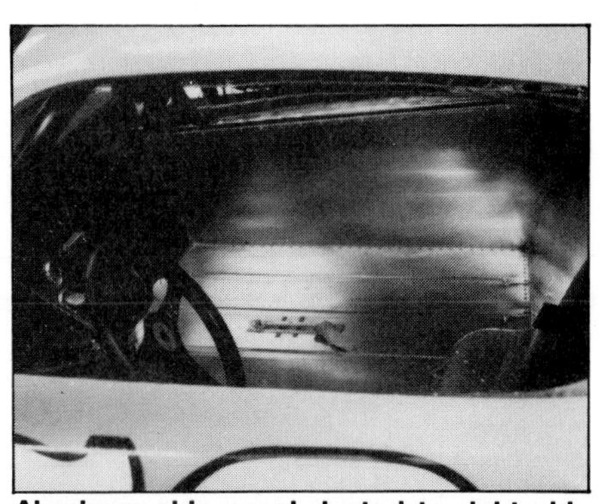
Aluminum side panel riveted to right-side door bar in this late-model stock car. Beads in floor panel are for strength. Steel dash panel is painted flat black to eliminate glare. Photo by Tom Monroe.

Side panels aren't as complicated as other interior panels. Virtually all side panels are flat, with the possible exception of a rolled edge at the top. Regardless, a good cardboard pattern is required to make a good piece.

The interior of a car is usually symmetrical like its exterior. The pattern for the side panel on one side can be used for the opposite panel. For example, the rear inside quarter-panel pattern can be used for both right and left sides. Door panels may need their own patterns because of different roll-cage configurations in the right and left door openings.

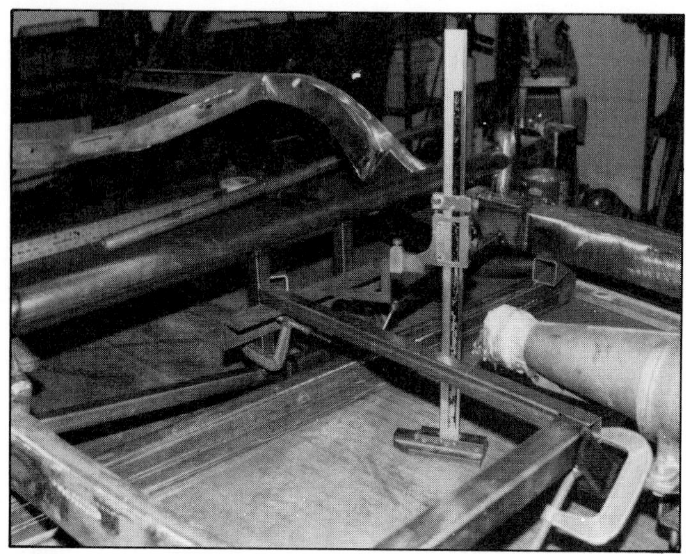

Adding tubes to existing structure makes a good floor foundation. Driver's-side floor had to be very strong for seat mounts.

Floor-pattern development is done after engine, transmission and all substructure is in place. Flange for transmission cover is part of this pattern.

Electric shears make quick work of cutting floor blank. Material is 0.040-in. mild steel. Foot area has already been beaded.

Some bends can't be made in a brake. Out come the C-clamps, angle iron and slapper. Presto, handformed bends.

Develop patterns for the side panels, transfer these patterns to sheet metal and cut out the blanks. Fit the panels one at a time. Remember to bead the panels before any forming is done.

Side-Panel Beading—There is a simple way to bead side panels. I roll the bead a fixed distance from the panel edge so it follows the outline of the panel. The result is beading which stiffens the panel, particularly at its edge. The bead also gives emphasis to the shape of the panel. A nicely shaped panel is enhanced. It makes your interior reflect the care you took to build it.

Forming—Except for mounting flanges, use large-radius bends rather than tight folds when forming side panels. This particularly applies to the top edges of door panels. It also applies to dashboards. Large-raduis bends give more of a professional look.

As you build each side panel, Cleco it in place. Each panel must fit to each other and to the car. Don't install anything permanently. When all side panels are Clecoed in place, go on to the floor.

FLOOR

Fabricating a floor is a giant job. It not only involves large panel/s, it includes the transmission cover and the drive-shaft tunnel. These two areas require special attention.

If a *live* axle is used—one that moves up and down with the wheels—the drive-shaft tunnel *must* have enough clearance at the rear for vertical drive-shaft travel. The transmission cover must include provisions for the shifter and the shifter boot. Therefore the shifter must be installed on the transmission. If the transmission cover fits closely around the transmission, the shifter *housing* will be a cube-like addition to the cover.

Floor Support—A floor should be strong and sturdy. It should have extra bracing immediately under the driver's feet. The floor will rest on flanges at the bottom edges of the front and rear fire walls. It will be supported at the sides of the car by one of several means. It may be bent up and riveted to the side panels. It might rest directly on top of an existing doorsill. The floor might even rest on an angle bracket you welded to the rocker-panel boxes. Or the floor may be attached directly to the frame-rail top or side. There are many possibilities.

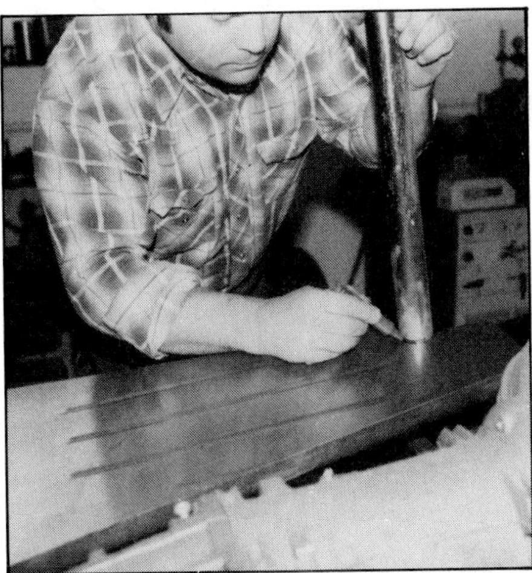

Using a felt-tip marker to indicate where trimming is necessary. Take your time when fitting panels.

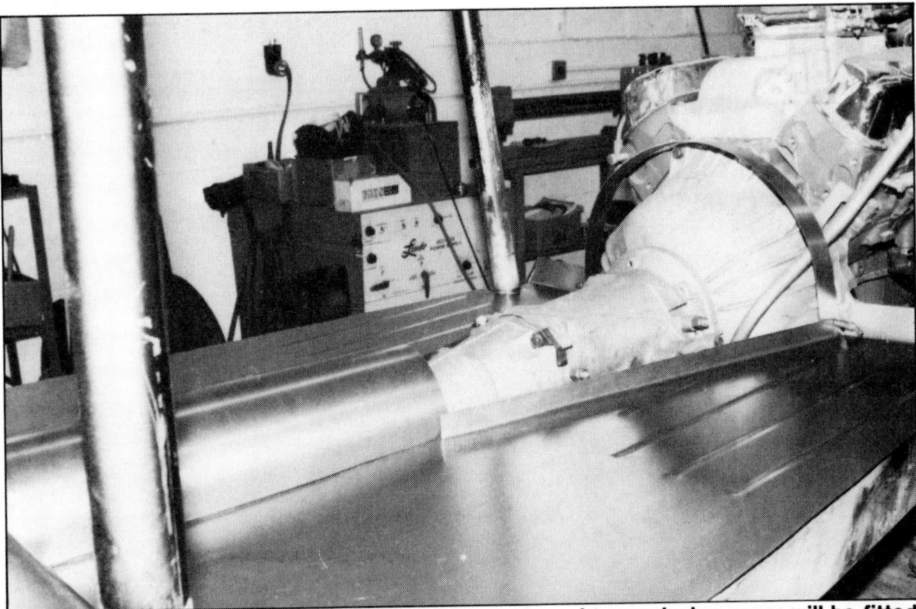

Hoop that's Clecoed in place is a "target." Cone-shaped transmission cover will be fitted to the hoop and drive-shaft cover.

The floor is an excellent place to use beading. It provides a great practical benefit by increasing the strength of the flat floorpan. This is done without increasing its weight one ounce. Although beading makes a panel more attractive, appearance is only a side benefit. A floor must be structurally sound so it will support the driver.

Floor Pattern—Right and left patterns are used to make floor panels. Start with the right side. You won't have to work around the seat mounts—if it's a left-drive single seater—and the pattern can be the basis for making the left-side pattern.

You may have the rare situation where you can fabricate the floor from a single piece of sheet metal because it doesn't have to be split its full length to clear the transmission or drive shaft. If this is the case, start with a piece of sheet metal at least as wide as the maximum interior width of the car. Find its center line and mark it so you'll have a reference when transferring right and left patterns to the blank.

After the right-side pattern is developed and the blank is marked, flop the pattern over and check it for left-side fit. You'll have to modify it to clear seat mounts and seat-belt anchors. Adjustments may also have to be made at the engine, bellhousing, transmission and pedal box. Also allow for overall-width differences.

There must be flanges for mounting the transmission cover and drive-shaft tunnel. These flanges can be added to the floor or the cover and tunnel. I recommend that you add flanges to the floor. They will provide additional stiffness to the floor. Add about 1-1/2 in. to the full length of the inside edge of each floorpan. Either way, flanges are needed. Your choice is *which* pieces get flanges and *how* the panels will interlock. There may also be flanges at other edges, depending on the design of your floorpan.

Once you have the floor blank/s cut out, bead the foot area similar to that pictured. *Then* bend flanges and do any other forming operations. Fit the panel/s to the car. Make changes to ensure the the floorpan fits perfectly. When you're satisfied with floorpan fit, Cleco the panel/s in place so you can develop the transmission cover and drive-shaft tunnel.

Drive-Shaft Tunnel—Start with the drive-shaft tunnel. It is relatively easy to build and you'll work to it with the transmission cover. You have a choice of two basic shapes—square or round. A square drive-shaft tunnel has three flat surfaces: the top and both sides. The second shape, and the one I prefer, is round. It has one large curved, or rolled surface, that may blend into flat sides if the tunnel is deep. I like a curved tunnel because it is stronger than a square tunnel and I believe it looks more professional.

Allow for maximum vertical drive-shaft travel in your drive-shaft-tunnel pattern. Give the drive shaft 1-in. of clearance at its vertical-travel limit. Simulate the cross section of a round cover by looping a tape measure over the drive shaft from one floorpan half

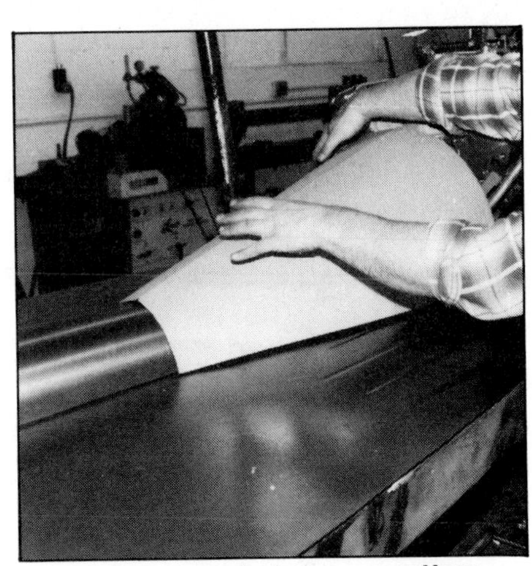

Developing transmission-cover pattern.

to the other. Make a 1- or 2-in.-wide metal test strip. Loop it over the drive shaft like you did with the tape, or put two breaks in the strip that are spaced the width of the top of the tunnel. It will be in the shape of a round or square U, depending on which tunnel shape you've chosen.

After you've established the cross section of the tunnel, determine its length. It will run from the transmission tail-shaft housing to the rear fire wall or rear-floor kickup. Make a cardboard pattern and check it for fit. Indicate flanges on the pattern if you didn't add flanges to the inner edges of the floorpan halves. Transfer the pattern to metal, cut out the blank, form it and Cleco the tunnel in place. You're now ready to make the last panel that will enclose the interior—the transmission cover.

Transmission Cover—Make a test strip like that pictured. Loop it over

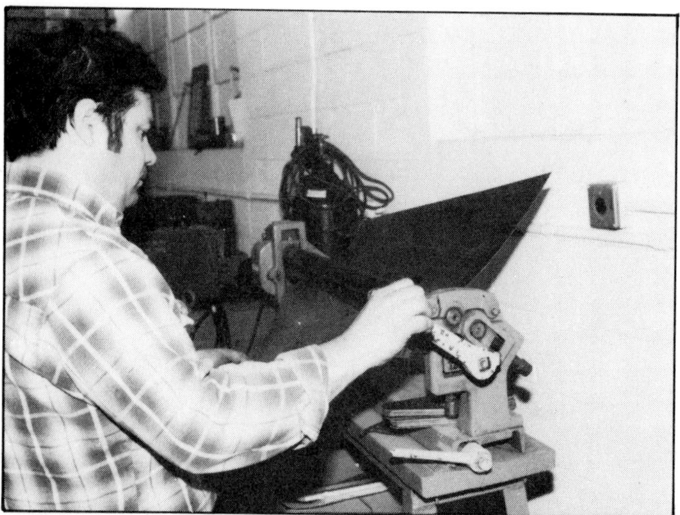
Rolling transmission cover requires that the roller be adjusted for more pressure at the small end to curve the blank into a cone.

Fit the transmission cover with care. It must fit tightly.

Checking transmission-cover fit after Clecoing it in place. It fits well along the floor and is tight against the drive-shaft tunnel. I'm pleased.

Box-like addition to transmission cover accommodates shifter and boot.

the transmission or bellhousing and provide at least a 2-in. clearance for the components underneath. Blend this section into the front of the drive-shaft tunnel. If the drive-shaft tunnel is round, the transmission cover will be cone-shaped. Using this test strip and the one for the drive-shaft tunnel, develop a cardboard pattern. Don't forget the hole for the shifter. Make a metal blank, cut it out and roll the cone following the procedure on page 36. Cleco the transmission cover in place.

Adding a *shifter box* to the transmission cover will complete the interior fabricating job. Determine the shape of the top of the box by matching it to the bottom of the shifter boot. The size of the hole in the top of the box is determined by how tightly the cover fits around the transmission and the *throw* of the shifter, both back-and-forth and from side-to-side. Build a cardboard box and tape it to the tunnel to check it for fit. When it's right, convert the box to metal. Bend the blank into a box using the cardboard pattern as your guide. Weld the box over the hole in the tunnel.

FASTENING PANELS

Unless you want them anodized, all you need to do with all those interior panels that are Clecoed in place is to secure them with permanent fasteners. Several methods can be used. How you fasten a panel depends on the material it is made from and how many times you plan on removing it: never, seldom, once in a while, or frequently.

There is always some question as to how to fasten interior sheet-metal panels. Several methods are available. Sheet-metal screws, *quick-release* fasteners or rivets are commonly used. Stock-car steel interiors are sometimes MIG or TIG welded in place. You cannot weld an aluminum panel unless the mating parts are aluminum. You need to know something about these fastening methods.

Screws—*Don't use sheet-metal screws for fastening car interiors.* They have the undesirable habit of loosening and backing out of their holes. And there's that nasty sharp point that sticks through the panel. If you've ever worked on a car that was full of sheet-metal screws, you should know what I'm referring to. You probably had your clothes ripped or your body cut by those sharp points. Avoid them! There are better ways of fastening panels.

Once in a while I use small machine screws—10-32 to be exact—to retain certain panels. I always use *nylon-insert*

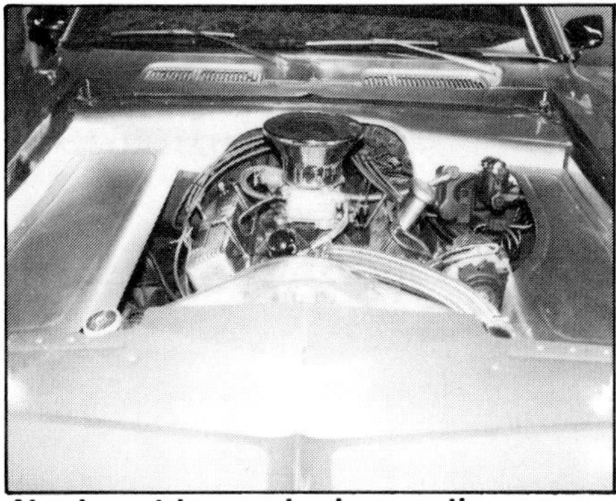
Aluminum trim panels clean up the appearance of this show/street-machine engine compartment. Anodizing and beading are an extra touch.

locknuts with them. This way the screws and nuts don't vibrate loose. Use these if you need to use screws for a particular panel. Don't use sheet-metal screws!

Pop Rivets—Pop rivets are good for securing interior panels if done right: Use the right *type* and *number* of rivets. Use steel or stainless-steel Pop rivets. Unlike everday aluminum rivets, they *stay* tight. Don't skimp on how many you use. The more rivets you use, the stronger and better sealed the interior will be. Don't space them more than 1-1/2 in. apart on long seams. Choose the correct grip length for the metal thickness you are joining. Refer to page 21 for determining rivet grip length. A rivet that's too long will not clamp panels tightly enough. Too short, and it won't hold at all. Be careful.

Quick-Release Fasteners—Unlike a screw that must be rotated several times to remove it, a *quick-release* fastener releases when it is rotated 90°, or 1/4-turn. This allows a panel to be removed very quickly. It can also be installed with the same relative speed. As a bonus, most quick-release fasteners stay with their panels when removed. So there is no danger of losing them.

There are two well-known quick-release fasteners. *Dzus* or *Southco* fasteners are used in panels that may need to be removed quickly, frequently or repeatedly. People seem to be fascinated by them—they are "trick." Consequently, they are overused. I use them only when they are needed.

Two popular Dzus fasteners are in use. The EHF-series Dzus fasteners are *self-ejecting.* One half is mounted on a plate riveted to the panel it retains. This assembly includes the plate, a spring and the fastener. The fastener projects through holes in the metal panels the Dzus joins. It hooks over a *lock wire,* or spring, that is riveted to the other panel. When rotated 1/4-turn counter-clockwise, the fastener unhooks from the lock wire and the spring pops the fastener out, releasing the panel.

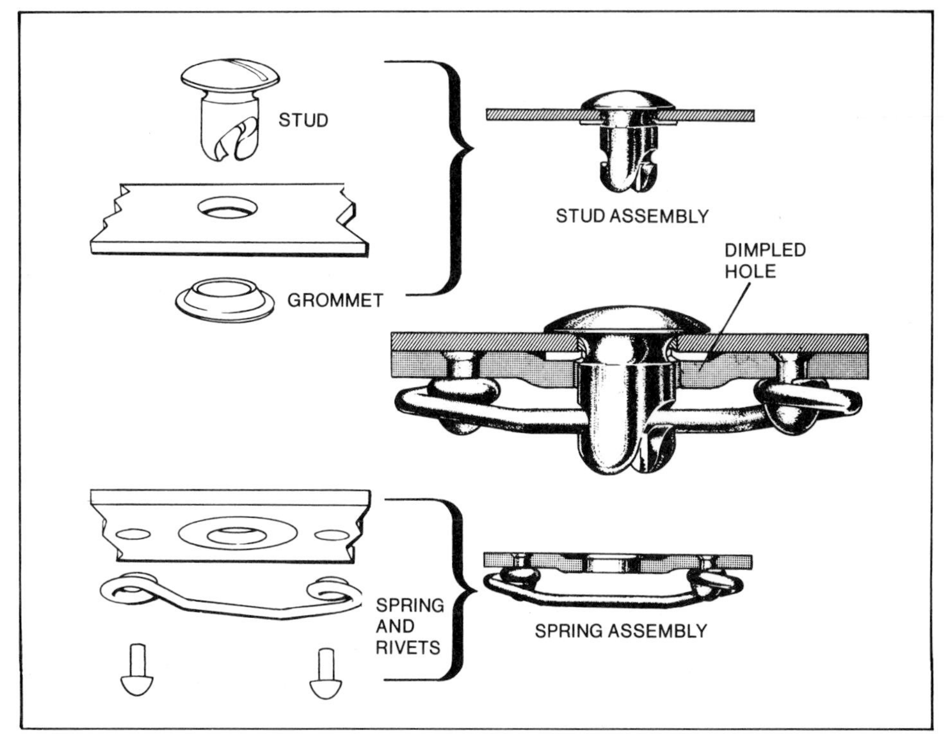

Details of AJ-series Dzus-fastener installation.

The AJ-series Dzus fastener works much the same way. But it doesn't pop up automatically when released. The AJ fastener is held to the panel by an aluminum grommet. It is a smaller, more compact fastener than the EHF series. The AJ series includes flat and domed-head fasteners. Both EHF and AJ series fasteners require *dimpling* the panel.

Dimpling—*Dimpling,* similar to *belling,* starts with a drilled hole. The edge of the hole is then *dimpled,* or bent down into a specific shape before installing the lock wire. This allows the mating panels to fit against each other. Dimpling kits are available from suppliers listed below. They include dimpling tools for installing several types of Dzus fasteners.

Dzus fasteners are available in different lengths for joining different thicknesses of metal. They are available directly from the manufacturer. They can also be ordered from Earl's Supply, Russell Performance Products or Moroso. Suppliers are usually helpful in determining the type or length you need if you'll explain what metals or parts you want to join.

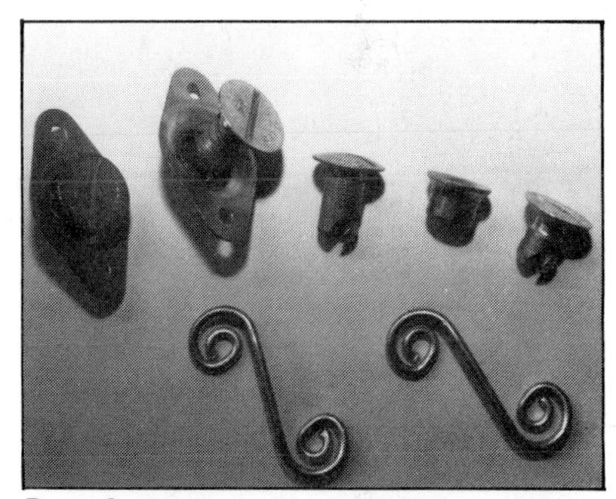
Dzus fasteners are worth their weight in gold! They hold panels securely and allow quick and easy removal. EHF self-ejecting fasteners at left. AJ series at right.

Welding—Steel stock-car interiors are sometimes welded in place. This makes an extremely strong, well-sealed interior. A welded interior even adds strength without adding much weight. The welded joints are well sealed and nearly impenetrable under most circumstances—including accidents. Of all the fastening methods, a welded interior makes the safest enclosure for the driver, offering great protection.

EQUIVALENT LENGTHS

1 Meter =
39.37 inches
3.2808 feet
1.0936 yards
1000 millimeters
100 centimeters

1 Centimeter =
0.3937 inch
0.0328 foot
10 millimeters
0.01 meter

1 Millimeter =
0.03937 inch
0.001 meter
0.10 centimeter

1 Inch =
0.0833 foot
0.0278 yard
25.4 millimeters
2.54 centimeters

1 Foot =
12 inches
0.3333 yard
0.3048 meter
30.48 centimeters

1 Yard =
36 inches
3 feet
0.9144 meter

Square Measure

Square In.		Square Ft		Square Yd
144	=	1	=	
1296	=	9	=	1

Cubic Measure

1728 cubic inches = 1 cubic foot
27 cubic feet = 1 cubic yard
231 cubic inches = 1 U. S. gallon
57.75 cubic inches = 1 quart
61.023 cubic inches = 1 liter
1000 cubic centimeters = 1 liter

1 cubic foot = 7.48 U.S. gallons
1 cubic inch = 0.004329 U.S. gallon
1 U.S. gallon = 0.13367 cubic foot
1 U.S. gallon = 231 cubic inches

Effect of Heat on Substances

Aluminum melts at 1256F (680C)
Cast iron melts at 3479F (1915C)
Glass melts at 2377F (1303C)
Magnesium melts at 1382F (750C)
Steel melts at 2500F (1371C)
Water boils at 212F (100C)

A-Pillar—structure of a car body that extends up from the door-hinge pillar and supports the roof at the front, doubling as the windshield frame at each side.

AK Steel—a steel alloy which has been "killed" with aluminum in the molten stage to refine its grain structure; a steel alloy with good ductility.

Alloy—a blending of metals; homogenous combination of metals in which the atoms of one replace or occupy positions between atoms of the other.

Aluminum Alloy—a range of metals combining aluminum with copper, magnesium, zinc, nickel and/or iron.

Anneal—with aluminum, to soften metal to make it more ductile by heating it to 640F (338C) and cooling rapidly to 450F (232C) until recrystalization occurs; in steel, a heating and cooling operation of steel in the solid state usually requiring slow gradual cooling.

Anodize—to coat a metalic surface electrolytically with a protective or colorative oxide.

Arc—a discharge of electric current crossing a gap between two electrodes.

B-Pillar—a vertical structure of a car body where the front door latch is located.

Baffle—an object designed to block or damp the movement of heat, air or liquid.

Base Station—in metalwork, the bottom section of a station buck that acts as a reference point.

Bell—to turn down the edge of a hole in sheet metal with a smooth curve.

Bead—in welding, a narrow half-round pattern where metal has been joined by heating; in metalworking, a decorative or structural half-round channel formed into the metal in a continuous line.

Bend—to cause something to assume a curved or angular shape; something which has been bent; a curve or angle.

Bending Sequence—the order in which several different bends are formed so that each successive bend is not blocked by a previous bend.

Blank—a cutout metal piece prepared for fabrication.

Brake—to bend a metal part to the proper radius and angle on a brake machine or by hand; the machine which forms angles or folds in metal.

Braze—a solder-like oxyacetylene joining method for steel, aluminum and cast iron using brass or bronze welding rod and flux; a bonding method, not a welding process.

Bung—a fitting in a liquid-holding container through which the container is filled or emptied.

Burnish—to make smooth or glossy as if by rubbing; polish, planish.

Butt-Weld—welding two metal pieces together at their butted edges.

Carburizing Flame—a gas-welding flame containing an excess amount of acetylene.

Chipboard—a lightweight flexible cardboard used to make patterns.

Chrome Moly—a steel alloy, designated SAE 4130, that is composed of 0.28—0.33% chromium and 0.15—0.25% molybdenum.

Collector—a metal assembly in an exhaust manifold that collects exhaust-gas flow from the primary exhaust tubes of an engine, funneling this flow into a single exhaust pipe.

Concave—curved like the inside of a bowl; hollow.

Convex—curved like the exterior surface of a bowl; curving or bulging outward.

Cowl—the body-structure assembly that connects the A-pillars between the windshield and the front fire wall.

Cross Section—a section formed by a plane cutting directly through an object.

Cut (of a file)—the direction and texture of the rows of cutting teeth arranged across the surface of a file.

Deburr—to remove uneven or jagged edges from the edges or surface of a metal piece.

Dimple—to turn the edge of a hole under or down to accept installation of a fastener; bell.

Direct Layout—marking metal according to dimensions needed for a given piece without first making a pattern or template; useful for making simple pieces.

Downforce—downward pressure created by air flow over a vehicle in motion.

Drag—the slowing force of air flow on a car in motion.

Dry Sump—an engine oil-supply system that utilizes a remote oil reservoir, or tank.

Ductility—capability of a metal to be shaped or hammered without cracking or breaking; flexibility.

Fillet-Weld—welding two metal pieces which join at an angle to each other; this weld usually fills the inner corner of the angle, sometimes the outer.

Fitting—a general term for any small part, such as plumbing, used in the structure of a component.

Flame Control—the ability to gas weld with the desired penetration and bead (with or without welding rod) in the exact location desired, time after time.

Flame Cutting—using an oxyacetylene welding system to melt and blow away metal with a burst of oxygen.

Flange—a protruding and angled edge used to strengthen or to attach one object to another.

Flux—fusible paste or powder used to facilitate brazing steel with brass or bronze welding rod; also a fusible paste or powder used in gas-welding aluminum or silver soldering. Flux may also be coated on a welding or brazing rod.

Fold—to bend, or brake; an angle formed in metal.

Grommet—a reinforced eyelet through which a fastener is attached.

Gusset—a triangular insert that fits across a corner for strengthening.

Heat Shield—a baffle intended to block and reflect heat.

Heat Treatable—metal alloys that can be hardened by heating; heat-treatable aluminum bears the letters H or T.

Heliarc—a kind of electric-welding named for the original shielding gas, helium, and the electric arc which is the heat source; also known as TIG-welding.

Hollow—in American metal-working usage, deeply indented or concave; a metal-shaping technique used to form a depression in metal; sometimes referred to as *raising* in British metal-working.

Jig—a tool or fixture holding a part upon which some operation is being performed.

Jounce—upward suspension movement.

Layout—in metal work, full-size outline of parts marked on sheet metal; for designer, a full-scale drawing of parts to be fabricated or assembled.

Main Hoop—the U-shaped bar of a roll-cage assembly that is mounted vertically at the floor immediately behind the driver and extends above his head.

Mandrel—a metal core over which material may be shaped.

Metal Finish—to smooth, polish and make even by hammering, grinding, filing or rubbing.

MIG—abbreviation for metallic inert gas; a welding method using a wire-fed electrode, a shielding inert gas, and an electric arc for the heat source.

Mild Steel—a low-carbon-steel alloy containing 0.30—0.50% carbon.

Neutral Flame—gas-welding flame with equal amounts of oxygen and acetylene; indicated by a bluish to orange flame with a white cone; high temperature of the flame is centered just past the end of the inner cone with the flame perpendicular to the metal surface.

Normalize—low- and medium-carbon steel heated 50—100F (10—38C) above critical temperature range and allowed to cool slowly in still air; used to refine the grain and to stress-relieve.

Oxides—a flaky film of metal impurities which sometimes occurs when welding.

Oxydizing Flame—a gas-welding flame containing more oxygen than acetylene; indicated by a purplish color and short inner cone.

Parallel—straight or curved lines which are equidistant from one another at every point.

Penetration—the depth reached by an object or force; the degree to which heat penetrates metal during welding.

Perpendicular—intersecting at right angles (90°); at a right angle to either horizontal or vertical.

Pi—the relationship between the diameter and the circumference of a circle, aproximately 3.1416.

Planish—to finish, make smooth or harden metal by pounding out irregularities; to smooth, toughen, flatten or polish metal by rolling or hammering; to burnish.

Plug-Weld—also known as a rosette weld, weld joins two overlaying metal pieces through a hole in the upper piece; weld fills the hole.

Pneumatic—operated or powered by compressed air.

Purge—to free of impurities or undesirable elements.

Radius Bend—in metal work, a curved fold as opposed to a straight angle fold.

Raise—a method of stretching metal by hammering it into a depression such as a hollowed wood base or shot bag; sometimes used to mean hollowing.

Reducing Flame—see "carburizing flame."

Rendering—to represent in detail in an artistic form; a precise scale picture of detailed construction.

Roll Bar—a U-shaped tube positioned so it protects a car's occupants in a rollover.

Roll Cage—an arrangement of several strong interrelated metal tubes intended to protect the occupants and strengthen a car's structure.

Rosette Weld—see "plug-weld."

Sand Bag—canvas bag filled with sand and sewn shut, used as a base over which metal is shaped; also used as a means to secure awkward parts during other operations.

Scale—in welding, a flaky oxide film of metal impurities sometimes formed on a metal, particularly iron alloys, when heated to high temperatures; in drawing, a method to represent the proportions of a measurement.

Semi-Circle—half a circle as divided by the diameter.

Shot Bag—a leather bag filled with very fine lead shot and sewn shut over which metal is shaped; #9 bird shot is most desirable for metal working.

Shrink—in metal work, to draw together and reduce the amount of metal area; usually resulting in an increase in metal thickness.

Side Loading—centrifugal force; the pressure outward from the center to the exterior of a circle as it is turned at speed; also called G-loading.

SK Steel—a steel alloy which a has been "killed" with silicon in the molten stage to refine its grain structure; a steel alloy with good ductility.

Slapper—long rectangular metal or wood tool used to curve, flatten or smooth a metal surface; sometimes a wooden slapper is covered with leather.

Spoiler—a shape arranged to project above the upper surface of the rear or the lower, front surface of a car body to disturb air flow, reducing aerodynamic lift.

Spring Back—the tendency of sheet metal or tubing to attempt to return to its original shape after bending or forming.

Station—wooden piece to represent a cross section of a metal construction at a given point.

Station Buck—wooden form representing the shape of a metal part; a device for ensuring accuracy in metal construction.

Step—to offset, or raise or lower a metal surface by forcing it through a die; usually the metal is raised or lowered the amount of the metal thickness itself.

Stretch—to lengthen or widen metal by hammering or rolling; to increase the area of sheet metal by hammering or rolling, resulting in thinning as a secondary effect.

Structural Integrity—structure which has the quality of being sound throughout.

Sump—depressed area in the engine oil pan of most cars, which serves as an oil reservoir.

Symmetrical—a shape identical in form and configuration on either side of a center line.

Tack-Weld—a welding method using small areas of weld spaced long a seam between adjoining surfaces; a temporary joining to prepare for final welding.

Temper—hardness of a metal that governs its strength and formability; to reheat metal after hardening to a temperature lower than critical range and then cooling.

Template—a pattern for laying out parts on a flat sheet of metal; a pattern showing bend lines, flanges or other special information about construction of the metal piece, including bending allowances.

TIG—abbreviation for tungsten inert-gas welding, using a tungsten electrode, and inert gas shield and an electric arc for the heat source; also known as heliarc.

Torsional Rigidity—resistance of a structure to twisting.

Unibody—frame and body of a car in one unit; composed of ribbed panels electrically spot-welded together, as opposed to a separate frame and body construction.

Witness Line—a reference mark for accurately fitting or joining two parts.

Work Harden—an increase in the hardness and strength of metal as it is formed, or worked; an undesirable over-hardening sometimes encountered when metal is shaped.

INDEX

OTHER BOOKS FROM HPBOOKS AUTOMOTIVE

HANDBOOKS

Auto Electrical Handbook: 0-89586-238-7
Auto Upholstery & Interiors: 1-55788-265-7
Brake Handbook: 0-89586-232-8
Car Builder's Handbook: 1-55788-278-9
Street Rodder's Handbook: 0-89586-369-3
Turbo Hydra-matic 350 Handbook: 0-89586-051-1
Welder's Handbook: 1-55788-264-9

BODYWORK & PAINTING

Automotive Detailing: 1-55788-288-6
Automotive Paint Handbook: 1-55788-291-6
Fiberglass & Composite Materials: 1-55788-239-8
Metal Fabricator's Handbook: 0-89586-870-9
Paint & Body Handbook: 1-55788-082-4
Sheet Metal Handbook: 0-89586-757-5

INDUCTION

Holley 4150: 0-89586-047-3
Holley Carburetors, Manifolds & Fuel Injection: 1-55788-052-2
Rochester Carburetors: 0-89586-301-4
Turbochargers: 0-89586-135-6
Weber Carburetors: 0-89586-377-4

PERFORMANCE

Aerodynamics For Racing & Performance Cars: 1-55788-267-3
Baja Bugs & Buggies: 0-89586-186-0
Big-Block Chevy Performance: 1-55788-216-9
Big Block Mopar Performance: 1-55788-302-5
Bracket Racing: 1-55788-266-5
Brake Systems: 1-55788-281-9
Camaro Performance: 1-55788-057-3
Chassis Engineering: 1-55788-055-7
Chevrolet Power: 1-55788-087-5
Ford Windsor Small-Block Performance: 1-55788-323-8
Honda/Acura Performance: 1-55788-324-6
High Performance Hardware: 1-55788-304-1
How to Build Tri-Five Chevy Trucks ('55-'57): 1-55788-285-1
How to Hot Rod Big-Block Chevys:0-912656-04-2
How to Hot Rod Small-Block Chevys:0-912656-06-9
How to Hot Rod Small-Block Mopar Engines: 0-89586-479-7
How to Hot Rod VW Engines:0-912656-03-4
How to Make Your Car Handle:0-912656-46-8
John Lingenfelter: Modifying Small-Block Chevy: 1-55788-238-X
Mustang 5.0 Projects: 1-55788-275-4
Mustang Performance ('79–'93): 1-55788-193-6
Mustang Performance 2 ('79–'93): 1-55788-202-9
1001 High Performance Tech Tips: 1-55788-199-5
Performance Ignition Systems: 1-55788-306-8
Performance Wheels & Tires: 1-55788-286-X
Race Car Engineering & Mechanics: 1-55788-064-6
Small-Block Chevy Performance: 1-55788-253-3

ENGINE REBUILDING

Engine Builder's Handbook: 1-55788-245-2
Rebuild Air-Cooled VW Engines: 0-89586-225-5
Rebuild Big-Block Chevy Engines: 0-89586-175-5
Rebuild Big-Block Ford Engines: 0-89586-070-8
Rebuild Big-Block Mopar Engines: 1-55788-190-1
Rebuild Ford V-8 Engines: 0-89586-036-8
Rebuild Small-Block Chevy Engines: 1-55788-029-8
Rebuild Small-Block Ford Engines:0-912656-89-1
Rebuild Small-Block Mopar Engines: 0-89586-128-3

RESTORATION, MAINTENANCE, REPAIR

Camaro Owner's Handbook ('67–'81): 1-55788-301-7
Camaro Restoration Handbook ('67–'81): 0-89586-375-8
Classic Car Restorer's Handbook: 1-55788-194-4
Corvette Weekend Projects ('68–'82): 1-55788-218-5
Mustang Restoration Handbook('64 1/2–'70): 0-89586-402-9
Mustang Weekend Projects ('64–'67): 1-55788-230-4
Mustang Weekend Projects 2 ('68–'70): 1-55788-256-8
Tri-Five Chevy Owner's ('55–'57): 1-55788-285-1

GENERAL REFERENCE

Auto Math:1-55788-020-4
Fabulous Funny Cars: 1-55788-069-7
Guide to GM Muscle Cars: 1-55788-003-4
Stock Cars!: 1-55788-308-4

MARINE

Big-Block Chevy Marine Performance: 1-55788-297-5

HPBOOKS ARE AVAILABLE AT BOOK AND SPECIALTY RETAILERS OR TO ORDER CALL: 1-800-788-6262, ext. 1

HPBooks
A division of Penguin Putnam Inc.
375 Hudson Street
New York, NY 10014